4th International Conference on Renewable Energy and Environment Engineering

IOP Conference Series: Earth and Environmental Science
Volume 897

Online
27 – 30 August 2021

ISBN: 978-1-7138-4879-0
ISSN: 1755-1307

Printed from e-media with permission by:

Curran Associates, Inc.
57 Morehouse Lane
Red Hook, NY 12571

Some format issues inherent in the e-media version may also appear in this print version.

Institute of Physics
Dirac House, Temple Back
Bristol BS1 6BE UK

Phone: 44 1 17 929 7481
Fax: 44 1 17 920 0979

techtracking@iop.org

Additional copies of this publication are available from:

Curran Associates, Inc.
57 Morehouse Lane
Red Hook, NY 12571 USA
Phone: 845-758-0400
Fax: 845-758-2633
Email: curran@proceedings.com
Web: www.proceedings.com

TABLE OF CONTENTS

IOP Conf. Series: Earth and Environmental Science **897** (2021) 011001 doi:10.1088/1755-1315/897/1/011001

Preface

2021 The 4th International Conference on Renewable Energy and Environment Engineering (REEE 2021) was successfully held online on August 27-30, 2021. The conference aims to provide an international platform for experts, scholars and business managers in the fields of Renewable Energy and Environment Engineering to share their latest research achievements.

The conference was scheduled to be held in Florence, Italy virtually. In view of long impact of COVID-19 and travel restrictions implemented by governments, we were forced to move the conference online. REEE 2021 used ZOOM as the platform holding online conference. Because there may be some emergencies and limitations existing, the test session was arranged to help authors learn some ZOOM basic functions and test their presentation slides or videos on the first day. Back-up videos were also collected in advance, if some presenter encountered the internet problem during presentation, the staff in zoom will help to play the video. On 28th and 29th, keynote & invited speeches and author parallel sessions were arranged. There were 6 speakers in plenary session, each presentation was 30/40 mins. They are: Prof. Mohamed Benbouzid from University of Brest, France (Speech Title: Microgrids energy management systems: Methods, solutions, and challenges); Prof. Javier Contreras Sanz from Universidad de Castilla-La Mancha, Spain (Speech Title: SiNGULAR Project: Smart and Sustainable Insular); Prof. Kai Strunz from Technical University of Berlin, Germany (Speech Title: Control of HVDC Overlay Networks: The OVANET Project and its Application to the German Power System Electricity Grids under Large-Scale Renewable Integration); Assoc. Prof. Teresa DONATEO from University of Salento, Italy (Speech Title: Reducing the environmental impact of a hybrid electric power system by optimal energy management: An application to Urban Air Mobility); Dr. Antonio Violi from University of Sannio, Italy (Speech Title: Dealing with the stochastic prosumanager problem with controllable loads) and Prof. Carlos Felgueiras from Polytechcnic of Porto, Portugal (Speech Title: Electronics in Higher Education for Sustainable Development). For author parallel session, each presenter has 15 mins to share the research ideas including Q&A part. After the presentation, the session chair and audience gave questions towards the presenter's presentation (2-3 mins). On the last day, the recorded presentation display is provided, attendees can come for the presentations they are interested in. 39 participants from Germany, France, Spain, Italy, Portugal, China, Thailand, Colombia, Ecuador, Brazil, United Arab Emirates, Croatia, Malaysia, Peru, Latvia, Ukraine, Indonesia, Japan, United States, Philippines, Romania, Belgium, Israel, Morocco, Lithuania attended the conference. Though the authors and speakers couldn't communicate face to face, the passion for involvement wasn't affected.

The proceedings is a compilation of the accepted papers and represents an interesting outcome of the conference. All the papers have been through rigorous review process to meet the requirements of international publication standard. At the end of the reviewing process, 24 papers were accepted in the conference proceedings-IOP Conference Series: Earth and Environmental Science (EES). The papers are divided into 4 chapters, they are Renewable Energy and Power Engineering; Energy Chemistry and Chemical Engineering;

IOP Conf. Series: Earth and Environmental Science **897** (2021) 011001 doi:10.1088/1755-1315/897/1/011001

Electric Power Engineering and Electronic Technology; Environmental Management, Energy Planning and Sustainable Development. Here we would like to thank all the technical committee members who made great efforts on paper reviewing.

The appreciation also goes to other conference committee members, conference chair, program chairs; keynote/invited speakers, session chairs and all the authors, thanks for their understanding and support in this special time. Hope all the participants had wonderful time during conference and got fruitful inspirations from the presentations delivered by speakers and authors.

Hopes you will enjoy the reading of this volume. Likewise, we are confident that the REEE 2021 proceedings captures the current state of the research in the related fields and hope it will motivate some of you in the longer term. We are looking forward to seeing you again in REEE next year.

Best regards,

Conference Chair

Prof. Elmar Bollin, Offenburg University of Applied Sciences, Germany

IOP Conf. Series: Earth and Environmental Science **897** (2021) 011001 doi:10.1088/1755-1315/897/1/011001

Statement of Peer Review

All papers published in this volume of IOP Conference Series: Earth and Environmental Science have been peer reviewed through processes administered by the proceedings Editors. Reviews were conducted by expert referees to the professional and scientific standards expected of a proceedings journal published by IOP Publishing.

IOP Conf. Series: Earth and Environmental Science **897** (2021) 011002 doi:10.1088/1755-1315/897/1/011002

Peer review declaration

All papers published in this volume of IOP Conference Series: Earth and Environmental Science (EES) have been peer reviewed through processes administered by the Editors.

All conference organisers/editors are required to declare details about their peer review. Therefore, please provide the following information:

- **Type of peer review:** Double-blind
- **Conference submission management system:** EasyChair System
- **Number of submissions received: 54** (46 Full papers + 8 Abstracts)
- **Number of submissions sent for review:** 46
- **Number of submissions accepted:** 24
- **Acceptance Rate (Number of Submissions Accepted / Number of Submissions Received X 100):** 52.17%
- **Average number of reviews per paper:** 2
- **Total number of reviewers involved:** 26
- **Any additional info on review process:**
  REEE holds the belief that the academic research work should be original and published only once. Each of selected paper was anonymously reviewed by two/three professional experts in the related subject area to ensure the final high standard and quality of each accepted submission. Authors have to revise according to reviewers' suggestions before submitting the final paper.
- **Contact person for queries:**
  Melanie Li Xiao (reee_conf@vip.163.com)
  Polytechcnic of Porto, Portugal

IOP Conf. Series: Earth and Environmental Science **897** (2021) 012001 doi:10.1088/1755-1315/897/1/012001

Aerodynamic improvement of Darrieus wind turbine

Oleg Goman, Andrii Dreus, Anton Rozhkevych and Krystyna Heti

Oles Honchar Dnipro National University, 49010, Haharina Ave., 72, Dnipro, Ukraine

Corresponding author's e-mail address: rozhkevych@ukr.net

Abstract. Until recently, vertical-axis wind turbines are less extensively developed in wind energetics. At the same time, there are a number of advantages in turbines of such type like their independence from the change of wind direction, lower levels of aerodynamic and infrasound noises, higher structural reliability (compared to horizontal engines), etc. With these advantages, vertical-axis wind turbines demonstrate promising capacities. Inter alia, the productiveness of such turbines can be refined through the aerodynamic improvement of the structure and comprehensive optimization of the rotor geometry. The main purpose of the presented paper is to aerodynamically improve vertical wind turbine in order to increase the efficiency of wind energy conversion into electricity. Within the framework of the classical theory of impulses, this article presents a study of the effect of variation in Reynolds number on the general energy characteristics of a vertical-axis wind turbine with two blades. The integral approach makes it possible to use a single-disk impulse model to determine the main specific indicators of the system. The power factor was calculated based on the obtained value of the shaft torque factor, which in turn was determined by numerically integrating the total torque generated by the wind turbine. To calculate the test problem, we used the classic NACA airfoils: 0012, 0015, 0018 and 0021. The proposed calculation algorithm makes it possible not to indicate the Reynolds number and corresponding aerodynamic coefficients at the beginning of the calculation, but to recalculate it depending on the relative speed, position of the airfoil and the linear speed of the airfoil around the circumference. Proposed modern design techniques can be helpful for optimization of vertical wind turbines.

1. Introduction

Wind energy is the fastest growing source of clean and renewable energy in the world. Development of this branch of energetics seems to be a priority for Ukraine – the country that is critically dependent on energy imports. Therefore, the development of alternative energy is a top objective of the modern energy industry. It requires ever-higher costs for green technologies of energy generation needed for sustainable development.

The development of the wind energy sector in Ukraine lags far behind the more developed markets. In addition to general economic factors, it should be noted that average annual wind speed in Ukraine is ranging from 4,5 to 5,5 m/s, which makes the use of major part of industrial wind turbines designed to operate at higher nominal wind speeds rather inefficient. The Central Geophysical Observatory of Ukraine expects increasing of average wind speed by 1-2 m/s in the next 30-40 years due to a climate change. Thus, the potential of wind energy industry will grow in future [8].

Wind power generation systems can be variable-speed or fixed-speed. Variable-speed ones are more attractive than fixed-speed ones because of the increased energy output. Until recently, it has been wrongly assumed that vertical-axis wind turbines cannot ensure any high efficiency level of energy

use. However, according to recent studies [1], Darrieus rotors can offer a specific speed of 6:1 and wind efficiency factor comparable to horizontal-axis turbines.
At the same time, among the advantages of such turbines, there is also the independence from the change of wind direction, lower level of aerodynamic and infrasound noises, higher structural reliability (compared to horizontal engines) that is beneficial in areas with turbulent wind flow such as rooftops, coastlines and cityscapes. Another advantage is the ability to place the generator directly on the turbine's foundation that results in a simpler transmission system. Such turbines operate better at a low capacity, which makes them very appealing to standalone and backup energy systems. With these advantages, vertical-axis wind turbines demonstrate promising capacities.
Characteristics of such turbines can be improved in particular through aerodynamic improvement of the structure and comprehensive optimization of the rotor geometry.
The general scheme of Darrieus rotor is showed in figure.1.
The Darrieus rotor is a research subject for many scientists [1–5]. To perfectly calculate the model, one should use the Navier-Stokes equations [7]. Numerical calculation and creation of the model for calculating these equations is a very time consuming process. For the mathematical calculation of Darrieus rotors we used aerodynamic coefficients of classic NACA airfoils: 0012, 0015, 0018 and 0021.

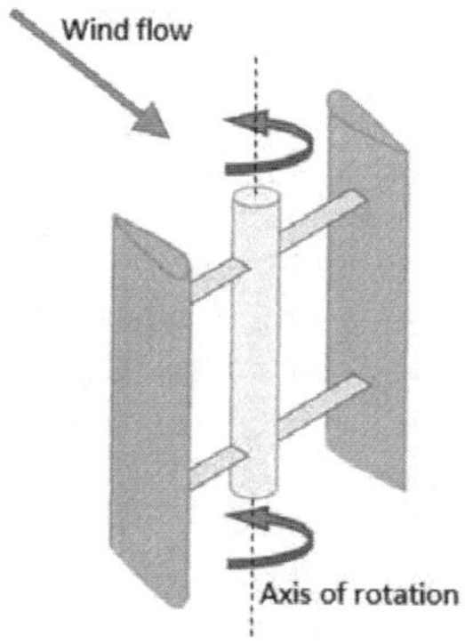

Figure 1. The scheme of a wind turbine with a vertical axis.

The proposed calculation algorithm, described in this article, uses a single-disk pulse theory with a refinement – taking into account the variation in Reynolds number, which, in turn, gives a more logical result than the calculation of the classical approach with a fixed Reynolds number.
The calculations show that the most effective airfoils for Darrieus wind turbine are NACA 0018 and NACA 0021.

2. Mathematical model

2.1 General characteristics of the mathematical model
The mathematical model is based on the classical single-disk impulse theory, the essence of which is that in the wind turbine the position of the "active disk" remains uncertain, but on this disk, the conservation laws in the integral sense are fulfilled.
In figure. 2 the distribution of wind flow for a single-disk calculation model is shown schematically.
In mathematical terms, the implementation of a single-disk impulse theory reduces itself to the following algorithm: forces acting on the blade and traverse, on the one hand, are expressed through aerodynamic coefficients of blade and traverse profiles (calculated by local angles of attack and local relative speed); on the other hand, the same forces are expressed by the momentum theorem. Mathematically, such a balance of forces is formalized in the form of a transcendental equation, the solution of which involves finding the main energy characteristics of the wind turbine – power and torque coefficients [1].

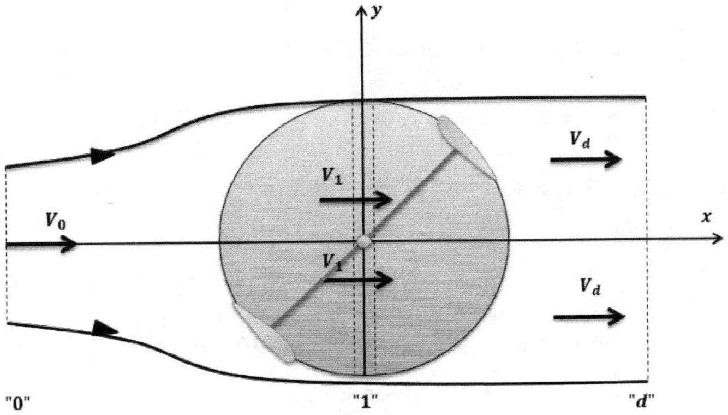

Figure 2. The scheme of a single-disc model.

The equation for finding of the value of the unknown speed in the "active disk" depending on the speed V_0 of the oncoming flow is:

$$\frac{V_0}{V_1} - 1 = F_1. \tag{1}$$

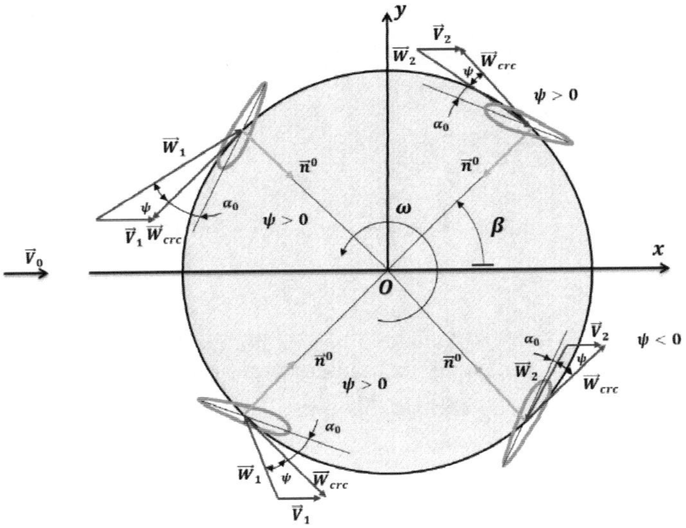

Figure 3. The scheme of relative velocities and angles of attack of the elements of the blade depending on the position of the azimuthal angle.

In equation (1), the dimensionless quantity F_1 is defined as follows:

$$F_1 = \mu_1 \int_0^{2\pi} \overline{W}_1^2 \left(c_x(\mathrm{Re},\alpha)\cdot \sin(\beta-\psi) - c_y(\mathrm{Re},\alpha)\cos(\beta-\psi)\right) d\beta. \tag{2}$$

The relative velocity W_1 of the incoming flow on the blade in the position characterized by the azimuthal angle β, is:

$$W_1 = \sqrt{V_1^2 + 2W_{crc}V_1 \sin\beta + W_{crc}^2}\,. \tag{3}$$

The determination of speed W_1 is necessary to find the coefficients of torque on the shaft and power of the wind turbine.

In equation (2):

- $\mu_1 = \dfrac{Nc}{16\pi Rk}$ – is a coefficient that takes into account the geometric parameters of the wind turbine;
- $\overline{W_1} = \dfrac{W_1}{V_1}$ – is the dimensionless relative velocity,
- V_1 – is a previously unknown speed in the active disc;
- N – is a number of blades;
- β – is an azimuthal angle,
- c – is a chord length of the airfoil;
- $c_x(\mathrm{Re},\alpha)$ and $c_y(\mathrm{Re},\alpha)$ – are aerodynamic coefficients of the selected blade profile depending on the local Reynolds number and the local angle of attack;
- Re – is Reynolds number;
- $\alpha = \alpha_0 + \psi$ – is a local angle of attack;
- $\psi = arctg\dfrac{-V_1\cos\beta}{R\omega + V_1\sin\beta}$ – is the angle between the relative velocity $\vec{W}$ and $\vec{W}_{crc}$;
- $\vec{W}_{crc}$ – is the linear velocity of the blade in a circumferential direction;
- α_0 – is the angle of installation of the blade relative to the tangent to the circle (considered positive if the leading edge of the profile deviates inward from the specified circle, and negative - outward).

2.2 The method of calculating the energy characteristics of the wind turbine depending on the variable Reynolds number

Rewriting the equation (1) in the form of:

$$V_1 = \frac{V_0}{F_1 + 1}. \tag{4}$$

The left part of equation (4) depends only on the speed V_1, so it is convenient to solve the functional connection in this form:

$$V_1 = f(V_1, V_0, \omega R), \tag{5}$$

Transcendental equation (5) can be solved by the method of successive approximations [10]. By specifying a range of numbers V_0, we can get a range of numbers V_1.

The total torque relative to the axis of the wind turbine, assuming that the center of pressure of the profile coincides with the middle of the chord, for one blade has the form of [1]:

$$M_{cp} = \frac{1}{2}\rho W^2 cR(c_y(\mathrm{Re},\alpha)\cdot\sin\psi - c_x(\mathrm{Re},\alpha)\cdot\cos\psi). \tag{6}$$

The average value of the total moment has the following form:

$$M_{cp} = \frac{N}{4\pi}\rho cHV_1^2\int_0^{2\pi} RW_1^2(c_y(\mathrm{Re},\alpha)\cdot\sin\psi - c_x(\mathrm{Re},\alpha)\cdot\cos\psi)d\beta. \tag{7}$$

The average power that is developed in one rotation is:

$$P_{cp} = M_{cp}\omega, \tag{8}$$

ω – is an angular speed of the wind turbine.

In turn, the coefficients of torque on the shaft and power of the wind turbine are in the formulas:

$$M_{cp} = c_m\frac{1}{2}\rho V_0^2 2R^2 H, \tag{9}$$

$$P_{cp} = c_m\frac{1}{2}\rho V_0^2 2R^2 H\omega = c_p\frac{1}{2}\rho V_0^3 2RH. \tag{10}$$

Using formulas (7), (9) and (10), we obtain the final form of the coefficients of torque on the shaft c_m and power coefficient c_p:

$$c_m = \frac{Nc}{4\pi}\frac{1}{(V_0 R)^2}\cdot\Phi(V_1,\omega R), \tag{11}$$

$$c_p = c_m\cdot\frac{\omega R}{V_0}, \tag{12}$$

$$\Phi(V_1,\omega R) = \int_0^{2\pi} R\left((R\omega)^2 + 2V_1 R\omega\sin\beta + V_1^2\right)\cdot\left(c_y(\mathrm{Re},\alpha)\cdot\sin\psi - c_x(\mathrm{Re},\alpha)\cdot\cos\psi\right)d\beta. \tag{13}$$

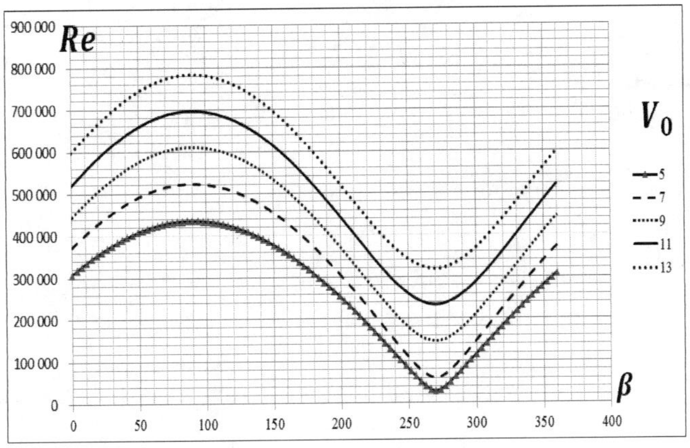

Figure 4. The dependence of the Reynolds number on the azimuthal angle β and the magnitude of the oncoming flow velocity V_0 for the profile NACA 0018.

The aerodynamic coefficients of the specified profiles for different Reynolds numbers in the range $Re = 10^4 \div 5\cdot 10^6$ [10] were used for the calculation.
The dependence of the Reynolds number on the azimuthal angle β and the magnitude of the oncoming flow velocity V_0 are shown in figure. 4 for the airfoil NACA 0018. For other airfoils the graphics are similar. The Reynolds number was calculated from the relative velocity (3) [6]:

$$Re = \frac{W_1 c}{\nu}. \tag{14}$$

The Reynolds number varies in the range from $2\cdot 10^3$ to $7{,}9\cdot 10^5$ at different speeds V_0 and a fixed linear velocity of the blade is around $W_{crc} = 5\ m/s$, and therefore this fact of the variation in Reynolds number is important in the calculation of the overall energy performance of the wind turbine.
Finally we get, that according to the method of variable Reynolds numbers, aerodynamic coefficients c_x and c_y are calculated depending on the magnitude of value (14) and relative speed (3) and are determined by linear interpolation [10] using the range of found and fixed Reynolds numbers, obtained experimentally for the indicated types of blades [9].

3. Results

To determine the main energy characteristics of the wind turbine based on the proposed mathematical model, an iterative numerical algorithm was built. The software module consists of two parts: in the first part the solution of the transcendental equation (5) is being calculated with the determination of the local Reynolds number (16), in the second – the calculation of the integral (9), with the determination of the torque coefficients on the shaft (13) and power (14). All subsequent graphs are presented depending on the number Z_0 – the tip-speed factor:

$$Z_0 = \frac{\omega R}{V_0}. \tag{15}$$

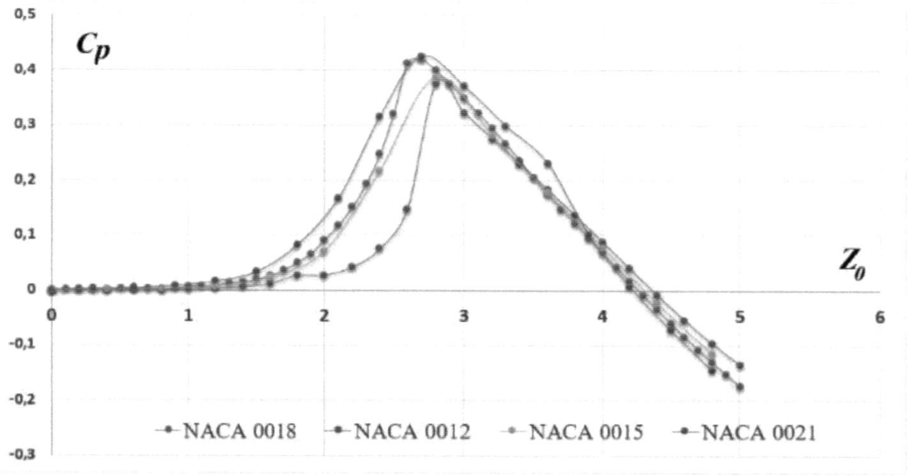

Figure. 5. The dependence of the power factor on the tip-speed factor at a wind speed $V_0 = 5\ m/s$.

The following geometric dimensions were taken for the calculation of the mathematical model:
- chord length of the blade - $c = 0.7\ m$;
- wind turbine radius - $R = 3.6\ m$;
- kinematic-viscosity coefficient - $\nu = 15.06 \cdot 10^{-6}\ m^2/c$.

Figure. 5 shows that the most optimal airfoils for wind speed $V_o = 5\ m/s$ are NACA 0018 and NACA 0021.

The calculations showed that the values of the power factors in some range of tip-speed factors are negative, which indicates that the rotation unit must consume energy from the outside and its operation is inefficient.

For this mathematical model of the wind turbine the working range of tip-speed factor is $0 < Z_0 < 4.2$.

4. Conclusion

The mathematical model for determination of the energy performance of a wind turbine with a vertical axis of rotation, based on a single-disk pulse theory is presented in this article. A numerical algorithm was created and parametric studies of the energy characteristics of the wind turbine model were performed. The proposed calculation algorithm differs from conventional algorithms, which are based on a single-disk theory, in the fact that it allows not to specify the Reynolds numbers before the calculation, but to calculate them directly depending on the relative velocity, the position of the airfoil and the linear velocity of the airfoil around the circumference.

The proposed mathematical model of calculation can be used for different types of Darrieus rotors, and the created software module allows varying the geometric parameters of the wind turbine. The numerical algorithm can be improved by using a two-disk impulse theory taking into account the variation in Reynolds number.

5. References

[1] Y. Abramovsky, "Aerodynamic theory of wind turbines. Study guide", Nauka i osvita, Dnipro, (2008).

IOP Conf. Series: Earth and Environmental Science **897** (2021) 012001 doi:10.1088/1755-1315/897/1/012001

[2] W. Tjiu, T. Marnoto, S. Mat, M.H. Ruslan, K. Sopian, "Darrieus vertical axis wind turbinefor power generation I: Assessment of Darrieus VAWT configurations", Renewable Energy, 75, pp. 50-67 (2015). https://doi.org/10.1016/j.renene.2014.09.038

[3] M. Scungio, F. Arpino, V. Focanti, M. Profili, M. Rotondi. "Wind tunnel testing of scaled models of a newly developed Darrieus-style vertical axis wind turbine with auxiliary straight blades". Energy Convers Manage. Vol.130, pp. 60-70 (2016). https://doi.org/10.1016/j.enconman.2016.10.033

[4] Omar S. Mohamed, Ahmed A.Ibrahim, Ahmed K. Etman, Amr. A. Abdelfatah, Ahmed M.R. Elbaz. "Numerical investigation of Darrieus wind turbine with slotted airfoil blades". Energy Conversion and Management: X. Vol.5, (2020). https://doi.org/10.1016/j.ecmx.2019.100026

[5] K. McLaren, S. Tullis, S .Ziada, "Measurement of high solidity vertical axis wind turbine aerodynamic loads under high vibration response conditions". Energy Conversion and Management. vol.130, pp. 60-70 (2016). https://doi.org/10.1016/j.jfluidstructs.2012.01.001

[6] M. S. Rahman, M. F. Tachie, "Reynolds number effect on flow characteristics of surface single and twin jets". Journal of Hydraulic Research. vol.57, (2019). https://doi.org/10.1080/00221686.2018.1522380

[7] M. S. Campobasso, J. Drofelnik, F. Gigante, "Comparative assessment of the harmonic balance Navier - Stokes technology for horizontal and vertical axis wind turbine aerodynamics", Computers & Fluids, vol.136, pp.354-370 (2016). https://doi.org/10.1016/j.compfluid.2016.06.023

[8] A. Rozhkevych, O. Goman, L. Knysh. "Estimation of the influence of the number of blades on the energy performance of a wind turbine with a vertical axis of rotation". Journal of the Oles Honchar Dnipro National University. Mechanics. vol.25. (2018). http://dx.doi.org/10.15421/371801

[9] J. Winslow, H. Otsuka, B. Govindarajan, I. Chopra, "Basic Understanding of Airfoil Characteristics at Low Reynolds Numbers (104 – 105)". JOURNAL OF AIRCRAFT, vol.55. (2018). https://doi.org/10.2514/1.C034415

[10] H. Safouhi. "A Generalized Technique in Numerical Integration". EPJ Web Conf. vol.173. (2018). https://doi.org/10.1051/epjconf/201817301011

IOP Conf. Series: Earth and Environmental Science **897** (2021) 012002 doi:10.1088/1755-1315/897/1/012002

Performance Evaluation of a Forward Swept Blade for Vertical Axis Wind Turbine through CFD Simulation

E Jr. J Golloy[1,3], and J Honra[2]

[1] School of Graduate Studies, Mapúa University, Intramuros Manila, Philippines
[2] School of Mechanical and Manufacturing Engineering Mapúa University, Intramuros Manila, Philippines
[3] Engineering Research and Development for Technology Graduate Scholar, Department of Science and Technology - Science Education Institute

ejgolloy@mymail.mapua.edu.ph

Abstract. With high power density wind farms, reduced environmental impact, and logistical requirements, vertical axis wind turbines offer a good value proposition for the challenging topography of the Philippines. In this study, the forward-swept wing was incorporated as a VAWT blade; its swept angle is defined along the axis of rotation and evaluated its performance through CFD simulation utilizing SST k-ω turbulence model with curvature correction function validated against experimental data. The forward-swept blade exhibited improved performance over a limited tip speed ratio (λ) with up to a 7.2% increase in peak performance occurring at $\lambda = 2.6$ and with the highest increase of 14 % in $\lambda = 2.387$. The increase is due to an increased peak torque output characterized by stall delay attributed to forward swept high angle of attack.

1. Introduction

Although traditional horizontal axis wind turbines are getting cheaper, the logistical component will still account for most of the developmental costs [1]. With the topographical makeup of the Philippines, this can be challenging. Alternatively, to address this logistical challenge and maximize the wind resource that can be harnessed, vertical axis wind turbines can be employed as these turbines are relatively small in comparison. What it lacks in individual turbine efficiency it compensates with higher power density. In a field experiment, a potential increase in wind farm power output by up to ten times its equivalent HAWT farms per land area was achieved by strategically clustering the turbines, synergistically enhancing performance. However, the turbines used in the experiment were reportedly unreliable, and they encountered many turbine-related failures [2].

It can be said that VAWT is lagging behind HAWT for it did not receive continuous development, unlike HAWT in the past [3]. Although relatively simpler, it is aerodynamically complex. VAWT dynamic loading was not well understood in the past, which led to poor design choices and premature failures [4]. The most common failures were fatigue failure from cyclic stress, which can be exacerbated by dynamic stall as it induces load fluctuations [3]. There are methods to mitigate these, both passive and active. However, the current study is particular about the Gorlov, also known as helical VAWT, which has been shown to alleviate dynamic stall effects resulting in the reduced magnitude of torque fluctuations [5]. However, due to reduced peak torque, helical designs have reduced start-up acceleration but have a more consistent output [6]. If the sweep angle and the blade

length are further increased, helical blades will be more susceptible to uneven blade loading, based on simulation results [5].
Forward swept wing has the characteristics that may offer an alternative solution. It has a high angle of attack, and the stall starts at the midplane [7]. Forward swept design was already investigated with traditional HAWT with mixed results, i.e., increased root bending stress that can be detrimental to pitched controlled turbines [8]. Another study achieved an up to 2.9% increase in power coefficient (C_p) in a CFD simulation [9]. Recently its application for VAWT was investigated and was termed inverted V-shaped blade, attaining a 15% increase in the coefficient of performance while retaining the desirable characteristics of helical design, but this is limited at a single tip speed ratio λ, and other flow characteristics were not evaluated yet [10]. A known problem with a forward-swept wing is its susceptibility to static divergence as per wind tunnel tests [11]. This issue occurs when the wing structure starts to yield under aerodynamic moments that originate from wingtips due to increasing lift, explaining the increased root bending stress observed in the HAWT blade application [8].

1.1 Swept blade and helical designs considerations

The airflow against a swept blade is divided into two components a velocity component perpendicular to the blade at a sweep angle called chordwise velocity component and the other called spanwise component that runs across the blade [12]. The increased spanwise flow may explain why helical design with steep angles tends to perform poorly at low wind speeds and angular velocity due to the blade stalling much earlier, where the peak C_p and torque is way below that of the straight-bladed design as the blade requires higher airflow to provide sufficient lift as seen on [6,13]. It was observed that as the sweep angle increases, its optimal powerband would further shift towards the higher windspeed and angular velocity, which coincides with [14,15] findings. There are consequences entailed in such operation. Higher sweep angles will inevitably experience uneven blade loading [5]. Also, inertial forces will be dominant at a high rotation speed, leading to the blade deflecting.

1.2 Observations on existing simulations and Numerical considerations

In turbulence modeling for VAWT for this particular case, it was observed that the SST k-ω [16] tends to result in an overestimated power coefficient values with fine meshes, especially with 2D mesh, particularly at high and low λ [17]. On the other hand, severe underestimation in moderate λ for coarse meshes but has good accuracy in low and high λ was observed [18]. Two factors are consistent with these observations: mesh density and the effect of isotropic assumption by RANS [19]. Due to local isotropic assumption, the model can have difficulty with flow features with high variations in mean strain rate, which is a known reason for the insensitive nature of eddy-viscosity models to curvature [20]. The model tends to over or underestimates turbulence effects under concave and convex curvatures, respectively, including inaccuracy with highly anisotropic flows. To account for this, a curvature correction function was utilized [21]. C_{curv} is the parameter that influences the strength of the correction and, by default, $C_{curv}=1$ [22]. However, for λ range where non-corrected SST k-ω exhibited accurate results, this may not be necessary.

2. Methodology

For conciseness, the straight and forward-swept blades were abbreviated to SB and FS-B, respectively. These were all conducted with ANSYS Fluent 2019 software in a workstation with AMD Ryzen 9 3950X 3.7Ghz 16-core processor and 64GB of RAM. Results are from the prescribed monitors of the instantaneous values of blade torque, and force components (axial, longitudinal, and transversal thrust). All data presented are sampled from the last revolution after reaching a time-periodic solution. The duration of a stable transient simulation for one revolution requires approximately 12 hours. Initial flow field calculation requires longer calculation time due to unsteady conditions; thus, mesh-to-mesh solution interpolation was incorporated in the methodology, substantially reducing the time required to obtain a periodic solution.

2.1 Numerical Model Validation Study

The model was based and validated against two experimental data [23,24]. The current simulation did not intentionally account for the struts or spokes to obtain an undisturbed flow for the straight blade so that the validation model can also be used in performance comparison. This approach is commonly employed in the literature to obviate additional cells imposed by the boundary layer mesh [10,15,18]. A blockage effect of approximately 10% and a correction coefficient of 1.5% were reported but were not applied in the experimental results. Therefore, the data presented in the experiment was also treated as is, including the reproduction of the wind tunnel dimensions to provide a direct comparison and preserve the effects of blockage in 3D for mesh independence analysis. The rest of the simulation was conducted at U = 9m/s with varying angular velocity, and the domain was enlarged to reduce the blockage effects as shown in figure 1.

2.2 Meshing

Two separate cell zones were generated from the whole fluid domain in ANSYS Fluent Meshing: a rotating sub-grid and a stationary one is shown in figure 2. The whole domain consists of polyhedral mesh, a first layer height of 0.025mm was specified for the boundary layer near the blade walls with 18 layers, and a growth rate of 1.25 was deemed sufficient to ensure a $y^+ \leq 1$.The polyhedral mesh was chosen as its quality is not severely affected by very high aspect ratio cells at the boundary layer and size change due to multiple faces compared to tetrahedral mesh and is much easier to generate than structured mesh [25].

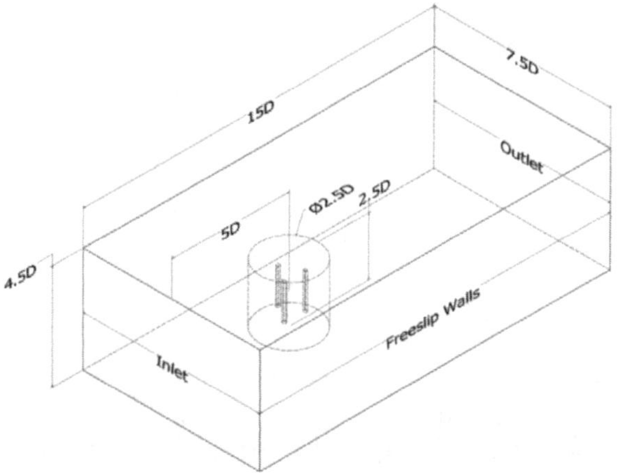

Figure 1. Turbine averaged instantaneous torque history of varying mesh density comparison.

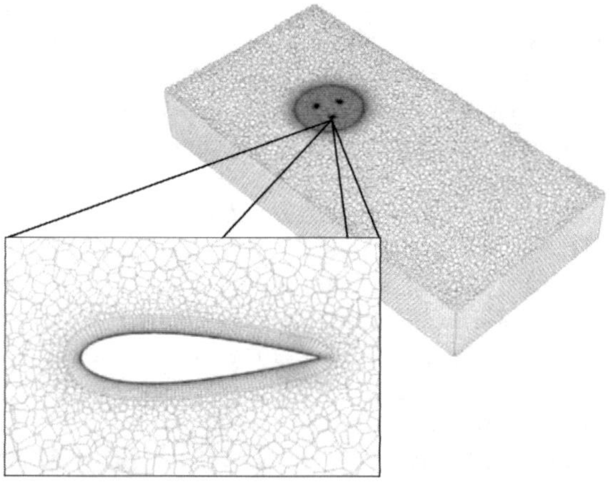

Figure 2. Generated mesh of the extended domain and boundary layer mesh near the wall.

2.3 Numerical model specification

The transient simulation was done in ANSYS Fluent, the flow conditions were assumed to be incompressible, and Shear-Stress Transport (SST) k-ω turbulence model [16] was employed in the study accompanied by curvature correction function [21] to obtain an accurate solution in known λ range that the model reportedly underestimates the results [18]. The default curvature correction value of C_{curv}=1 was applied.

2.4 The computational domain

The blade walls are non-slip, while the domain walls have free-slip conditions applied. Specified air density is ρ= 1.155 kg/m^3 inlet and outlet were set to velocity and pressure, respectively, with 1% turbulence intensity [23]. The wall zones between the rotating and stationary domain were interfaced together, prescribed with a matching option to account for non-conformal boundary walls for the sliding mesh method. The sliding mesh assumes a rigid mesh and is extensively used as this is more efficient and less demanding compared to dynamic mesh [10,15,18]. Blade chord Reynolds number considered for this study ranges from $\mathrm{Re_c}=1\times10^5 \sim 1.62\times10^5$.

2.5 Wind turbine calculations

Time step size Δt can be obtained by dividing the azimuthal angle increment $\Delta\theta$ with angular velocity ω in radians:

$$\Delta t = \frac{\Delta\theta}{\omega} \tag{1}$$

The force and torque are calculated every timestep and used to monitor convergence aside from residuals to ensure accuracy. The total force $\mathbf{F_a}$ are calculated from the blade face and is given by the sum of the dot product of a specified force vector $\vec{\mathbf{a}}$ with pressure $\vec{\mathbf{F_P}}$ and viscous $\vec{\mathbf{F_v}}$ forces, while torque is the sum of the cross product of these forces with moment vector $\vec{\mathbf{r}}$. $\vec{\mathbf{r}}$ is defined from the moment center to the specified blade wall [26].

$$\mathbf{F_a} = \vec{\mathbf{a}} \cdot \vec{\mathbf{F_P}} + \vec{\mathbf{a}} \cdot \vec{\mathbf{F_v}} \tag{2}$$

$$\vec{\mathbf{T_b}} = \vec{\mathbf{r}} \times \vec{\mathbf{F_P}} + \vec{\mathbf{r}} \times \vec{\mathbf{F_v}} \tag{3}$$

The coefficients are expressed as:

$$C_X = \frac{F_X}{0.5\rho A_s U^2} \tag{4}$$

$$C_Y = \frac{F_Y}{0.5\rho A_s U^2} \tag{5}$$

$$C_q = \frac{T}{0.5\rho U^2 A_s r} \tag{6}$$

$$C_P = \frac{T\omega}{0.5\rho U^3 A_s} \tag{7}$$

2.6 Solution settings

PISO exhibited better performance than SIMPLE for pressure-velocity coupling in this study. Iterations per timesteps take longer but reach convergence much faster and is stable even at higher under-relaxation values. Only the neighbor correction is enabled to shorten the iteration time. A value of 4-5 is sufficient to provide stable simulation from initialization with second-degree accuracy. After reaching a stable solution, the neighbor correction factor was reduced to 1. Spatial discretization and transient formulation are all set to second-order accuracy. Convergence criteria based on residuals was set to 1×10^{-4} [18], and a maximum of 20 iterations per time step was deemed sufficient [10]. For $\lambda = 2.035$ due to considerable flow variations, residuals fluctuate between $1\times10^{-3} \sim 1\times10^{-4}$.

2.7 Mesh independence analysis

All simulations were run until a stable instantaneous torque was obtained, dictated by the relative change in adjacent peaks (less than 1%) of up to 4-5 rotations after a periodic solution was achieved [18]. The analysis was run at U = 8m/s, 400rpm, and $\mathrm{Re_c}=1.18\times10^5$ with an initial 0.5° azimuthal increment where the highest performance C_P as recorded [23]. The mean coefficient of performance measured in the experiment at 8m/s was $C_P = 0.285$, with an error at around $\pm 8\%$. The medium mesh was selected as it is well within error and accounts for the increase in cell count for the low blockage case (increase in domain size).

The azimuthal increment was increased to 2° [10], and a temporal analysis was carried out to further ascertain accuracy. Only two increments were compared Table 2 since the slight increase in peak torque has similarity to Rezaeiha et al. [27]. This implies that a small azimuthal increment does not readily equate to better accuracy and will still depend on the conditions.

Table 1. Mesh independence analysis.

	Minimum specified scoped size at the wall	Total number of cells generated	Simulated C_P	Relative Error %
Fine	0.8 mm	18×10^6	277	-2.81
Medium	0.9 mm	10×10^6	271	-4.91
Coarse	1 mm	7×10^6	258	-9.47

Table 2. Temporal resolution analysis at 8m/s for medium mesh.

	Increment	Simulated C_P	Relative Error %
Case 1	0.5°	0.271	-4.91
Case 2	2°	0.277	-2.81

2.8 Model validation at varying operating condition

For the enlarged domain, the SB was divided in half through symmetry to decrease the cell count. Due to the reduced blockage effect, a decrease in power coefficient was observed compared to the experimental but is well within error. The validation results at varying λ with low blockage are shown in figure 4 and are in line with Su et al.[10]. The observed overestimation is attributed to significant gradients near-wall from the increased angular velocity at high λ and dynamic stall at lower λ. Note that curvature correction was disabled in lower λ as it causes overestimated results, and the uncorrected SST k-ω is sufficiently accurate in this range.

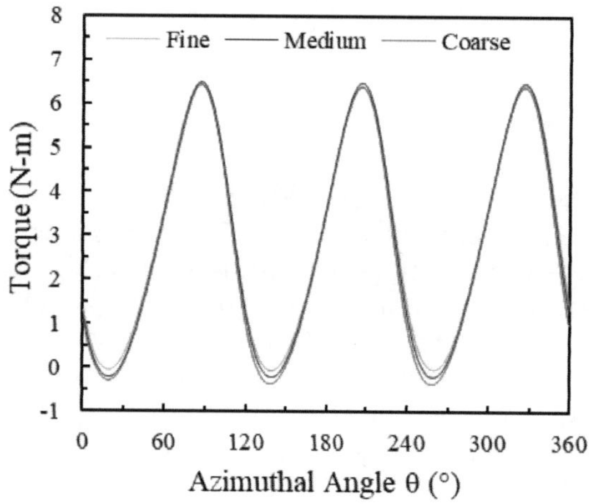

Figure 3. Turbine averaged instantaneous torque history of varying mesh density comparison.

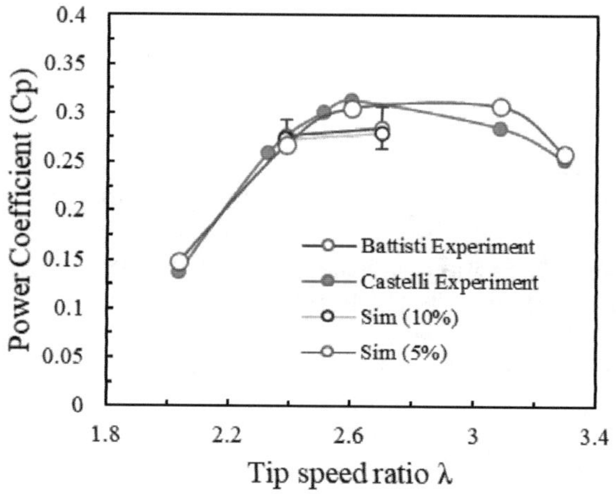

Figure 4. Comparison of simulated C_P against two experimental data [24,23]. The percentage indicates blockage.

2.9 Simulation of the Forward Swept Blades

Symmetry condition for FS-B was not applied similar to Su et al., (2020), for there was an observed in-plane interaction of the vortices due to spanwise flow. The symmetry plane acts like a fence prohibiting the propagation of dynamic stall vortex, which significantly affects results. An initial simulation was done involving FS-B in two forward swept configurations $\theta_{FS} = 6°$, $\theta_{FS} = 10°$ and SB (from validation) at peak performance in single λ based from experiment [23] to characterize its effects. This comparison was made due to the excellent accuracy exhibited by the model in this condition, ensuring confidence in the results. $\theta_{FS} = 6°$ is approximately equivalent to 0.6c inverted V-shaped blade of Su et al.[10] shown in figure 5 $\theta_{FS} = 6°$ will be referred to as FS-B 6° which will be used in further analysis at varying λ range and compared it with the simulated SB based from Castelli et al. [24]. The initial simulation will use the experiment-based domain, and the varying λ simulation will use the enlarged domain to decrease any potential interference in wake due to high blockage.

2.10 Forward swept blade schematic

The general dimensions of the model are the same as the experiment [23]. The forward-swept is defined by the circular phase shift of the blade tips in a leading forward position. The shape emulates a diverging forward-swept wing and a mirrored Gorlov/helical blade with a low sweep as shown in figure 5 superimposed on the inverted V-shaped blade [10] in addition to the conceptual turbine and its potential support configuration.

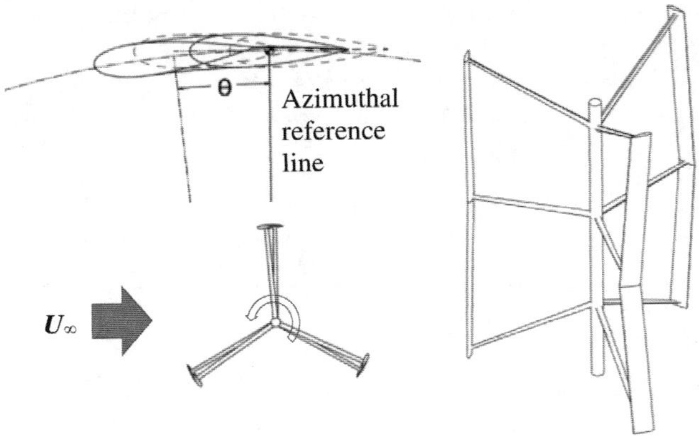

Figure 5. Forward swept blade superimposed on an inverted V-shaped blade with the reference azimuthal line & swept angle θ top left) as shown. A conceptual turbine illustration (right), with airflow & rotation indicated by the arrow (bottom left).

3. Performance evaluation and analysis of the results

The results are illustrated in torque and thrust as these are the most straightforward in evaluating the performance characteristics of the blade. The initial simulation consists of a comparison between FS-B having two different sweep angles against SB. One revolution is represented by subdividing it into four sub-regions: windward ($315° \leq \theta < 45°$) upwind ($45° \leq \theta < 135°$) leeward ($135° \leq \theta < 225°$) and downwind region ($225° \leq \theta < 315°$).

3.1 Sweep effects on performance

Simulation of two sweep angles was done to observe sweep effects on the performance shown in figure 6, which reveals increased peak torque at low sweep angle in the upwind region but lower torque at downwind region as seen from the instantaneous moment coefficient contrary to Su et al.[10] inverted V-shaped blade, but has the similar trend in reduced peak torque as the sweep angle increases but has a wider band. This observed characteristics with increasing sweep angle resemble helical blades in a simulation conducted by Alaimo et. al [13], particularly at the downwind region but will require further evaluation as this is only done for a single λ and two angles.

3.2 Vortex formation

In figure 7, FS-B skin friction lines originating from the leading-edge representing attached flow take a chordwise path which may look horizontal but obliquely shifts towards midplane (inboard) at the line of separation (vertical line) after that, exhibits a more pronounced shift upon separation. Due to the stall primarily occurring at the midplane seen in figure 8, the peak aerodynamic forces are also situated there. Considering only the half of the blade, representing a helical counterpart, the load is biased towards the trailing end of the blade similar to Scheurich et al.[5] findings. This was also reflected in the skin friction lines in figure 7 where separation line (vertical) is not visible indicating full separation. From the trailing edge, the lines representing the reverse flow are highly skewed compared to the angle of the blade, indicating a longer travel path thus, causing a delay in separation which increases with sweep angle, as seen in 120° post-stall in figure 7, this is reflected in prior figure 6 showing a higher torque retained for FS-B 10°. The increase in peak torque output at the windward region is characterized by the delay of separation as described, while torque in the downwind region has a slight reduction. Increasing the sweep angle further delays the separation, which should reflect a higher peak torque; however, it was reduced instead, shown with the FS-B 10° case figure 6. The likely cause is the variation in the angle of attack (AoA) of the blade section similar to Scheurich et al. [6]. Observe the increase in the incidence of reverse flow in figure 7 for FS-B 10° which is slightly larger compared to low sweep angle of 6°; this interferes with the propagation of dynamic stall vortex due to the increased spanwise flow consequently reducing its strength thus the reduction in peak torque as observed at the current λ. For the helical equivalent, this could mean that it is convected away from the blade, reducing its interaction which explains the higher degree of angle of attack at the trailing section from the spanwise flow and the high degree of wake asymmetry originating from the trailing section observed by Scheurich et al. [6].

Table 3. Average torque and coefficient of performance for SB and FS-B with two sweep angles.

Blade	Average Torque [N-m]	C_P	ΔC_P[%]
SB 0°	2.932	0.277	-
FS-B 6°	3.163	0.299	7.942
FS-B 10°	3.152	0.298	7.581

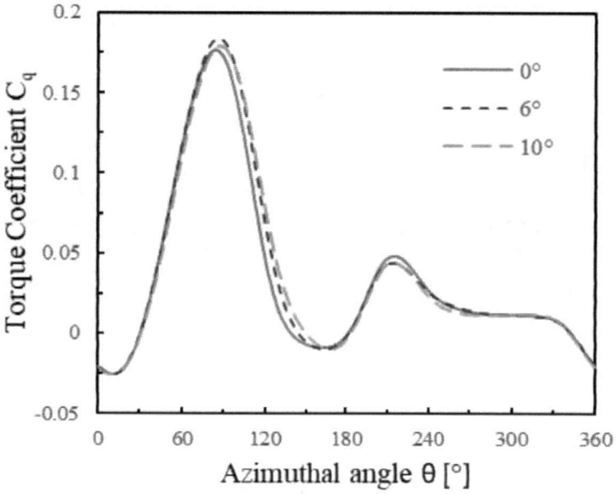

Figure 6. Instantaneous torque coefficients of blades with sweep angles of 0°, 6°, 10° at 8m/s vs azimuthal angle position within one revolution.

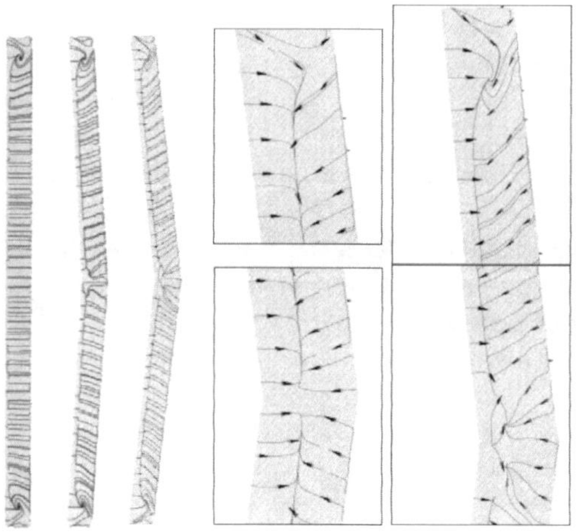

Figure 7. Skin friction lines at 120° left group). Post-stall close-up at 90° and 120° of FS-B 10° right group).

3.3 Performance at varying tip speed ratios λ

For the proceeding simulations, only the FS-B 6° was continued and was compared with the simulated experimental equivalent, SB. The averaged values in figure 9 of torque and power suggest an increase with λ until peak efficiency then declines afterward. In terms of power coefficient, the highest increase is at λ = 2.387 (14%) and for peak efficiency at λ = 2.6 (7.2%). The overall effect of instantaneous loading asymmetry can be seen from the average thrust values. At λ = 2.035 comparing the two blades, a slight reduction of transversal thrusts was observed for FS-B but had consistently higher longitudinal thrust for the rest of λ range. In figure 10 & 11 of the instantaneous plots, FS-B 6° exhibited increased peak torque in a limited λ range from λ = 2.387 to λ = 3.082. At λ = 2.035 where the dynamic stall is more pronounced, FS-B has slightly reduced peak torque and averaged torque. At the downwind region, the magnitude of blade torque and force fluctuation compared to SB was reduced. The most significant increase in peak torque and loads was observed at λ = 2.387, equating to a much higher average torque output and efficiency. Due to a relatively low sweep FS-B, instantaneous torque and thrust are almost similar with SB at λ = 3.082 to λ = 3.292. Peak longitudinal thrust continually increases with λ in the upwind region caused by the increasing pressure gradients as the angular speed increases.

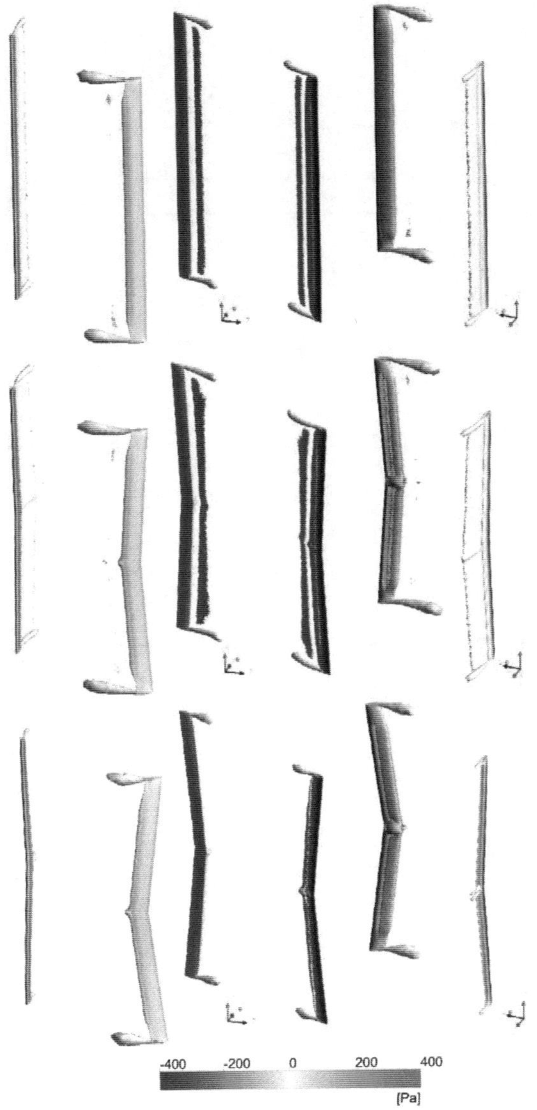

Figure 8. Q-criterion viewed from upwind (left) & downwind (right). SB (top), FS-B 6° middle), & 10° bottom).

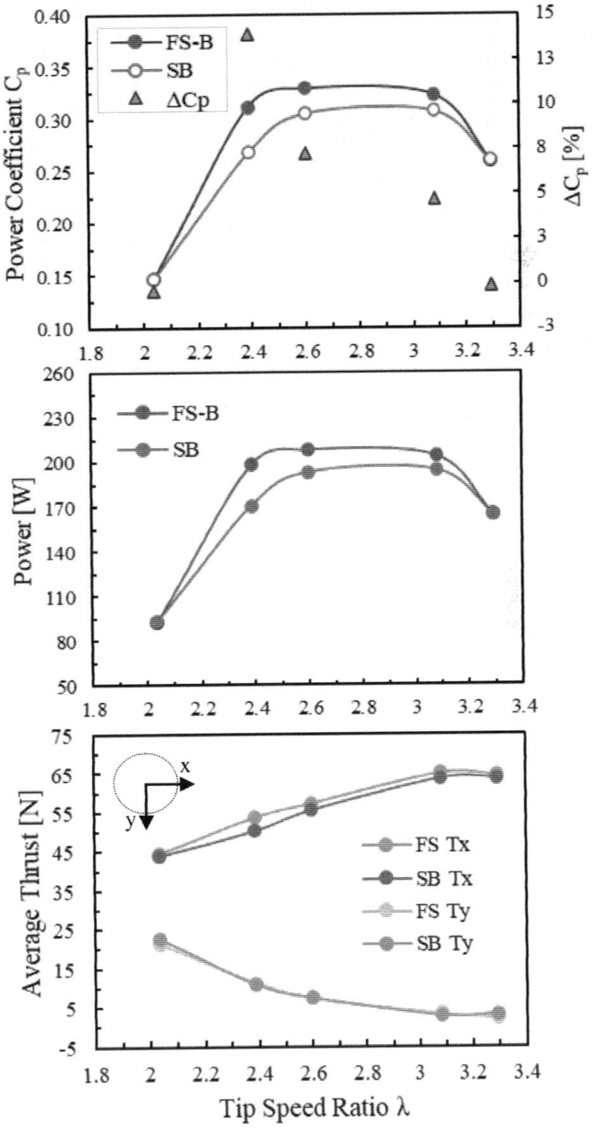

Figure 9. Comparison of averaged power coefficient (top), power (middle), & thrust force (bottom) at varying tip speed ratios λ.

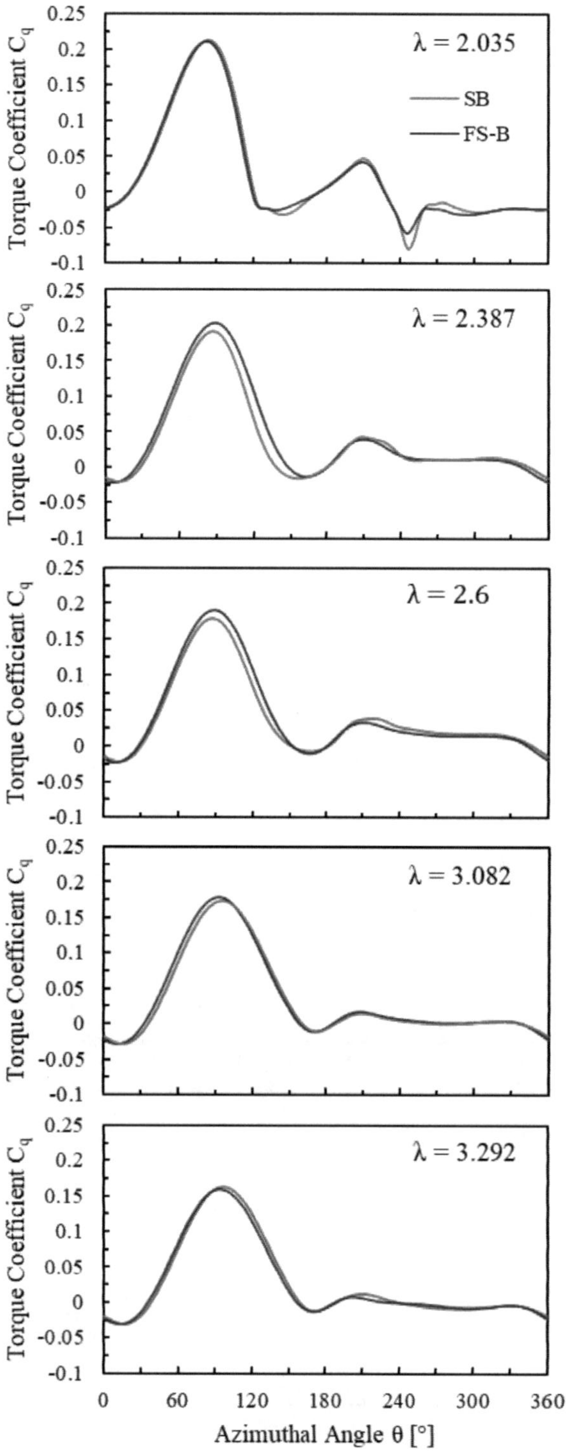

Figure 10. Comparison of blade instantaneous torque coefficients C_q at varying tip speed ratios λ

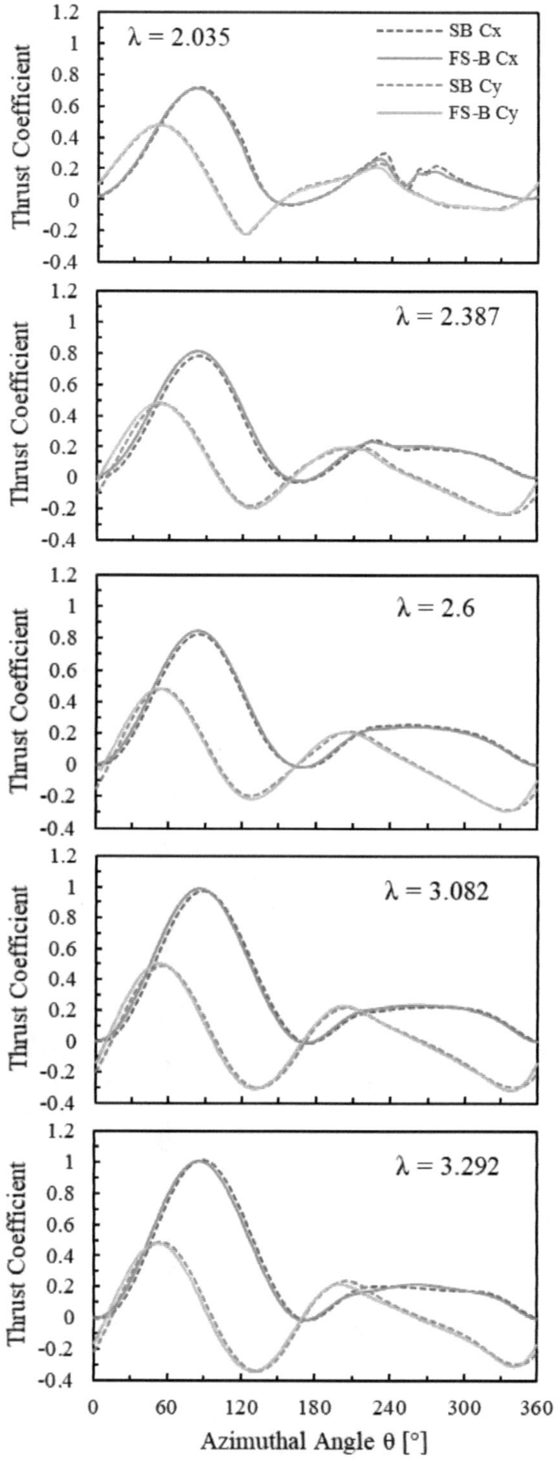

Figure 11. Comparison of blade instantaneous longitudinal C_X & transversal C_Y thrust coefficients.

3.4 Blade section loads

Based on the analysis of flow visualizations, it is apparent that loads on the blade are uneven. Further investigation was done to determine the magnitude of these load variations to understand loading conditions and mitigate these in the design stage. The analysis requires the blade face to be subdivided into different sections as illustrated in figure 12 which is from the half portion of the blade. This was conducted in two tip speed ratios where there is a significant difference in loading conditions. The results were not normalized to give the reader a better outlook of the magnitude of the loads. In figure 12, in line with the previous observation from vortex formation, FS-B peak load is situated at the midplane section (c), which is even higher than SB in $\lambda = 2.035$ while producing lower total blade torque. A noticeably reduced tip loading for FS-B can be seen at section (a), a large difference from section (c) compared with SB. The section (b) and (c) loads for SB are similar but have a more pronounced difference at $\lambda = 2.035$ leeward and downwind regions. The favorable reduction of the fluctuation magnitude at the downwind region for FS-B previously observed in figure 10 can be attributed to reduced tip stall at section (a) while SB undergoes a higher magnitude of a secondary stall. In $\lambda = 2.387$ section (a) of FS-B has good stall characteristics where the torque load is distributed at a much wider band due to sweep effects. It can be said that the reduction is caused by the induced blade tip vortices, which hinder reverse flow as shown in prior figure 7 & 8, which has a similar effect to a vortex generator as demonstrated by Choudhry et al. [28].

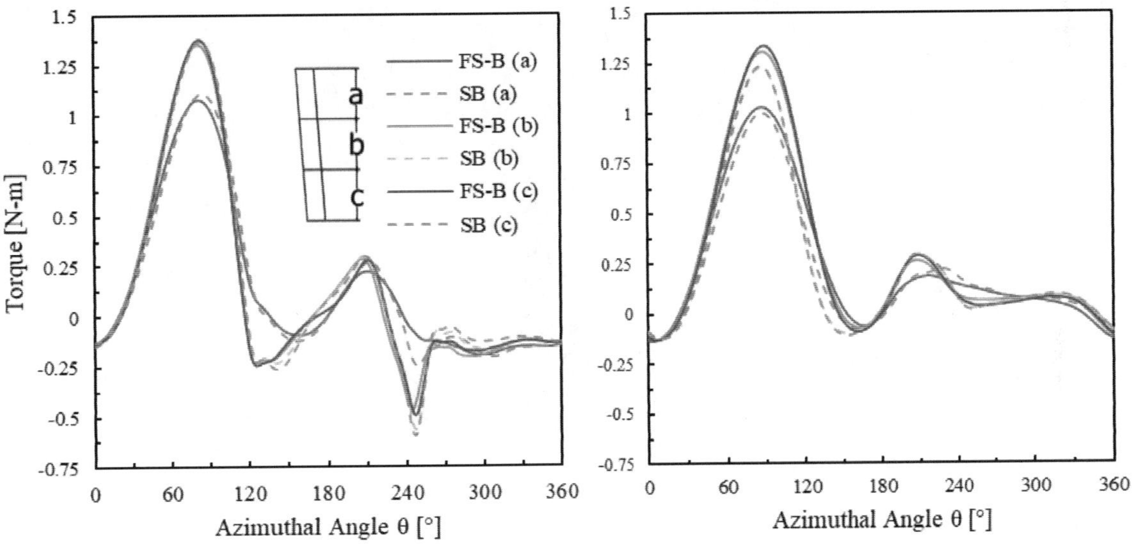

Figure 12. Blade section torque loads for at two tip speed ratios $\lambda = 2.035$ (top) & $\lambda = 2.387$ (bottom).

3.5 Wake interference

Analysis of near wake was done to assess the influence of wake shed, particularly at the downwind region. Prior results suggest a clear relationship between the magnitude of dynamic stall upwind in peak torque and torque recovery at the downwind, consistent with results from other research [10,17]. Vorticity contours from FS-B at $\lambda = 2.035$ were taken due to a much higher observed dynamic stall at midplane to visualize and analyze this. Keep in mind that this represents the current λ and some wake feature may or may not be present and varies with other cases. It can be seen from figure 13 that the lifting face switched from inside to outside, which is also reflected in prior figure 8 (middle figure). The increase in torque output in sections (b) and (c) in the leeward region previously shown in figure 12 is due to the secondary stall indicated by the formation of C2. A1 shed by the previous blade appears to interact with the stall vortex in full separation at 240° C2 and A1 are counter-rotating and dissipate slowly. The source of A1 is from the separated dynamic stall vortex as indicated by A2 in the leeward region and A3 at upwind from the following blades. It is likely that at higher λ the development of stall vortex (e.g., C2) is stunted due to much lower torque values that can be seen from instantaneous torque at downwind seen in figure 10 as λ increases.

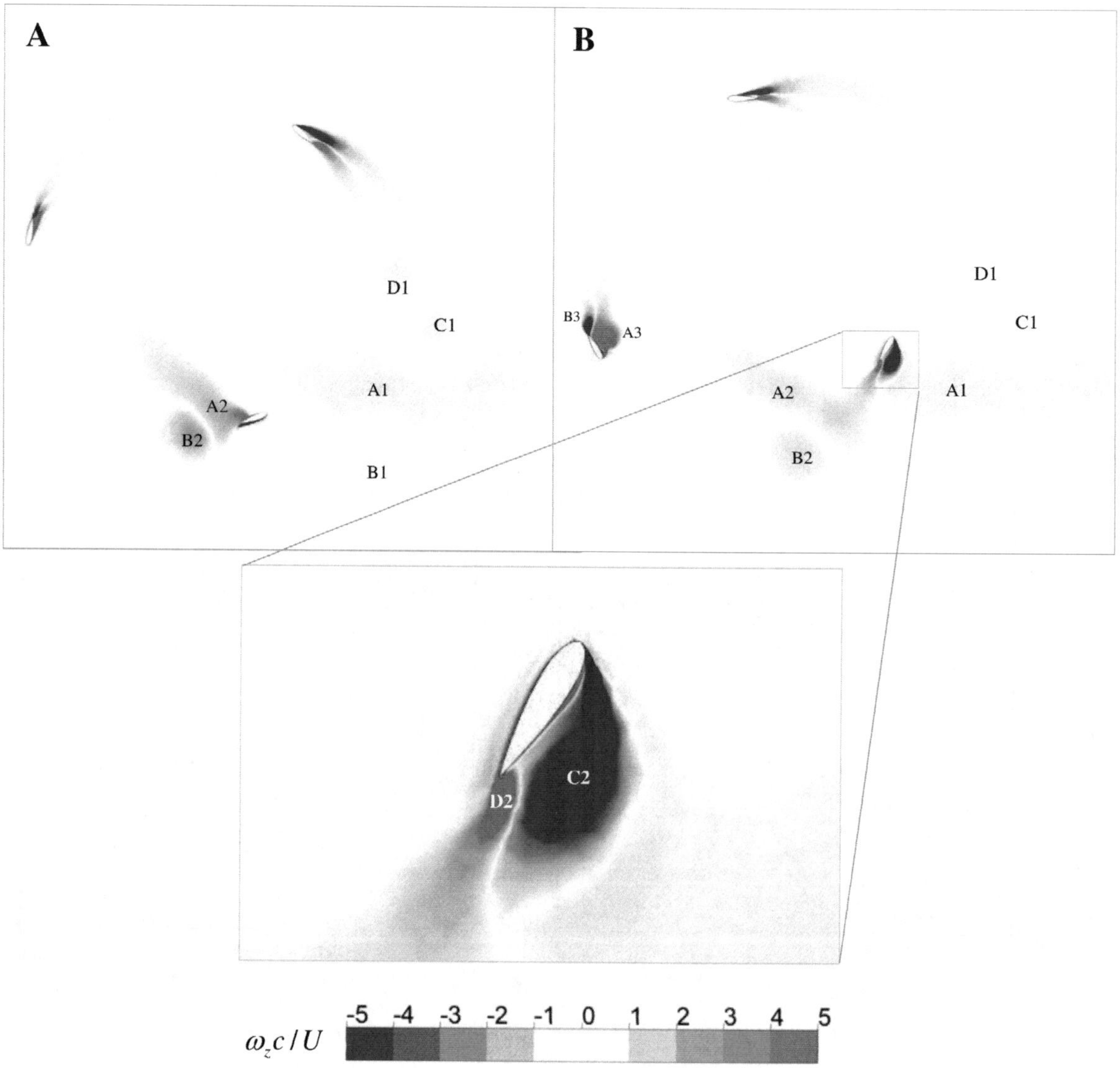

Figure 13. Non-dimensionalized vorticity contours revealing switched lifting face at 200° (A) then wake interaction at 240° (B). Vortex shed history are labeled alphanumerically to track its origin and formation.

4. Conclusion

The forward-swept blade design performance was evaluated through computational fluid dynamics (CFD) simulation. The results are represented by plots of power, torque, and thrusts in addition to flow patterns and vortex formation. The forward-swept blade increased peak torque output and efficiency of up to 7.9% at 8m/s, λ = 2.695. In 9m/s at λ = 2.387 and λ = 2.6 a 14% and 7.2% increase was achieved, respectively. Peak loads are situated midplane where stall predominantly occurs, which is higher than the straight blade. Flow separation is delayed; thus, the torque curve of the forward-swept blade is much wider. Dynamic stall formation is influenced by the spanwise flow induced by the swept blade, and its intensity is reduced with increasing sweep and higher λ. Inhibiting dynamic stall formation reduces peak torque; its magnitude at the upwind can influence torque output downwind from wake interference.

5. References

[1] IRENA 2020 Renewable power generation costs in 2019 *International Renewable Energy Agency, Abu Dhabi.* 144
[2] Dabiri J O, Greer J R, Koseff J R, Moin P and Peng J 2015 A new approach to wind energy: Opportunities and challenges *AIP Conference Proceedings* **1652** 51–7
[3] Möllerström E, Gipe P, Beurskens J and Ottermo F 2019 A historical review of vertical axis wind turbines rated 100 kW and above *Renewable and Sustainable Energy Reviews* **105** 1–13
[4] Ashwill T D, Sutherland H J and Berg D E 2012 *A retrospective of VAWT technology.*
[5] Scheurich F, Fletcher T M and Brown R E 2010 The influence of blade curvature and helical blade twist on the performance of a vertical-axis wind turbine 29th ASME Wind Energy Symposium (Orlando, Florida)
[6] Ferrari L, Bianchini A and Magnani S 2011 Start-Up Behavior of a Three-Bladed H-Darrieus VAWT: Experimental and Numerical Analysis 10
[7] Gudmundsson S 2014 The Anatomy of the Wing *General Aviation Aircraft Design* (Elsevier) pp 299–399
[8] Verelst D R and Larsen T J 2010 *Load Consequences when Sweeping Blades - A Case Study of a 5 MW Pitch Controlled Wind Turbine* (Danmarks Tekniske Universitet, Risø Nationallaboratoriet for Bæredygtig Energi)
[9] Kaya M N, Kose F, Ingham D, Ma L and Pourkashanian M 2018 Aerodynamic performance of a horizontal axis wind turbine with forward and backward swept blades *Journal of Wind Engineering and Industrial Aerodynamics* **176** 166–73
[10] Su J, Chen Y, Han Z, Zhou D, Bao Y and Zhao Y 2020 Investigation of V-shaped blade for the performance improvement of vertical axis wind turbines *Applied Energy* **260** 114326
[11] Diederich F W and Budiansky B 1948 Divergence of swept wings
[12] Vos R and Farokhi S 2015 Aerodynamics of Swept Wings *Introduction to Transonic Aerodynamics* Fluid Mechanics and Its Applications vol 110 (Dordrecht: Springer Netherlands) pp 427–511
[13] Alaimo A, Esposito A, Messineo A, Orlando C and Tumino D 2015 3D CFD Analysis of a Vertical Axis Wind Turbine *Energies* **8** 3013–33
[14] Han D, Heo Y, Choi N, Nam S, Choi K and Kim K 2018 Design, Fabrication, and Performance Test of a 100-W Helical-Blade Vertical-Axis Wind Turbine at Low Tip-Speed Ratio *Energies* **11** 1517
[15] Divakaran U, Ramesh A, Mohammad A and Velamati R K 2021 Effect of Helix Angle on the Performance of Helical Vertical Axis Wind Turbine *Energies* **14** 393
[16] Menter F R 1994 Two-equation eddy-viscosity turbulence models for engineering applications *AIAA Journal* **32** 1598–605
[17] Rezaeiha A, Kalkman I and Blocken B 2017 Effect of pitch angle on power performance and aerodynamics of a vertical axis wind turbine *Applied Energy* **197** 132–50
[18] Lei H, Zhou D, Bao Y, Li Y and Han Z 2017 Three-dimensional Improved Delayed Detached Eddy Simulation of a two-bladed vertical axis wind turbine *Energy Conversion and Management* **133** 235–48
[19] Syawitri T P, Yao Y F, Chandra B and Yao J 2021 Comparison study of URANS and hybrid RANS-LES models on predicting vertical axis wind turbine performance at low, medium and high tip speed ratio ranges *Renewable Energy* **168** 247–69

[20] Elliott K J, Savory E, Zhang C, Martinuzzi R J and Lin W E Analysis of a curvature corrected turbulence model using a 90 degree curved geometry modelled after a centrifugal compressor impeller 8

[21] Smirnov P E and Menter F R 2009 Sensitization of the SST Turbulence Model to Rotation and Curvature by Applying the Spalart–Shur Correction Term *Journal of Turbomachinery* **131** 041010

[22] Ansys® Fluent 2019 Curvature Correction for the Spalart-Allmaras and Two-Equation Models *ANSYS Help Viewer* Help system (ANSYS, Inc.)

[23] Battisti L, Persico G, Dossena V, Paradiso B, Raciti Castelli M, Brighenti A and Benini E 2018 Experimental benchmark data for H-shaped and troposkien VAWT architectures *Renewable Energy* **125** 425–44

[24] Castelli M, Ardizzon G, Battisti L, Benini E and Pavesi G 2010 Modeling Strategy and Numerical Validation for a Darrieus Vertical Axis Micro-Wind Turbine *Volume 7: Fluid Flow, Heat Transfer and Thermal Systems, Parts A and B* ASME 2010 International Mechanical Engineering Congress and Exposition (Vancouver, British Columbia, Canada: ASMEDC) pp 409–18

[25] Sosnowski M, Krzywanski J, Grabowska K and Gnatowska R 2018 Polyhedral meshing in numerical analysis of conjugate heat transfer ed P Dančová *EPJ Web Conf.* **180** 02096

[26] Ansys® Fluent 2019 Computing Forces, Moments, and the Center of Pressure *ANSYS Help Viewer* Help system (ANSYS, Inc.)

[27] Rezaeiha A, Kalkman I and Blocken B 2017 CFD simulation of a vertical axis wind turbine operating at a moderate tip speed ratio: Guidelines for minimum domain size and azimuthal increment *Renewable Energy* **107** 373–85

[28] Choudhry A, Arjomandi M and Kelso R 2016 Methods to control dynamic stall for wind turbine applications *Renewable Energy* **86** 26–37

Acknowledgments

The authors would like to thank the Department of Science and Technology - Science Education Institute -Engineering Research and Development for Technology (ERDT) program for funding this research.

IOP Conf. Series: Earth and Environmental Science **897** (2021) 012003 doi:10.1088/1755-1315/897/1/012003

Design, modelling and optimisation of a small-scale Solar Organic Rankine Cycle system for rural power generation

L F Patiño[1], U Azimov[1*], C P Tavera-Ruiz [2], J M Castellanos [3], P Gauthier-Maradei[2] and L Castro[2] *

[1] Faculty of Engineering and Environment, Northumbria University, Newcastle Upon Type, NE1 8ST, UK

[2] Grupo de Investigación en Tecnologías de Valorización de Residuos y Fuentes Agrícolas e Industriales para la Sustentabilidad Energética - INTERFASE, Escuela de Ingeniería Química, Universidad Industrial de Santander – UIS, Carrera 27, Calle 9 Ciudad Universitaria, Bucaramanga, Colombia.

[3] GEMATECH S.A.S., Bucaramanga, Colombia.

[*] Corresponding author: licasmol@uis.edu.co; ulugbek.azimov@northumbria.ac.uk

Abstract. This research study develops the design and model of a Solar Organic Rankine Cycle (SORC) coupled to a bio-digester for small-scale generation in rural areas, in Betulia, Colombia. Moreover, the model is optimised employing a Genetic Algorithm with the software Matlab and the thermodynamic library CoolProp. The objective variables were the mass flow rate of the working fluid, the pressure and temperature of the expander inlet, the solar collectors' type and the temperature of the water circuit for the bio-digester. The results indicate an overall efficiency between 8.42 and 9.45% with a Levelized Cost of Energy (LCE) between 3.85 and 5.63 £/W. Additionally, the power output is directly related to the mass flow rate of the working fluid. Likewise, increasing the scale of the SORC decreases the LCE. Finally, the results suggest that a superheated fluid reduces the efficiency and the LCE and can deliver more heat to the bio-digester. It is advisable the utilisation of a scroll expander and a counter-flow plate exchanger with a Direct Vapour Generation configuration. The model is a flexible tool capable of integrating more equations and components, with the evaluation of different fitness functions.

1. Introduction

The high consumption of resources such as coal, oil, natural gas and other fossil fuels coincides with the increase of environmental problems, especially the global warming, water contamination and solid waste [1]. Several renewable technologies have been improving during the last decade to relieve this harm to the environment. The main renewable alternatives currently promoted are wind turbines, photovoltaic solar panels, biomass digesters and geothermal systems [2]. However, some of them can be expensive for small scale and remote areas, where exist a lack of energy that requires attention, using simple but efficient technology.

In Colombia, according to reports from the Mining and Energy Planning Unit (UPME), around 66% of the national territory does not have public electricity service through the National Interconnected System (SIN). This is mainly due to the fact that in remote locations with difficult access it is costly and not feasible to carry out civil interconnection works. Also, some areas access is restricted by armed groups.

Due to the lack of electricity in the non-interconnected areas in Colombia, the population has opted for alternative energy supply routes, mainly using diesel generation plants. Although this type of plant has become the most common option in the rural population, its implementation implies a high cost of operation and maintenance, not to mention that it generates environmental impacts by producing greenhouse gases and noise pollution, directly affecting the quality of life of the peasant population.

Solar-thermal power systems have shown a solution and potential source of clean energy for rural communities. These systems are an engaging alternative since they contribute to the advantage of providing thermal energy and electrical power with the same equipment [3]. The Organic Rankine Cycle (ORC) is an independent and flexible system that employs heat source to electrical and thermal power. According to Quoilin *et al.* [4], the ORC in small-scale systems is suitable since they work with low working temperature, they show low maintenance requirements, and the systems are relatively simple. These advantages convert ORC technology attractive and suitable for this purpose and can be considered one of the most common and competitive cycle to generate electricity from thermal energy [5]. It is composed of a pump, an evaporator, a turbine or expander and a condenser.

Two main designs are considered in the literature: the Heat Transfer Fluid design (HTF) and the Direct Vapour Generation design (DVG). The HTF design is the most popular design in research since it is less sensitive to the heat source fluctuations. This type of system collects the heat from the source and flows in a closed-loop using a specific fluid, and using three heating exchange (HE): the evaporator, the regenerator and the condenser [6].

The DVG design uses solar collectors as evaporator. Therefore, it requires only one HE in the condensation process, as well as a regenerator. This configuration is more suitable for the solar heat source. It reduces the costs, the irreversibility, and the complexity of the system and increases the thermal efficiency with fewer components [6, 7]. Furthermore, the DVG design is less common in literature.

Considering that SORC is an efficient and environmentally friendly alternative, is projected as an attractive solution for non-interconnected areas. Moreover, there is no research dedicated to developing a Direct Vapour Generation SORC system that uses its waste heat to improve another low-temperature system. As an example, the production of biogas using a small bio-digester, which is suitable for rural areas, is raised by the heating of the waste at low temperatures.

The aim of this research study is to develop a model of a Solar Organic Rankine Cycle (SORC) system to produce small-scale power in rural areas. The design considers a bio-digester that absorbs the waste heat from the SORC to raise the production of biogas. This model is optimised to find the most suitable sizes of each component, pursuing the best efficiency and the lowest cost. The location of the project is Betulia, Colombia, where bio-digester is a real project led by the Industrial University of Santander. In this work the optimisation of the thermodynamic parameters of the SORC taking into account the efficiency and levelized cost of energy (LCE), as well as the integration of solar collectors, bio-digester, heat transfer model and ORC, taking into account the location.

2. Methodology

Initially the type of ORC design to work was selected. In this research the DVG design was selected due to this system has lower thermal inertia than the HTF system, making it faster to start [8]. Finally, the simplicity of DVG design will allow having easier maintenance with fewer components.

The SORC proposed (Figure 1) use as heat source the solar collectors, and contemplate three added components: the pump, the expander, and the heat exchanger. The type of collector is a variable that is optimised in the model, being dependent on the efficiency and the price. Since it collects heat with solar collectors, it will work during daylight. The bio-digester has a volume of $6m^3$ and it is developed by Universidad Industrial de Santander.

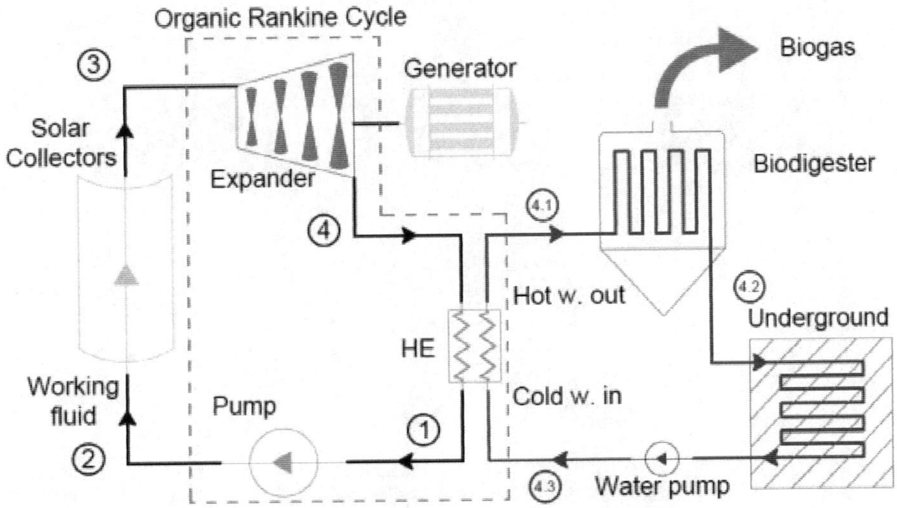

Figure 1. Schematics of the Solar Organic Rankine Cycle system coupled with Bio-digester.

The proposed system uses the heat waste from the SORC to increase the temperature in the bio-digester and accelerate the process of biogas generation. Since it is a need to decrease the temperature of the working fluid in point 1, the water circuit must cool after the bio-digester. For this, the pipes pass through the underground since the temperature is constant. This passive method gives stability to the system. The hourly ground temperatures were provided by Universidad Industrial de Santander from data measured for fifteen days, obtained a mean value of 21.084 ºC.

The fluid selected was R-245fa since it requires smaller equipment is a worldwide-cheap refrigerant, is available in Colombia and has a low critical temperature, which benefits the performance of the ORC. Aditionally, this fluid has low toxicity and non-flammable, make it safe for the rural people. Regardless of its high GWP, the lifetime of the fluid in the atmosphere is 7.2 years [9]. Furthermore, its non-ODP, make the fluid friendly with the environment.

About the solar collector type, the optimisation procedure will find the best collector taking into account the objective function. Currently, the model can evaluate the efficiency and the levelized cost of energy. However, with more develop of the model, other factors will apply, such as the availability.

To select a software some characteristics was considered, such as flexibility, licence availability, complexity and others. The properties of the main software available were compared, considering five aspects for the final selection: the price and the availability in the University, the robustness and prestige, the thermophysical library, popularity, and previous knowledge of the software. Examining this, Matlab was selected [10].

For the calculation of thermodynamics of the fluid was chosen CoolProp, due to it is an open-source and a recognised library developed by the National Institute of Standards and Technology of U.S. Department of Commerce [11, 12]. This tool can calculate the state thermodynamic properties of a fluid by knowing another two properties of state using high-accuracy Helmholtz energy formulation and cubic equation of state and fitting algorithms [13].

For the optimisation method the model uses the Genetic Algorithm (GA) for two main reasons. The first one is related to previous knowledge acquired during the studies of the author. The GA can find different solutions for problems with nonlinear constraints, searching in several regions of a space at the same time. Moreover, the fitness function implemented in the code involves nonlinear equations. GA can solve problems when the objective function is stochastic, discontinuous, and highly nonlinear [14]. The diagram of optimisation approach is shown on Figure 2.

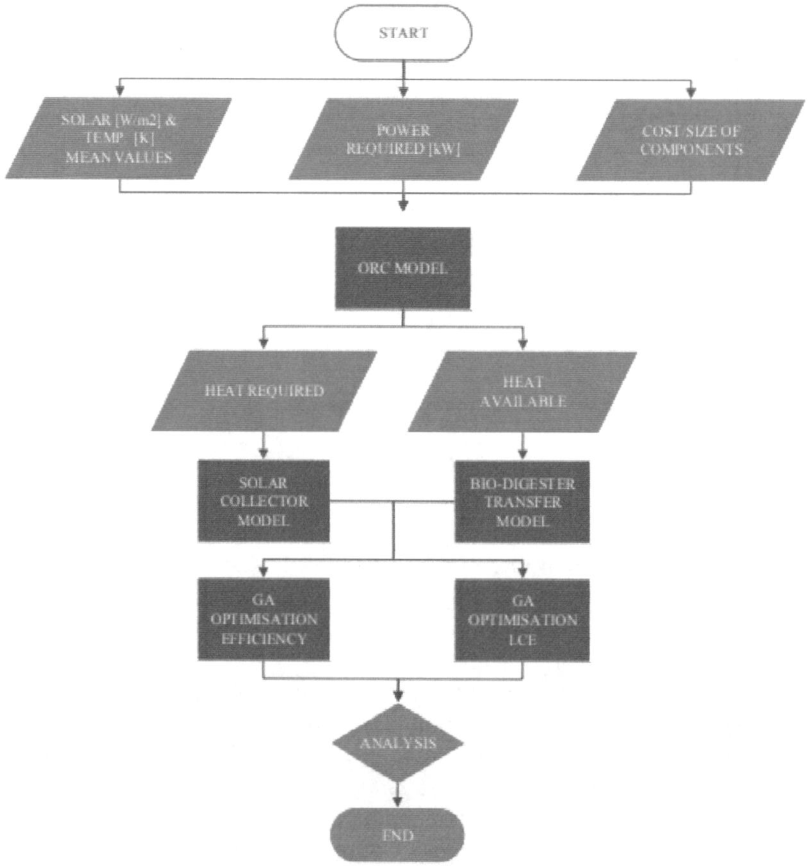

Figure 2. Diagram of processes of the SORC

The ORC system is based on equations that mostly all authors from literature review apply in their RC models. On the other hand, it is necessary to mention that the model assumes the value of some variables called 'optimisation variables': mass flow rate of the working fluid, the pressure in point 2, the temperature in point 3, the type of solar collectors and the temperature of water in point 4.

Three main inputs start the model. For one side, the location gives the solar radiation, ground temperature and ambient temperature. Table 1 gives these values. On the other hand, the power required to lead to the size of the SORC components. In this case, this value is between 1000 and 1100 Watts. Finally, the cost of each component is an input that could vary according to the location. However, these values are derived from the literature review.

Table 1. Local Data of Betulia, Colombia.

Variable	Value	Unit
Ambient temperature	294.84	K
Ground Temperature	294.234	K
Blackbody sun temperature	5800	K

The final step of the model is optimisation. As it was noted, the optimisation variables are the values that genetic algorithm changes and mutates to optimise the system to find the minimum or maximum value of the objective function. Furthermore, two different functions are assessed separately: the efficiency of the system and the Levelized Cost of Energy (LCE).

3. Results and Analysis

The model begins with the calculation of thermodynamic parameters of point 1 (Figure 1). Here, the temperature of the working fluid is similar to ground temperature. Also, its state is saturated (quality of vapour=0). Using CoolProp library were calculate all parameters in this point, such as pressure P_1, enthalpy H_1, entropy S_1, density D_1, viscosity V_1 and others. Then, the ideal entropy of point 2 (S_2) is assumed as the same as S_1. Using the optimisation variable P_2, it is calculated the ideal enthalpy H_2.

Different analysis can be done with the model. In this case, two main results are reviewed. To begin with, the efficiency optimisation gives a result of 9.45% of the overall system as the objective function. Also, the LCE is 5,63£/W. Figure 3 displays the T-S diagram of the SORC.

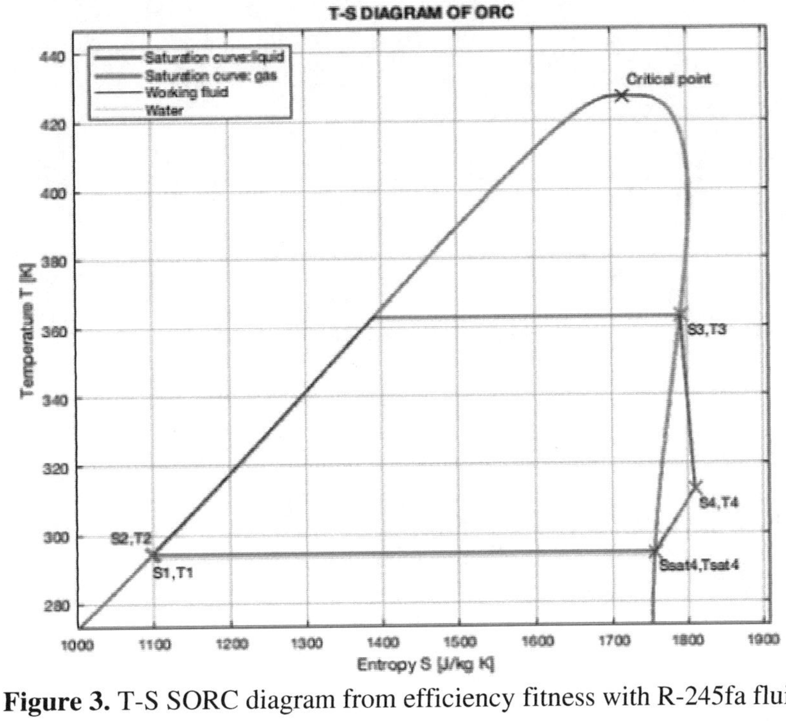

Figure 3. T-S SORC diagram from efficiency fitness with R-245fa fluid.

Due to the algorithm is looking for the maximum efficiency, the model sets the minimum heat required from solar collectors to obtain an output of 1000 Watts.
Consequently, the fluid is heated until the fluid achieves the saturated vapour point. Looking at the constraints, temperature T3 must be higher or equal as the saturated temperature, to not damage the expander with two-phase fluid. Also, the model selects the EFCP collectors, since it presents the best efficiency at temperature T3.
On the other hand, this configuration is available to give a small quantity of heat to the bio-digester, since the temperature in point four is 39.26 °C. According to the second law of thermodynamics, this is the maximum temperature that the bio-digester circuit can achieve. The UIS recommends a temperature of 80 °C for the bio-digester. Nevertheless, the main aim of the project is the generation of electrical power. Moreover, any temperature gained in the bio-digester increases the efficiency of it.

The second analysis is done by using LCE optimisation. The best result shows an efficiency of 8.42% and an LCE of 3.85 £/W. Figure 4 shows the T-S diagram of the SORC for this case.

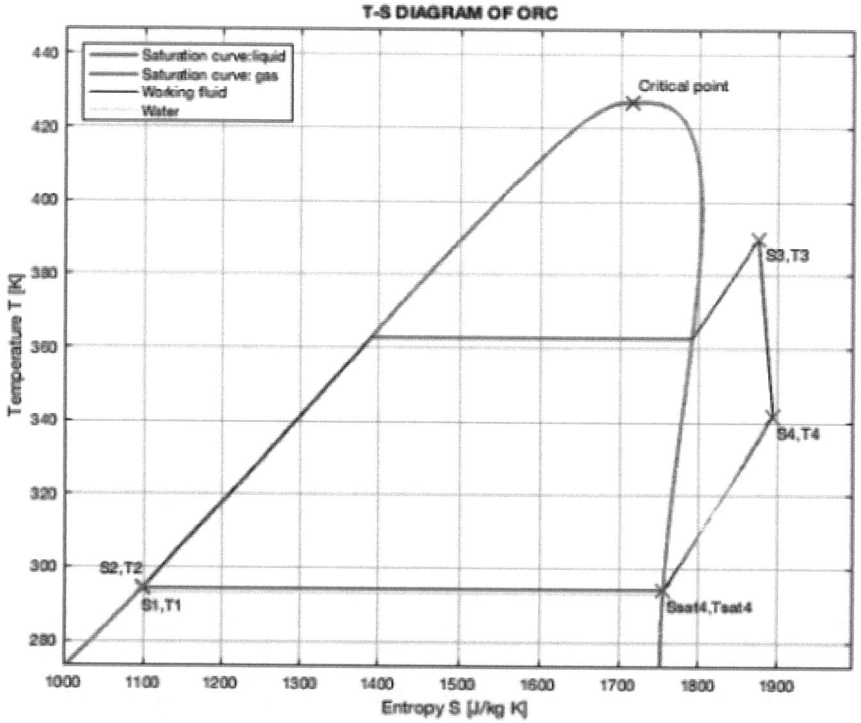

Figure 4. T-S SORC diagram from LCE fitness with R-245fa fluid

In this case, the algorithm searches for the lowest cost. As a consequence, two main changes appear. Firstly, the fluid achieves a superheated temperature in point three. Different factors explain this behaviour. Firstly, a higher temperature permits the system to get a smaller pump, reducing the consumption and costs. At the same time, it also enables getting a higher power output from the expander. Thus, these factors reduce the LCE, due to the costs are not altered in a significant way. In other words, delivering more power with the expander and a similar pump consumption decreases the LCE. For this, it is necessary to achieve super-heated fluid. Secondly, the type of solar collector differs from the previous result.

Additionally, the heat available for the bio-digester is larger than the preceding configuration, achieving a maximum temperature in point 4 of 66 °C. Table 2 displays the comparison of the results of both fitness functions.

Firstly, the LCE function tends to get the highest power (upper constraint of 1100 Watts). The efficiency function gives the opposite result (lower constraint of 1000 Watts). Also, the LCE function permits to give around 800W more of thermal power to the bio-digester with greater temperature, improving the production of gas. Moreover, the difference in final cost between both results is more than 1400 £, which is a considerable amount of budget, taking into account the conditions of the real project. The difference is mainly related to the cost of the collectors.

The most suitable configuration is the LCE function results considering the input data, the lower budget and the simplicity of the collectors. It is important to mention that costs such as pipelines, structure and control systems are not included.

Firstly, the output power of the SORC varies the LCE result. As the power output is increased, the LCE decreases. For instance, a simulation with no constraint regarding the 1000-1100 watts limit,

IOP Conf. Series: Earth and Environmental Science **897** (2021) 012003 doi:10.1088/1755-1315/897/1/012003

gave an LCE of 2.44 £/W, with an output power of 330.3 kW and an efficiency of 8.56%. The reason for this is due to the cost ratio increase slower as the power output ratio.

Table 2. Comparison of output data for the SORC, using Efficiency fitness and LCE fitness

Variable	Efficiency Result	LCE Result	Unit
T3	363	390	K
T4	312.41	341.71	K
P1	128280	128280	Pa
P3	1000000	998821	Pa
Qcond	-6695.1	-7357.3	W
Q available	507.11	1343.9	W
Qcoll	7700.1	8456.1	W
Wturb	1029.3	1122.3	W
Wpump	24.23	23.51	W
eff	9.45	8.42	%
lce	5.63	3.85	£/W
Acoll	26.77	32.85	m^2
Solar collector	EFPC	FPC	
Costexp	780	780	£
Costpump	288.32	287.93	£
Costhe	904	504.14	£
Costcoll	3686.5	2661	£
Total cost	5658.82	4233.07	£
Woutput	1005.07	1098.79	W

For this case, the upper limit for power is the mass flow rate. Therefore, the power output depends directly on the flow rate. This result is similar to Xu *et al.* [7]. Their work suggests that "to maintain a reasonable outlet temperature under higher solar radiation, larger flow rate is needed. Therefore, higher solar radiation and a larger flow rate can provide more power and more efficiency" [7]. The second aspect is that efficiency is not affected by the output or the mass flow rate. According to the model, the efficiency does not vary changing the overall power output. Based on the results, an approach of real components can be done.

During the developing of the project, some deviations appeared concerning the research design. For instance, at the beginning it was considered just an optimisation of solar collectors. Also, the Rankine Cycle was considered just as a mathematical model. However, the research found that it is necessary to optimise the RC and the heat source at the same time. Therefore, the model is more flexible. In contrast, the combination between the SORC and the bio-digester was assumed as a black box. However, the heat exchanger model is more complex than it was expected. Its development required a longer time.

Nevertheless, a first approach of the SORC coupled with the bio-digester was settled successfully, using the software and data input considered in early stages. More research must be considered regarding the real-life project, such as the costs of the pipeline and physical structure, and a deeper heat-transfer model between the HE and the bio-digester. Also, a next step must be considered regarding control, since the SORC should adapt itself to the variations of the heat source. Therefore, it is meaningful to investigate and integrate control strategies [15]. For this, an hourly input performance must be developed.

IOP Conf. Series: Earth and Environmental Science **897** (2021) 012003 doi:10.1088/1755-1315/897/1/012003

The parameters of the model can change depending on the objectives and constraints. For instance, it is possible to prioritize thermal generation for the bio-digester with a simple modification to the code. Also, it is possible to add new solar collectors, expanders and other components. The model can evaluate the changes quickly. Furthermore, it can also change the working fluid.
For instance, the model must improve the calculation of costs, regarding pipelines, accessories and structure, maintenance, and lifetime. Also, this work does not contain the limit temperature of the collectors, which could affect the selection of it.

4. Conclusions

The present study describes the development of an optimisation model of a small-scale Solar Organic Rankine Cycle focused on rural areas, coupled with a bio-digester. This project was carried out with the Industrial University of Santander (UIS). The final aim is to continue the improvement of the model, to build a real project in Betulia, a village located in Colombia. It was used the Matlab software and the thermodynamic library CoolProp. The optimisation model considered as optimisation variables the mass flow rate of the working fluid, the work pressure, the maximum temperature, the type of collectors and the maximum temperature of water for the bio-digester circuit. With this, the sizes of the components were calculated.
The results give an important approach of the real behaviour that the system could get, giving the most suitable parameters, depending on the location, the budget and real characteristics of the components. Moreover, some considerations have been proved and must be considered in the future. For instance, the output of the SORC (which depends on the mass flow rate) vary the LCE. As the power output increases, the cost per Watt decrease. This factor does not affect the overall efficiency. Similarly, a superheated fluid decreases efficiency but improves the LCE. Additionally, more heat can be released to the bio-digester.
Some challenges are expected to solve in the future. The main one consists of the optimisation of the system using hourly data. With this, the model analysis will be more realistic. Moreover, this will help to generate a control system to get the most suitable configuration of the components since the input conditions are changing constantly, such as solar radiation and the ambient temperature. Also, the implementation of new objective functions can expand the possibilities of final configurations.

5. References

[1] Li, P., Han, Z., Jia, X., Mei, Z., Han, X., & Wang, Z. 2019 Analysis and comparison on thermodynamic and economic performances of an organic Rankine cycle with constant and one-dimensional dynamic turbine efficiency *Energy Conversion and Management* **180** p 665–679.

[2] U.S. Energy Information Administration 2019 Renewable Energy Sources – Energy Explained, Your Guide To Understanding Energy *U.S. Energy Information Administration.* Available at: https://www.eia.gov/energyexplained/?page=renewable_home (Accessed: 4 August 2019).

[3] Markides, C. N. 2015 Low-Concentration Solar-Power Systems Based on Organic Rankine Cycles for Distributed-Scale Applications: Overview and Further Developments Frontiers in Energy Research. *Frontiers* **3** p 47.

[4] Quoilin, S., Orosz, M., Hemond, H., & Lemort, V. 2011 Performance and design optimization of a low-cost solar organic Rankine cycle for remote power generation. *Solar Energy* **85(5)** p 955–966.

[5] Tzivanidis, C., Bellos, E. and Antonopoulos, K. A. 2016 Energetic and financial investigation of a stand-alone solar-thermal Organic Rankine Cycle power plant *Energy Conversion and Management* **126** p 421–433.

[6] Aboelwafa, O., Fateen, S. E. K., Soliman, A., & Ismail, I. M. 2018 A review on solar Rankine cycles: Working fluids, applications, and cycle modifications *Renewable and Sustainable Energy Reviews* **82** p 868–885.

[7] Xu, G., Song, G., Zhu, X., Gao, W., Li, H., & Quan, Y. 2015 Performance evaluation of a direct vapor generation supercritical ORC system driven by linear Fresnel reflector solar concentrator *Applied Thermal Engineering* **80**, p 196–204.

[8] Marion, M., Voicu, I. and Tiffonnet, A. L. 2014 Wind effect on the performance of a solar organic Rankine cycle *Renewable Energy* **68** p 651– 661.

[9] Honeywell 2007 HFC-245fa *Product Stewardship Summary* Available at: http://www51.honeywell.com/sm/common/documents/Public-Ris-Summary-HFC-245fa.pdf (Accessed: 25 August 2019).

[10] Mathworks 2019. *Matlab Software*. Available at: https://www.mathworks.com/products/matlab.html.

[11] NASA POWER 2019 Data Access viewer Available at: *https://power.larc.nasa.gov/data-access-viewer/* (Accessed: 24 July 2019)

[12] National Institute of Standards and Technology 2019 REFPROP Available at *https://www.nist.gov/srd/refprop* (Accessed: 01 September 2019).

[13] Caldiño-Herrera, U., Castro, L., Jaramillo, O. A., Garcia, J. C., Urquiza, G., & Flores, F. 2017. Small Organic Rankine Cycle Coupled to Parabolic Trough Solar Concentrator *Energy Procedia* **129** p 700–707.

[14] Konak, A., Coit, D., & Smith, A. 2006 Multi-objective optimization using genetic algorithms: A tutorial *Reliability Engineering and System Safety* **91** p 992- 1007.

[15] Hu, D., Zheng, Y., Wu, Y., Li, S., & Dai, Y. 2015 Off-design performance comparison of an organic Rankine cycle under different control strategies *Applied Energy* **156** p 268–279.

Acknowledgments

The authors are especially grateful to the Industry-Academia Partnership Program (IAPP) of the Royal Academy of Engineering for the funding and support (Grant No. IAPP18-19\269). As well as to the Faculty of Engineering and Environment of the University of Northumbria in the UK and the School of Chemical Engineering of the Universidad Industrial de Santander in Colombia.

Claudia Tavera-Ruiz thanks the postdoctoral program of the Ministry of Science, Technology and Innovation of Colombia for the funding.

IOP Conf. Series: Earth and Environmental Science **897** (2021) 012004 doi:10.1088/1755-1315/897/1/012004

Wind Energy Potential Assessment Based-on WRF Four-Dimensional Data Assimilation System and Cross-Calibrated Multi-Platform Dataset

Nurry Widya Hesty[1], Dian Galuh Cendrawati[1,*], Rabindra Nepal[2], and Muhammad Indra al Irsyad[1]

[1] Research and Development Center for Electricity, Renewable Energy, and Energy Conservation Technologies, Ministry of Energy and Mineral Resources, South Jakarta, Indonesia 12230
[2] School of Business, Faculty of Business and Law, University of Wollongong, New South Wales, Australia

* dian.cendrawati@esdm.go.id

Abstract. Indonesia has a target of achieving 23% of renewable energy share in the total energy mix in 2025. However, Indonesia does not have accurate and comprehensive data on renewable energy potentials, especially wind energy. This article aims to assess the theoretical potential of wind speed and to visualize the wind speed by province for the entire Indonesia. Our assessment relied on the Weather Research and Forecasting (WRF) model using Four-Dimensional Data Assimilation technique, also known as Nudging Newtonian relaxation. The robustness of our analysis is confirmed by using high-resolution data from the National Centers for Environmental Prediction – Final (NCEP - FNL) and Cross-Calibrated Multi-Platform (CCMP) Reanalysis satellite data. This study shows the WRF method is a feasible option to estimate wind speed data.

1. Introduction

Wind farm is one of feasible option to decarbonize electricity sector in Indonesia [1]. Wind power resource in Indonesia has been measured by various institutions [2]. However, further reliable and high-resolution assessments of wind energy resources are crucial to improve the quality of wind energy potential data in Indonesia.

This study proposes an assessment methodology that assimilates two weather data sources (i.e., Cross-Calibrated Multi-Platform/ CCMP and the National Centers for Environmental Prediction – Final/ NCEP-FNL) to produce more accurate and higher-resolution data. NCEP data is a common data resource for wind speed assessment due to its accuracy but has a lower resolution than that of CCMP. Meanwhile, CCMP has higher spatial and temporal resolutions but has lower accuracy in high-wind speed (i.e., > 15 m/s) and rainy conditions. Previous studies mostly used data from NCEP only [3-5]. An exception is Hesty and Hadi [6] who assimilated CCMP and NCEP-FNL but their wind energy assessment was only for a specific site (microscale). The novelty of this study is the assimilation of those two-weather data to estimate wind energy in mesoscale with Indonesia as a case.

2. Literature Review

IOP Conf. Series: Earth and Environmental Science **897** (2021) 012004 doi:10.1088/1755-1315/897/1/012004

The improvement of the wind energy assessment would not have been as successful without the use of the numerical weather prediction (NWP) model. Micro-level analysis commonly uses models of MM5, Wind Atlas Analysis and Application Program (WAsP), and WindSIM [7-9]. MM5 and WAsP may produce comparable results, but MM5 has a critical advantage in that it only needs reanalysis data without requiring wind measurement data [10]. Reanalysis data is useful for wind resource assessments in a case when observational data is not available. NWP model, a software to describe atmospheric processes and changes, along with reanalysis data is the main tool to construct historical climate data in a regional grid by integrating various past observation and measurement systems. NWP models are useful to downscale reanalysis data sets while adding physical phenomena, due to their smaller spatial and temporal time scales, including the consideration of local topographical features. The most accurate data for the wind energy simulation is ERA-Interim reanalysis for the onshore area and NCEP-R2 reanalysis for the offshore area [3, 11]. Hesty and Hadi [6] assimilated CCMP and NCEP-FNL to increase data resolution from 27 km into 3 km for wind speed assessment in West Java coast, Indonesia.

The MCP method also uses long-term reference data derived from the NWP model and the atmospheric reanalysis data set [12]. One of the most widely used NWP models is the Weather Research and Forecasting (WRF) model, which provides relatively accurate wind estimates for analysis on flat and homogeneous flat terrain [13, 14]. For higher terrain complexity, WRF requires more detailed terrain data [11]. As a mesoscale model, NWP models are commonly coupled to microscale wind flow model to obtain a higher spatial resolution and accuracy [4].

Our resource assessment lays on an assessment model built from WRF Four-Dimensional Data Assimilation System (WRF-FDDA) FNL and CCMP dataset. These two approaches could provide more accurate and higher resolution data without requiring huge computation resources. The finding of our study calls for more discussions about opportunities of wind energy investments in Indonesia as detailed out in section 5.

3. Methodology

We estimated wind energy potentials by constructing an atmospheric mesoscale model based on a regional wind map. We improved the model by measured wind resource data coupled with a data assimilation technique. For this purpose, we used the atmospheric mesoscale WRF model with a spatial resolution of 5 x 5 km to map wind resources at 50 meters (m) above ground level (agl). The outer domain has a horizontal resolution of 27 km and the resolution for the inner domain is 5 km. It has 35 vertical levels and the lowest crucial levels are at around 10, 30, 52 and 97 m agl. Initial and boundary conditions are from FNL datasets with a spatial resolution of 1 x 1°.

We assimilated ocean surface wind data provided from the CCMP satellite data by coupling numerical model with the Newtonian relaxation technique. The CCMP data contains high-resolution wind data generated from the integrations of wind measurements from Remote Sensing Systems (REMSS) satellites and Variational Analysis Method (VAM). The CCMP surface winds dataset contains 0.25° gridded ocean surface wind. We then added prognostic equations nudging the predicted variables toward available observation data (interpolated in each model grid). The nudging equation:

$$\frac{\partial \emptyset_m}{\partial t} = \frac{(\emptyset_{obs} - \emptyset_m)}{\tau} \quad (1)$$

where $\emptyset_m$ is the variable of prognostic model, $\emptyset_{obs}$s is the measured variable, and τ is the time scale of a relaxation. The spatial variation of nudging is:

$$f(r) = e^{-r/r_0} \quad (2)$$

where r is the distance from the measuring point, and r_0 is a reference distance representing the nudging range. The weight of the nudging is obtained by multiplying Equations (1) and (2). As a result, the selected optimal relaxation time scale is one hour and the selected nudging radius is 25 km. We then convert the estimated wind speed data into wind energy potentials by assuming that a 1 MW wind

IOP Conf. Series: Earth and Environmental Science **897** (2021) 012004 doi:10.1088/1755-1315/897/1/012004

turbine is for one hectare of land with wind speed above 6 m/s and a 100-kW wind turbine is for one hectare of land with wind speed between 4 – 6 m/s.

4. Results

Fig. 1 shows the modelled annual mean wind speed at 50 m agl by using the Geographic Information System (GIS). We classified the wind speed into eight speed classes from below 3 m/s (i.e., the lowest class) represented by green colour to higher than 9 m/s (i.e., the highest class) represented by red colour. Wind resources in coastal areas are extremely high exceeding 5 m/s, which is the average cut-in wind speed for many commercial wind turbines [15, 16]. Furthermore, Fig. 1 shows that the wind speed in several west parts of Indonesian offshore areas is over 7 m/s.

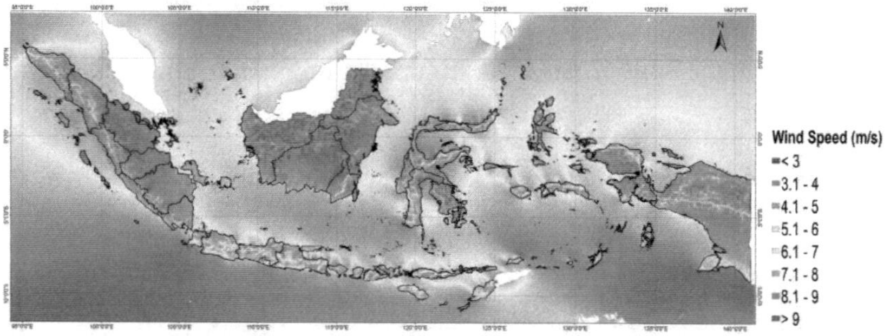

Fig.1. Indonesia global wind speed at 50 m height - resolution 5 km.

We evaluated the model by using wind measurement data from three meteorological masts (met masts) in University of Cendrawasih (Uncen) - Jayapura, Tamanjaya - West Java, and Bantaeng - South Sulawesi. The met masts observed climatological and weather conditions at a height of 50 m agl. Table 1 summarize the sites of met mast and the measuring period. Fig. 2 compares the average hourly data between the measured data (i.e., blue line) and the modelling results (i.e., red line). The average measured wind speeds are 2.33 m/s in Jayapura, 6.70 m/s in Sukabumi, and 4.66 m/s in Bantaeng, Meanwhile the modelled wind speeds are 2.78 m/s in Jayapura, 7.20 m/s in Sukabumi, and 5.44 m/s in Bantaeng.

Table 1. The coordinate of met mast and measuring periods

Site	Coordinate		Measuring periods
Jayapura	-2.58249	140.6575	October 2005 to December 2006
Sukabumi	-7.2688	106.5288	January to December 2008
Bantaeng	-5.5825	120.0475	June to December 2006

Table 2 summarizes statistical parameters of bias, root mean square error (RMSE), and correlations between the modelled- and observed- wind data. The bias is below 0.5 m/s for all sites or lower than the maximum bias (i.e., + 0,5) proposed by Emery, Tai and Yarwood [17]. The RMSE for Bantaeng is the highest among the other sites while the correlations for all sites are between 0.6 and 0.7. The WRF model produces overestimated results at low wind speeds and underestimated results at wind speed above 9 m/s for all sites.

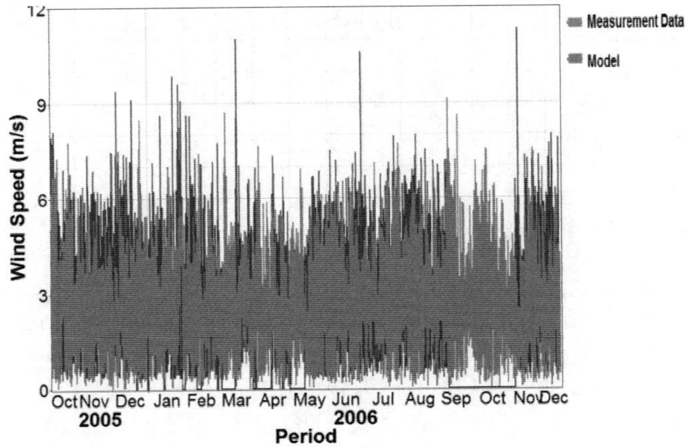

(a) Jayapura site

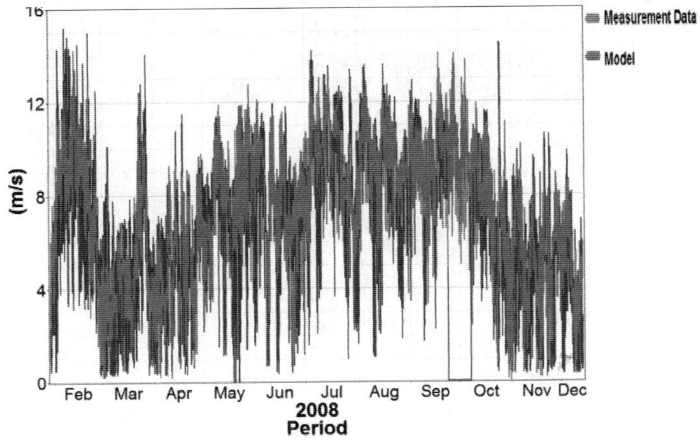

(b) Sukabumi site

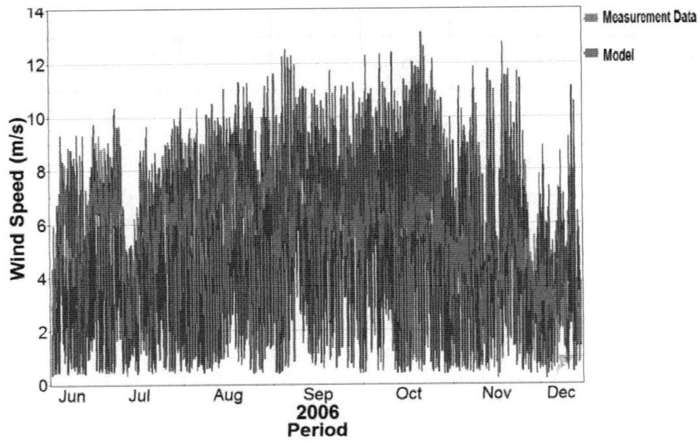

(c) Bantaeng site

Fig 2. Measured wind speed (blue color) and modelled wind speed (red color) at 50 m agl in three sites

Table 2. Bias, RMSE and Correlation of Model Results and Actual Data

Site	Bias	RMSE	Correlation
Jayapura	0.324	2.300	0.734
Sukabumi	0.368	2.373	0.736
Bantaeng	0.324	2.870	0.667

Comparing geographical characteristics on different sites can identify terrain effects on wind distribution data. Wind speed distributions and their Weibull-fits of the observed and modelled wind speed at 50 m agl for each mast are different for each location. Sukabumi data has the widest range of wind speeds while the lowest range of wind speed is in Jayapura site. The channeling in Sukabumi is clearly visible; however, the modelled wind speeds are lower than in the observations. One of the possible causes is that terrain model is too smooth. In Bantaeng, most observed data is low wind speed while the modelled data is mostly high wind speed. The WRF model cannot precisely replicate wind directions in Bantaeng even though the terrain is flat. The observation data has higher wind speeds and more northerly component than the modelled data. The deviation is due to the inaccurate representations of surface roughness elements like forests and buildings. Therefore, the representation in numerical weather models should be improved for this specific site. The application of a roughness length alone is not enough to characterize the interaction of atmospheric flows with the surface. The modelled wind speed for Jayapura is stronger than the observed data, but the wind direction in both modelled and observed data is similar. The modelled data appears to capture the directional wind distribution of observation data quite well.

5. Policy Implications

We convert the average wind speed data into wind energy potentials. The potential data is then used by RUEN [18]. The largest potentials are in electricity systems of Java, Madura, and Bali (JAMALI) (24,011 MW) and East Nusa Tenggara (10,188 MW). RUEN has expected to build wind farms with total capacity for 716 MW and 266 MW in those regions respectively [18]. Yet, the feasibility of wind farm should consider at least two other factors that are grid flexibility and electricity tariffs.

The JAMALI system is the largest electricity grid system with total electricity supply 149,9 TWh in 2019. Currently, the main electricity supply in the JAMALI system is mainly coal-fired power plants and combined cycle gas turbine (CCGT) for 70.3% and 22.6% of respectively [19]. The JAMALI system does not have neither solar farms nor wind farm yet; however, PV rooftop in the system is growing with installed capacity 4,849 kWp in December 2019 [20]. Moreover, RUEN [18] also expects that the JAMALI system will have wind farms with total capacity 716 MW by 2025. However, wind farm investment in JAMALI system is less interesting since the average generation cost of the State-owned Electricity Company (PLN) in the JAMALI system is only 6.91 ¢US$/ kWh, which is the ceiling price for PLN to buy renewable-based electricity from independent power producers (IPP). The regulation of Minister of Energy and Mineral Resources (MEMR) No. 53 /2018 [21] and No. 50 /2017 [22] sets PLN regional generation cost in previous year as the ceiling price for regions with generation costs lower than the PLN's average national generation cost, i.e., 7.86 ¢US$/ kWh in 2019. In contrast, East Nusa Tenggara systems have average regional generations costs (i.e., 17.58 ¢US$/ kWh) higher than the average national generation cost. East Nusa Tenggara actually has several separated grid systems with different generation costs. The largest system (and the generation cost in 2019) are Sumba system (20.81 ¢US$/ kWh), Timor system (18.17 ¢US$/ kWh), West Flores (17.58 ¢US$/ kWh), and East Flores (21.28 ¢US$/ kWh). The generation costs for smaller systems in Nusa Tenggara reached 21.34 ¢US$/ kWh in 2019. For such regions, MEMR [21] and MEMR [22] allow PLN to buy renewables-based electricity at maximum 85% of regional electricity production costs. It means that the ceiling tariff will be between 14.94 and 18.14 ¢US$/ kWh, which are higher than average levelized cost of energy (LCOE) of wind power plants around the world, i.e., 4.6 to 9.9 ¢US$/ kWh in 2019 [23]. However, by assuming VRE only can supply 20% of total electricity productions, East Nusa Tenggara only can take 200 GWh electricity generated from wind turbine. If

IOP Conf. Series: Earth and Environmental Science **897** (2021) 012004 doi:10.1088/1755-1315/897/1/012004

the capacity factor of wind turbine is 35% [23], total capacity of wind farms that could be installed in East Nusa Tenggara is around 65 MW. Beyond the attractive ceiling tariff, the Indonesia government offers three incentives that are:

a. Import duty exemptions for two years that can be extended for one year [24];
b. Tax holiday up to 20 years [25];
c. Tax allowance [26].

6. Conclusions

This study uses WRF and GIS models to estimate theoretical wind energy resources in Indonesia. The modelled data is then validated by using empirical data measured by three met masts in Jayapura, Bantaeng and Sukabumi. As results, the WRF model is reliable to estimate mean wind speeds in all Indonesia provinces. The wind speeds, presented in a GIS map, are useful information for wind energy planning in national and regional levels. It is the first map for Indonesia context and it also has been used officially by the General National Energy Plan [18]. Our WRF model validated by three measurement data is the initial stage to provide more robust wind maps for the entire of Indonesia. In general, the modelling slightly overestimates the wind speed and the deviations are related to local topographical feature and at low- high wind speed.

This analysis emphasizes that the model outputs have realistic meteorological patterns. Once more data is available, future studies should conduct spatial analysis by points by using other prominent methods such as Inverse Distant Weight and Kriging in GIS environment. Moreover, wind energy potential data should be extended to offshore wind energy potentials

7. References

[1] Al Irsyad M I, Halog A, Nepal R and Koesrindartoto D P 2020 Economical and environmental impacts of decarbonisation of Indonesian power sector *J. Environ. Manage.* **259** 109669

[2] Martosaputro S and Murti N 2014 Blowing the wind energy in Indonesia *Energy Procedia* **47** 273-82

[3] Carvalho D, Rocha A, Gómez-Gesteira M and Santos C S 2014 WRF wind simulation and wind energy production estimates forced by different reanalyses: Comparison with observed data for Portugal *Appl. Energy.* **117** 116-26

[4] Beaucage P, Brower M C and Tensen J 2014 Evaluation of four numerical wind flow models for wind resource mapping *Wind Energy* **17** 197-208

[5] Romanic D, Hangan H and Ćurić M 2018 Wind climatology of Toronto based on the NCEP/NCAR reanalysis 1 data and its potential relation to solar activity *Theor. Appl. Climatol.* **131** 827-43

[6] Hesty N W and Hadi T W 2015 Reanalisis weather research and forecast-four dimension data assimiliation (WRF-FDDA) untuk meningkatkan akurasi estimasi potensi energi angin di daerah lepas pantai (studi kasus: pantai selatan Jawa Barat) *Jurnal Sains Dirgantara* **13**

[7] Buhr R, Kassem H, Steinfeld G, Alletto M, Witha B and Dörenkämper M 2021 A Multi-Point Meso–Micro Downscaling Method Including Atmospheric Stratification *Energies* **14** 1191

[8] Sharma P K, Warudkar V and Ahmed S 2019 A comparative analysis of wind resource parameters using WAsP and windPRO *Int. J. Green Energy.* **16** 152-66

[9] Mentis D, Siyal S H, Korkovelos A and Howells M 2016 A geospatial assessment of the techno-economic wind power potential in India using geographical restrictions *Renew. Energy.* **97** 77-88

[10] Feng C, Cui M, Hodge B-M and Zhang J 2017 A data-driven multi-model methodology with deep feature selection for short-term wind forecasting *Appl. Energy.* **190** 1245-57

[11] Carvalho D, Rocha A, Gómez-Gesteira M and Santos C S 2014 Offshore wind energy resource simulation forced by different reanalyses: comparison with observed data in the Iberian Peninsula *Appl. Energy.* **134** 57-64

[12] Weekes S, Tomlin A, Vosper S, Skea A, Gallani M and Standen J 2015 Long-term wind resource assessment for small and medium-scale turbines using operational forecast data and measure–correlate–predict *Renew. Energy.* **81** 760-9
[13] Mughal M O, Lynch M, Yu F, McGann B, Jeanneret F and Sutton J 2017 Wind modelling, validation and sensitivity study using Weather Research and Forecasting model in complex terrain *Environ. Model. Softw.* **90** 107-25
[14] Westerhuis S, Fuhrer O, Bhattacharya R, Schmidli J and Bretherton C 2021 Effects of terrain-following vertical coordinates on simulation of stratus clouds in numerical weather prediction models *Q. J. R. Meteorol. Soc.* **147** 94-105
[15] Akour S N, Al-Heymari M, Ahmed T and Khalil K A 2018 Experimental and theoretical investigation of micro wind turbine for low wind speed regions *Renew. Energy.* **116** 215-23
[16] Li H and Chen Z 2008 Overview of different wind generator systems and their comparisons *IET Renewable Power Generation* **2** 123-38
[17] Emery C, Tai E and Yarwood G 2001 Enhanced meteorological modeling and performance evaluation for two Texas ozone episodes. (California, US: ENVIRON International Corporation) p 235
[18] GOI 2017 Presidential Decree Number 22 of 2017 on the National Energy General Plan (Jakarta, Indonesia: Government of Indonesia (GOI))
[19] DGE 2020 Statistik Ketenagalistrikan Tahun 2019. (Jakarta, Indonesia: Directorate General of Electricity (DGE) - Ministry of Energy and Mineral Resources (MEMR)) p 122
[20] DGE 2020 Kebijakan Pemanfaatan Pembangkit Listrik Tenaga Surya (PLTS) Atap. (Jakarta, Indonesia: Directorate General of Electricity (DGE) - Ministry of Energy and Mineral Resources (MEMR)) p 19
[21] MEMR 2018 Peraturan Menteri Energi dan Sumber Daya Mineral Nomor 53 Tahun 2018 tentang Perubahan atas Peraturan Menteri Energi dan Sumber Daya Mineral Nomor 50 Tahun 2017 tentang Pemanfaatan Sumber Energi Terbarukan untuk Penyediaan Tenaga Listrik. (Jakarta: Ministry of Energy and Mineral Resources (MEMR))
[22] MEMR 2017 Peraturan Menteri Energi dan Sumber Daya Mineral Nomor 50 Tahun 2017 tentang Pemanfaatan Sumber Energi Terbarukan untuk Penyediaan Tenaga Listrik. (Jakarta: Ministry of Energy and Mineral Resources (MEMR))
[23] IRENA 2020 Renewable Power Generation Costs in 2019. (Abu Dhabi, UAE: International Renewable Energy Agency (IRENA)) p 160
[24] MoF 2015 Peraturan Menteri Keuangan Nomor 66/PMK.010/2015 tentang Pembebasan Bea Masuk atas Impor Barang Modal dalam Rangka Pembangunan atau Pengembangan Industri Pembangkit Tenaga Listrik untuk Kepentingan Umum. (Jakarta, Indonesia: Ministry of Finance (MOF))
[25] MoF 2020 Peraturan Menteri Keuangan Nomor 130/PMK.010/2020 tentang Pemberian Fasilitas Pengurangan Pajak Penghasilan Badan. (Jakarta, Indonesia: Ministry of Finance (MOF))
[26] MoF 2020 Peraturan Menteri Keuangan Nomor 11/PMK.010/2020 tentang Pelaksanaan Peraturan Pemerintah Nomor 78 Tahun 2019 tentang Fasilitas Pajak Penghasilan untuk Penanaman Modal di Bidang-bidang Usaha Tertentu dan/ atau di Daerah-Daerah Tertentu. (Jakarta, Indonesia: Ministry of Finance (MOF))

Acknowledgement
N.W.H and D.G.C. acknowledge funding supports from the Research and Development Center for Electricity, New-Renewable Technology and Energy- Ministry of Mineral and Energy Resources, Republic of Indonesia. The authors also wish to thank those who aided this research, particularly Luthfi Imanal Satrya (Bandung Institute of Technology) and Michael Hölling (University of Oldenburg) who provided insightful comments.

IOP Conf. Series: Earth and Environmental Science **897** (2021) 012005 doi:10.1088/1755-1315/897/1/012005

Optimizing the size of a hybrid renewable energy system using various types of storage technologies

Noé Fernando Proaño[1], Mayra Salazar Román [1], Lata-García Juan[1]

[1]Universidad Politécnica Salesiana, Electrical Department, Guayaquil, Ecuador

nproano@est.ups.edu.ec

Abstract. Hybrid renewable energy (HRE) systems without connection to the grid have a series of drawbacks due to the lack of reliability in the predictability of the climate, for this reason it is necessary to achieve the optimal sizing of the system components together with reliable storage systems to ensure the correct energy supply. At present commercially there are several storage technologies, lead acid batteries are the most used however they are losing space compared to lithium ion in stationary applications despite the uncertainties for cost and profitability, another technology is pumped hydro, appears as a clean solution unlike the previous ones. In this document the hybrid system composed of photovoltaic panels, a biogas generator and the three types of storage individually by means of technological tools is optimized, the result of the techno-economical simulation indicates that the system with pumped hydro has the lowest energy cost (COE) with 0.299 $/kWh, while the system with acid lead batteries has the highest energy cost at 0.401 $/kWh but has the lowest initial project cost 1.05 M$, lithium ion batteries reach a longer life time.

1. Introduction

The Paris agreement signed on December 12, 2015 by 197 countries, seeks to substantially reduce greenhouse gas emissions and limit the increase in temperature to 2 ° C in the coming years [1], during the discussion of the agreement consensus was reached between scientists and politicians on the depletion of non-renewable resources and to continue with the rate of environmental pollution will reach catastrophic consequences in the coming decades. One of the main pollution sectors is energy, therefore short-term actions must be taken to mitigate the impact of energy consumption even more when evidence reveals that current consumption rates are not sustainable in the long term [2].

On the other hand, in isolated communities the use of generators based on fossil fuels is becoming more common, which is not viable due to environmental issues, in addition to the volatile costs of fuels and the difficulty in transport leads to be unattractive. The renewable energies are a solution to the energy problem and to reduce environmental pollution, in recent years with the advancement of technology in renewable generation systems has been promoted the increase in the implementation of solar plants, wind farms and biomass-based generators with marginal energy costs, but for small isolated hybrid systems the cost is not yet competitive due to the price of the storage components [3].

The storage system has two functions, one is to store energy of the renewable generators when the consumer does not need and the second is to deliver the stored energy when the generators cannot meet the total demand in the charge, additional has features such as smoothing the instabilities of the network due to the intermittency of several meteorological variables and improve the quality of the energy that is directly linked to the sizing of the storage bank.

IOP Conf. Series: Earth and Environmental Science **897** (2021) 012005 doi:10.1088/1755-1315/897/1/012005

One of the first challenges in the implementation of generation systems is the technical dimensioning due to the intermittency of renewable sources so it becomes a hard task, because the low predictability and sudden changes of the weather, in the literature of high impact has been proposed methodologies to follow, optimization algorithms together with specialized software's to achieve the optimal sizing [4, 5].

In the reference [6] the authors indicate that in a micro-grid is the right sizing of the battery bank therefore the system will have a better performance and efficiency unlike non-optimized renewable energy systems, the authors propose a new methodology to found the optimal size where storage technology, life time and degradation of the equipment are involved, the proposed model uses linear programming of mixed integers (MILP), the results are verified by a numerical simulation with historical meteorological and load data.

Mulleriyawage [7] in his research describes that residential battery banks are in except because of the high market price, the optimal size can reduce costs so it can be economically reasonable, the aim of the work is to minimize the price of energy, including the total price of installation based on the maximization of self-consumption, the case study was carried out in Australia, the results show that the implementation is profitable with current market prices as long as the current government incentives are maintained according to the capacity of the SF, a tariff scheme for time of use is more convenient than a self-consumption system , with a better return on investment, lower net current cost and reducing the emission of polluting gases.

Fareeha [8] presents his research work where he aims at the economic technical design of a HRE system for remote areas, the optimization problem has been solved by means of the HOMER software depending on the management of demand and the type of storage, in the document the performance of four cases is evaluated for the solution of the problem posed, the resulting system will result in the lowest overall cost of the system increasing the efficiency and performance of the same.

Various software's are used around the world which are a powerful tool for the sizing, analysis, feasibility study of hybrid systems [9–11].

According to the above, the main objective of the work is the optimal sizing of the hybrid system with different types of storage taking advantage of the resources of the locality, the comparative analysis of the different scenarios based on the price of energy, initial cost, reliability and energy balance. The article is structured as follows: the second section presents the description of renewable sources and the model of the hybrid system. The third section describes the system optimization for each of the selected technologies. The fourth section discusses the results obtained. The final part includes the conclusions of the work

2. Description of the system

The hybrid system without connection to the grid will meet the energy needs of the loads and offices of the solid waste dump of the Iguanas in the city of Guayaquil, being the main place of collection of urban solid waste in the country.

The characteristic curve is of commercial type, this is presented in Figure 1, the estimated power is 438.9 kW, during the night the load is low, in the day it is where there is the greatest demand.

The work aims to optimizer the components of the hybrid system of renewable energies using different storage technologies to know the competitive and economic advantages of the systems, to achieve the objective three types of technologies are used, in the first analysis acid lead batteries are used which are the most used, the second analysis is by lithium ion batteries, batteries widely used in portable devices and the third analysis is by means of hydraulic pumping, which is a technology that makes its way in the market has the advantage of not emitting polluting's.

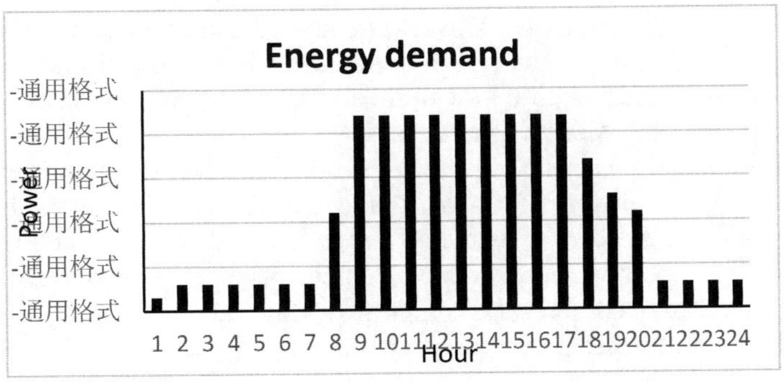

Figure 1. Characteristic demand curve.

2.1. Solar radiation

The solid waste dump of the Iguanas is located at coordinates -2.083, -79.954, solar irradiance values are from NASA, the site under study has an important solar resource with an average of 4.58 kW/m2/day. The monthly irradiance data are shown in Figure 2.

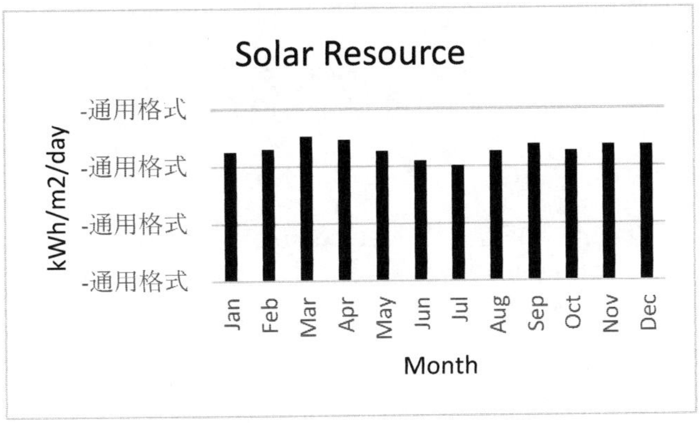

Figure 2. Monthly solar resource values

2.2 Modelling of the photovoltaic system

Photovoltaic solar panels convert solar energy into electricity through the photoelectric effect. The output energy of the photovoltaic field can be calculated using equation 1 [5]:

$$E_{PV} = E_{RPV} \times PR_{PV} \times \left(\frac{G_T}{G_{T,STC}}\right) \times \left[1 + \beta_P \times \left(T_C - T_{C,STC}\right)\right] \tag{1}$$

Where; E_{RPV} is the nominal power of the photovoltaic generator (kW), PR_{PV} is the reduction factor (%), G_T is the solar irradiance over the receiving plane (kW/m2), $G_{T,STC}$ is the solar irradiance under standard test conditions (STC), βp is the temp coefficient (%), T_C is the temp of the photovoltaic cell, $T_{C,STC}$ is the temp under STC.

2.3 Biomass

Municipal solid waste (MSW) is garbage or waste generated by urban areas, waste is composed of organic waste from food, paper, wood and inorganic waste such as glass, plastic and metals, depending on the recycling processes implemented by local governments, certain household waste arrives at the landfill, these wastes are not classified as hazardous.
There are two types of landfills according to the collection of waste, the organized landfill horizontal pipes are installed for the collection of biogas, while the vertical pipes are placed at the time of accumulating the landfill, the gas can be extracted between 8 and 12 years. While in the unorganized landfill the filling is placed without horizontal or vertical pipes, which have to be placed when the capacity of the landfill is completed, the top is filled with concrete, the biogas extraction time is 3 to 7 years. Fuel gas can be used in internal combustion engines [12], [13].
Agreements between caf and the municipality of Guayaquil will allow the use of methane gas from the Iguanas dump, the project consists of the design of incorporation of 65 wells for the extraction of gas reducing 260,000 tons of carbon dioxide equivalent per year [14], the city of Guayaquil has a production of 3419 tons / day of solid waste [15].

2.4 Modeling of gasifying biomass
Biogas is produced by bacteria until it finish the process of biodegradation of matter in the absence of air or anaerobic, this process is natural in landfills controlled by urban solid waste, methane is the main component of the gas, the energy value total of biomass is dependent by the concentration of gas methane. The power generated is obtained by equation 2.

$$P_{MTg} = \frac{Q_{BM} \times \eta_{BMG} \times CV_{BM} \times 1000}{DOH_{BMG} \times 365 \times 860} \tag{2}$$

Where, Q_{BM} is the amount of annual biomass measured in tons, ηBMG is the throughput of the biomass generator, the calorific value of the matter is CV_{BM}, DOH_{BMG} is the daily hours of operation of the generator and 860 is the conversion value from kcal to kWh.

2. 5 Types of storage technologies
The correct energy storage allows the flexibility and reliability of the system, several technologies have been developed to drive the transition to decarbonized electrification. Each one offers certain advantages regarding the design and type of function to be performed.

2.5.1 Lead acid batteries. This type of batteries have been widely used for HRE systems both in the energy backup and the stability of the grid, these elements are composed of two electrodes immersed in an electrolyte so that chemical reactions occur, currently according to the technology used they can be divided into several categories.
Lead-acid batteries are the oldest rechargeable and widely used due to the low cost and a considered number of operating cycles but among the disadvantages is the low energy density, a low lifetime compared to another type of battery and the explosive gases it emits.
The loading process can be calculated using equation 3.

$$E_{Bat}(t) = (1 - \alpha) \times E_{Bat}(t-1) + \left(E_G(t) - \frac{E_L(t)}{\eta_{Conv}}\right) \times \eta_{CC} \times \eta_{rbat} \tag{3}$$

2.5.2 Hydraulic Pumping. It is one of the clean storage technologies, and the principle of operation of hydraulic pumping is to drive water from a lower tank to a higher one when you have an excess of energy, when the load requires energy the water of the upper tank falls moving a microturbine producing

IOP Conf. Series: Earth and Environmental Science **897** (2021) 012005 doi:10.1088/1755-1315/897/1/012005

electrical energy. It is a mature technology which does not involve environmental pollution [16], to calculate the energy needed to pump the fluid from the lower tank equation[16] (4) is used, the volume of fluid is V, h is the distance from the upper tank, ηp is the throughput of the pumping system while the energy generated by the microturbine can be calculated using equation (5), where ηg is the generation t throughput.

$$E_{pumping} = \frac{\rho \,.\, g \,.\, h \,.\, V}{\eta_p} \tag{4}$$

$$E_{generator} = \rho \,.\, g \,.\, h \,.\, V \,.\, \eta_g \tag{5}$$

2.5.3 Lithium batteries. Lithium batteries is the result of technological development (LiFePO4), batteries have some advantages over their competitors such as not suffering from the memory effect and being able to charge at any time of the state of charge, they have a higher density of capacity and storage which has as a consequence a longer life time reaching up to 15 years. The maintenance is almost zero without emitting polluting gases by not presenting self-discharge.
The energy total storage by the battery bank can be calculated using equation 6. Where $B_{bat.0}$ is the initial charge of the battery bank, V_{bat} is the battery voltage, I_{bat} is the battery current.

$$B_{bat} = B_{bat.0} + \int V_{bat} .\, I_{bat}\, dt \tag{6}$$

3. Simulation and results

The different configurations according to the selected technologies are simulated in HOMER to achieve optimal sizing, the result of each of the systems is evaluated by three parameters 1) price of energy for kWh, 2) the cost of capital starts l which are the economic resources that are needed to start the project, 3) NPC, is the present value of all component costs over the life of the project.

3.1 System proposal

The selected solar panels are from the Jinko eagle per 60/300w with an efficiency of 18.33%. The biogas generator is from the manufacturer Jenbacher JGC 420 B81 with an electrical power of 1,416 kW. The inverter is from the manufacturer Fronius Symo with a power of 20 kW per unit.

3.1.1 PV, Biogas, Lead Acid Battery system. The scheme of the first proposed system to be studied is shown in Figure 3, the resulting optimal system is conformed by a photovoltaic matrix of 123 kW, 1 generator Jenbacher J420,493 batteries of the manufacturer Trojan SSIG 12 120 has a capacity of 118 Ah and 3 converter of 20 kW mark Fronius.

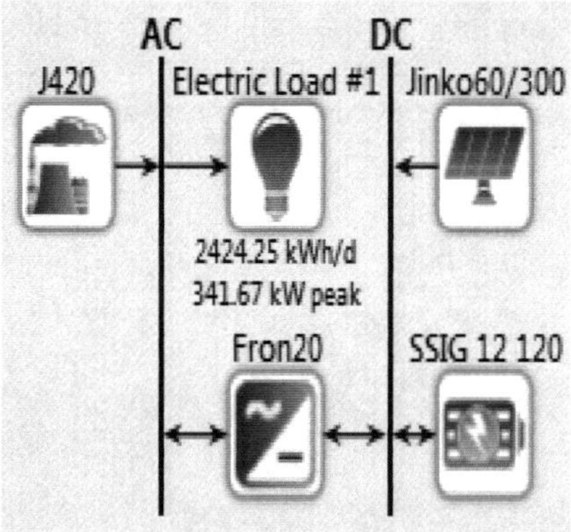

Figure 3. Proposal of the hybrid Pv, Biogas, Pb system

The photovoltaic array has an average output of 19. 6 kW, with a maximum output of 116 kW, the total production is 172,021 kWh/yr, the cost of energy is 0.0959 $/kWh, the biogas generator has an electricity production of 1.738,294 kWh/yr starting 386 times a year, the operational life time is 8.20 yr, the efficiency of the equipment is 13. 9%, the operating cost is 18.6$/hr and the marginal operating cost is 0.0759 $/kWh. El battery bank is composed of 493 units with a nominal capacity 698 kWh, the storage system has a range of5.53 h, the life expectancy is low with 1.78 years, delivering 193,256 kWh over the operating time, the energy stored in the battery is 1 21,192 kWh/yr, the losses in the bank are 24,278 kWh/yr.

The optimization achieves the following economic results, the total cost of installation is 4. 58M USD (NPC), initial capital is 1. 05M USD, the annual operating cost is 273,188 $/yr and the levelized price of energy (COE) is 0.401 USD/kWh.

During the entire time of operation of the project, the element that has a higher cost is the biogas generator of 3,211,192.24 $ followed by the group of batteries with a total cost of 1,136,952. 30$ and also with the cost per operation and maintenance with 127,465.31$

3.1. 2 PV, Biogas, Lithium system. The system is made up of a photovoltaic generator of 120 kW, 1 biogas generator Jenbacher J420, 375 lithium batteries with a capacity of 100 Ah, 5 inverter of 20 kW Fronius, in figure 4 the configuration of the system is shown. The total energy production of the system is 1,873,911 kWh/yr, has an excess energy of 982,900 kWh/yr, the unsatisfied load is 0% which guarantees the energy supply.

The photovoltaic generator has an average output of 19.1 kW during the day, the maximum output of the generator is 113 kW, the total production is 167.426 kWh/yr, the energy cost of0.0959 $/kWh with a yield of 15.9%, the biogas generator has an electricity production of 1.706,566 kWh/yr with an average output power of 356 kW, the equipment starts 407 times a year, the life time of the generator is 8.36 yr, the efficiency of the equipment is 13. 7% and the marginal cost of operation is 0.0759 $/kWh. The lithium battery bank is composed of 375 units with a nominal capacity of 480 kWh, the battery bank has a range of 4.75 h, the life expectancy is high with 16 years, delivering 1,680,000 kWh during the operation of the project, the energy stored in the battery is 106,970 kWh /yr, the losses in the bank are minimal compared to another type of storage with 4.284 kWh /yr.

The total price of the components where it includes the initial capital, change of equipment and maintenance is 3,925,872.12 dollars, the initial capital is 1.29M operating cost is 203,513 dollars and the price of energy is 0.343 $/kWh. The equipment that has the highest cost is the J420 generator with 3,157,197.94 followed by the battery bank with 531,485.86, the equipment that has a higher cost per maintenance operation is the biogas generator with 61,884.02 dollars.

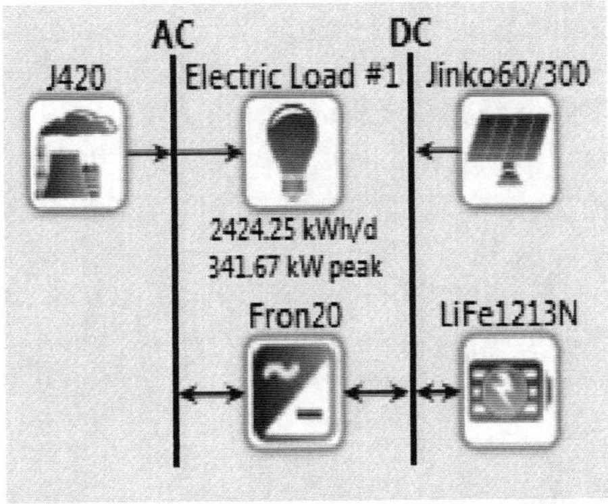

Figure 4. Proposal of the hybrid Pv, Biogas, Li-ion system

3.1.3 PV, Biogas, Pumped hydro system. The third system consists of a photovoltaic generator of 135 kW, 1 biogas generator Jenbacher J420, 5 pumped hydro 245 kWh, 5 inverter of 20 kW Fronius, the scheme is presented in Figure 5. The total production of the system is 1,741,495 kWh /yr of which 1,552,914 kWh corresponds to the biogas generator, the excess energy is 824,753 satisfying the load at 100%.

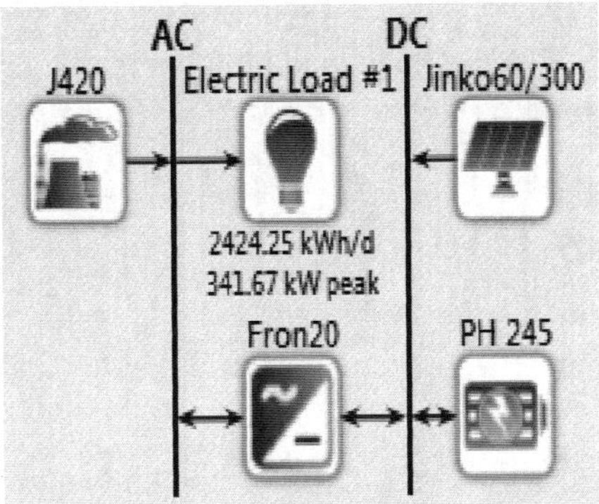

Figure 5. Proposal of the hybrid Pv, Biogas, Ph system

Table 1. Results of different systems hybrid

Systems	**COE ($)**	**NPC ($)**	**Operating cost ($)**	**Initial capital ($)**	**Renewable Production (kWh)**	**Losses Storage (kWh)**	**Autonomy Storage (h)**	**Annual Throughput Storage (kWh)**	**Expected life Storage (yr)**
PV/Biogas/B Lead Acid	0.401	4.58 M	273,188	1.05 M	1,910,315	24,278	5.53	108,722	1.78
PV/Biogas/B Lithium	0.343	3.93 M	203,513	1.29 M	1,706,566	4,284	4.75	105,060	16
PV/Biogas/Pump Hydro	0.299	3.42 M	182,863	1.06 M	1,741,495	29,511	12.6	11,309	7

The photovoltaic system has an average of 21.5 kW the total production is 188,581 kWh/yr, the energy cost of0.0959 $/kWh with a yield of 15.9% running for 4,404 hrs/yr, the biogas generator has an electricity production of 1,552,914 kWh/yr starting for 443 times a year with a yield of 12.4%, the average output power of the J420 is 356 kW, the operating life time is 9.18 yr, while the marginal cost of operation is 0.0759 $/kWh.

IOP Conf. Series: Earth and Environmental Science **897** (2021) 012005 doi:10.1088/1755-1315/897/1/012005

The storage system is composed of 5 hydraulic pumping units of 245 kWh, the system has 12.6 hr of autonomy, the nominal storage capacity is 1,271 kWh, the energy delivered is 979,982 kWh for the 7 years of operation, the energy lost in the system is 29,511 kWh /yr.
The optimization achieves the following economic results, the total cost of installation is 3.42M USD (NPC), the initial capital is 1.06M USD, the annual operating cost is 182,863 $/yr and the level cost of energy (COE) is 0.299 USD/kWh.
The total cost of the project is 3,424,427 dollars, the element that has a higher cost is the biogas generator of 242,569.93 $ followed by hydraulic pumping but with a large difference of about 90% with a total cost of 1,136,952.30 $ but the highest cost is in the area of operation and maintenance with 129,275.17 dollars.

4. Economic result of the systems

In the Table 1 presents the result of the different systems analysed, the scheme formed by hydraulic pumping achieves the best results with the lowest cost of energy and a considerable life of the system, the state of charge (SOC) indicates the energy stored in the storage system, the indicator is presented in percentage, in the Figure 6 the SOC of the pumped hydro is shown, when it is at 100% the upper reservoir is completely full, when it reaches 0% the upper reservoir is empty therefore it has no energy backing in case of needing the load.

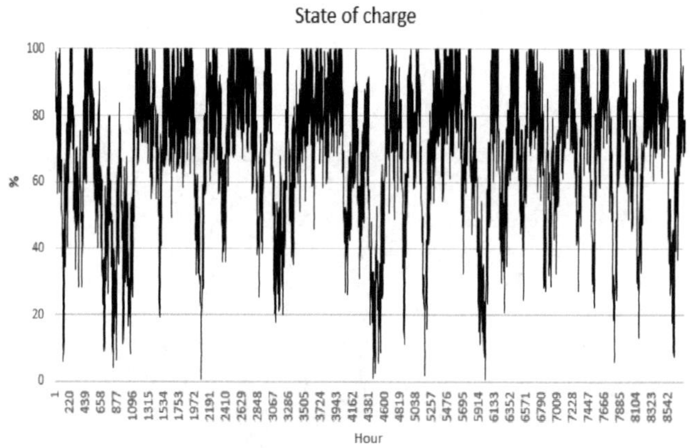

Figure 6. Generic 245 kWh Pumped Hydro State of Charge

In few occasions the load consumes 100% of the energy stored in the reservoir, while in the year the stored energy is at 100%, figure 6 shows in positive the power input that enters the reservoir of the system and in negative the power that is delivered by the reservoir.

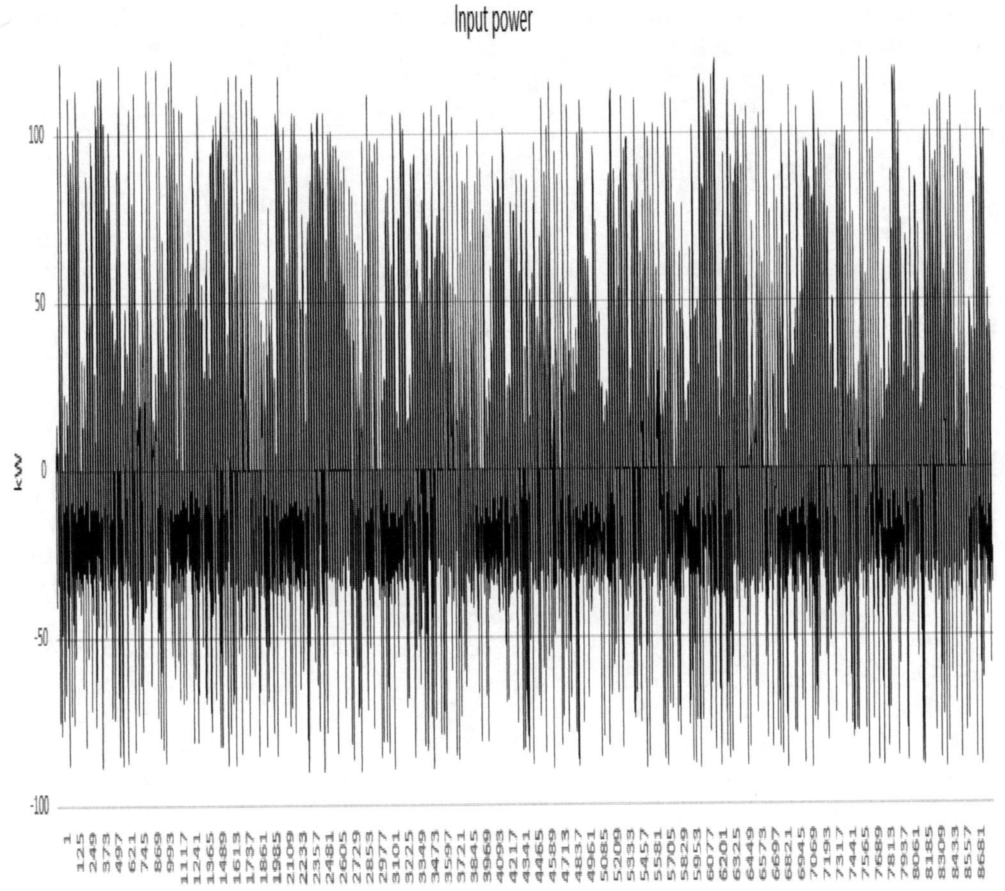

Figure 7. Generic 245 kWh Pumped Hydro Input power

4.1 Comparison of greenhouse gas emissions

The first system made up of lead acid batteries emits 850,969 kg / yr of carbon dioxide, 2,830 kg / yr of carbon monoxide, 79.8 kg / yr of particulate matter and 5,937 kg / yr of nitrogen oxides into the environment. The second system made up lithium of lithium batteries emits 835,437 kg / yr of carbon dioxide, 2,778 kg / yr of carbon monoxide, 78.3 kg / yr of particulate matter and 5,829 kg / yr of nitrogen oxides. The third system made up of pumped hydro emits 760,218 kg / yr of carbon dioxide, 2,528 kg / yr of carbon monoxide, 71.3 kg / yr of particulate matter and 5,304 / yr of nitrogen oxides. The last system analysed has the best indicators regarding the gases emitted into the environment

5. Conclusions

In the article the optimization of the hybrid system taking advantage of the generation of biogas from municipal solid waste is analyse, the behaviour of three types of storage technology is compared and evaluated, the first analysis was carried out with lead acid batteries having the highest cost of energy with 0.401 $ / kWh, a lower expectation in the life time with 1.78 years, however it presents the lowest initial cost of the system so it remains one of the preferred ones, it also has very high energy losses with 24278 kwh/yr ,the second technology analysed to are lithium batteries, the leveled cost of energy is 0.343 $/kWh , has the highest initial cost of the three systems studied with 1.29 M USD, but also has the highest life expectancy with 16 years.

The best result has the hydraulic pumping system with a photovoltaic generator of 135 kW, a biogas generator J420, 5 hydraulic pumping systems, the cost of energy is 0.299 $/kWh.

The size of the photovoltaic generator varies according to the type of storage system, with lithium batteries the smallest PV size is achieved.

6. References

[1] Nations U El Acuerdo de París | Naciones Unidas
[2] Cook J, Oreskes N, Doran PT, et al (2016) Consensus on consensus: A synthesis of consensus estimates on human-caused global warming. Environ Res Lett 11:. https://doi.org/10.1088/1748-9326/11/4/048002
[3] (2017) Renewable Capacity Statistics 2017. /publications/2017/Mar/Renewable-Capacity-Statistics-2017
[4] University of Zaragoza iHOGA software. https://ihoga-software.com/versiones/. Accessed 26 Apr 2017
[5] HOMER - Hybrid Renewable and Distributed Generation System Design Software. http://www.homerenergy.com/. Accessed 18 Apr 2017
[6] Amini M, Khorsandi A, Vahidi B, et al (2021) Optimal sizing of battery energy storage in a microgrid considering capacity degradation and replacement year. Electr Power Syst Res 195:107170. https://doi.org/10.1016/j.epsr.2021.107170
[7] Mulleriyawage UGK, Shen WX (2020) Optimally sizing of battery energy storage capacity by operational optimization of residential PV-Battery systems: An Australian household case study. Renew Energy 160:852–864. https://doi.org/10.1016/j.renene.2020.07.022
[8] Akram F, Asghar F, Majeed MA, et al (2020) Techno-economic optimization analysis of stand-alone renewable energy system for remote areas. Sustain Energy Technol Assessments 38:100673. https://doi.org/10.1016/j.seta.2020.100673
[9] Elkadeem MR, Wang S, Sharshir SW, Atia EG (2019) Feasibility analysis and techno-economic design of grid-isolated hybrid renewable energy system for electrification of agriculture and irrigation area: A case study in Dongola, Sudan. Energy Convers Manag 196:1453–1478. https://doi.org/10.1016/j.enconman.2019.06.085
[10] Khan MJ, Yadav AK, Mathew L (2017) Techno economic feasibility analysis of different combinations of PV-Wind-Diesel-Battery hybrid system for telecommunication applications in different cities of Punjab, India. Renew Sustain Energy Rev 76:577–607. https://doi.org/10.1016/J.RSER.2017.03.076
[11] Cordero PA, Lata García J, Jurado F Optimization of an Off-Grid Hybrid System Using Lithium Ion Batteries
[12] Briones-Hidrovo A, Copa J, Tarelho LAC, et al (2021) Environmental and energy performance of residual forest biomass for electricity generation: Gasification vs. combustion. J Clean Prod 289:125680. https://doi.org/10.1016/j.jclepro.2020.125680
[13] Khan EU, Martin AR (2015) Optimization of hybrid renewable energy polygeneration system with membrane distillation for rural households in Bangladesh. Energy 93:1116–1127. https://doi.org/10.1016/j.energy.2015.09.109
[14] Nuevo aporte a Guayaquil en el tema de residuos sólidos | CAF. https://www.caf.com/es/actualidad/noticias/2019/12/nuevo-aporte-a-guayaquil-en-el-tema-de-residuos-solidos/. Accessed 10 Jun 2021
[15] Cartografía de los residuos sólidos en Ecuador 2020 – Alianza Basura Cero Ecuador. http://www.alianzabasuraceroecuador.com/3d-flip-book/cartografia-de-los-residuos-solidos-en-ecuador-2020/. Accessed 10 Jun 2021
[16] Kapila S, Oni AO, Kumar A (2017) The development of techno-economic models for large-scale energy storage systems. Energy 140:656–672. https://doi.org/10.1016/J.ENERGY.2017.08.117

IOP Conf. Series: Earth and Environmental Science **897** (2021) 012006 doi:10.1088/1755-1315/897/1/012006

Use of renewable energies as part of the strategy to increase the access of energy to remote communities. A Colombian case: Isla Múcura

Luis Miguel León Gil[1], Francisco Javier Amórtegui Gil[1], and Cristian David Rodríguez Reyes [1]

[1]Carrera 30 #45-03, Bogotá, Colombia
Universidad Nacional de Colombia: sede Bogotá

Lumleongi@unal.edu.co, fjamoreguig@unal.edu.co, crdrodriguezre@unal.edu.co

Abstract. Use of renewable energies in remote places with no access to main electrical infrastructure becomes fundamental to guarantee equity as well as affordable and clean energy, impacting the United Nations Sustainable Development Goals (UN SDGs). There is also fundamental correct and effective capacitation to operators, to avoid expected-lifetime reductions due to malfunctioning or maloperation. In this paper, the case of Isla Múcura is studied: Photovoltaic (PV) system and storage diagnosis, recovery and community capacitation activities were performed in a fishing and tourism-based community with no knowledge about electrical infrastructure. It was detected that correct capacitation and reliable information devices are fundamental to improve maintenance, operation, and management of the overall generation system

1. Introduction

Use of self-generating power plants became a suitable solution for remote locations where electrical energy is required, but geographical or economic conditions are not appropriate to implement a physical splice with interconnected electrical networks. Colombia, due to its complex geography and remarkable population concentrations in main cities, focused the coverage of electrical service through National Interconnected Grid (SIN by its Spanish translation) for almost 97% of citizens [1], located in approximately 50% of Colombian territory.

Aiming to increase the access of nationals to electric fluid, Colombia created government entities as the institute for Planification and Promotion of Electrical Solutions in Non-Interconnected Zones (IPSE by its Spanish translation), which is intended to identify, promote, foment, develop and implement energy solutions for non-interconnected zones; improving life conditions of settlers, building peace and equity in the country, as well as pushing forward the use of renewable energies [2].

Isla Múcura is a 30ha island part of the Colombian territory, located in the influence area of the "Corales del Rosario" National Natural Park. This place and its inhabitants keep a fishing and tourism-based economy. Life conditions there are characterized by the lack of provision in public services, however, electricity supply is provided by a 30 kWp PV array + storage based on lead acid battery cells donated from Japan Government to IPSE and the Island, as well as a 145 kVA Diesel self-generating power plant.

IOP Conf. Series: Earth and Environmental Science **897** (2021) 012006 doi:10.1088/1755-1315/897/1/012006

Most importantly, the energy system management oversees the local community, who were not properly trained to operate and maintain the PV system in optimal conditions, this has led to detriment in the capacity of the PV plant and storage system. On the other hand, due to lack of consumption information, there is not a fair economic recognition system for used energy.
To determine the status of the overall system, energy generation and storage profiles were obtained as shown in section 2. Batteries recovery tests were studied, the results can be found in section 3. Energy consumption measurement system was implemented to monitor the usage of energy resources, as mentioned in section 4. Electrical recommendations to improve the functionality of the PV system are in section 5. Finally, social impact and conclusions are in section 6 and 7, respectively.

2. Preliminary Diagnosis

Several field inspections were performed to establish the multi-line diagram and the status of each one of the components, part of the system. In general terms, the 30kWp PV plant is composed of 3 clusters which are two parallel arrays of 20 PV modules each one. The general characteristics of the array are shown below. Each cluster is connected independently to a 10kW DC/AC power inverter, which are attached to a single AC bus.

Table 1. PV modules array electrical characteristics.

	PV Plant
PV Power	30 kWp
Voc	748 V
Isc	52.98 A
Vmpp	602 V
Impp	49.86 V

The electrical system also has a DC storage stage connected directly to the AC general bus, which is composed by two independent AC/DC inverters in configuration Master (1) – Slave (2). Energy storage is composed of 96 lead acid cells working at 2V and 2500Ah (C120) each one, distributed in two sets controlled by power inverters. Two strings are connected in parallel composed by 48 series arranged cells, conforming a 48V and 10kAh storage system.
As energy back-up, a 145kVA and 416VAC diesel generator is used altogether with a reduction transformer to match the AC bus voltage. The diesel generator is oversized compared to the overall connected load and operates when storage level is critical. The automatic transference is not working and so the process is performed manually. From the general inspection, a multi-line electrical diagram was sketched (Fig. 1) considering energy generation sources, the storage system, and main loads.

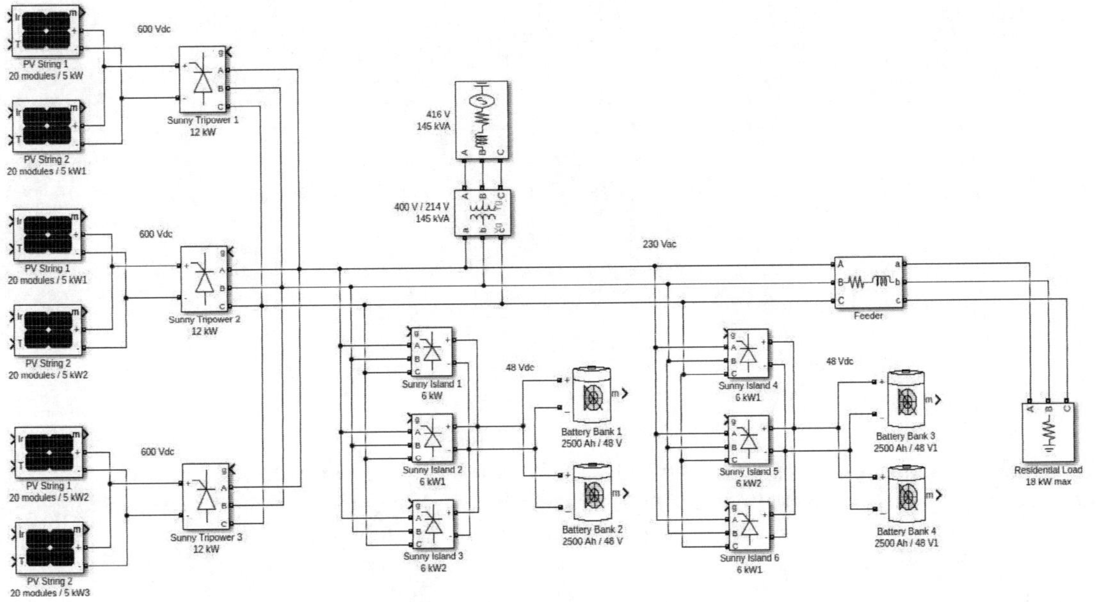

Figure 1. Electrical diagram of Isla Múcura, including PV and diesel generation, as well as the storage system and the main loads connected by final users.

2.1. PV Modules

From visual inspection to the PV array, it could be determined that cleaning maintenance is not easily possible without fall risk, complementary to this critical situation, several birds and animals step up over the PV modules leaving excretions and shading several cells, generating localized heating and expected-lifetime reduction. To determine the generation capacity of the array, a silicon reference cell was installed and compared with the generation profile of the plant.

Considering the measured irradiance at a selected point (Fig. 2), the corresponding output power from the three installed clusters, and the ambient temperature operation of PV modules, it is possible to obtain the overall efficiency for the generation plant. As result of the analysis, the total generated power of the plant at 15:00 hrs with an irradiance value of 620 W/m2 was 15.3 kW measured at the output of cluster connection. From measured irradiance profile and thermal measurements, it is possible to determine that the expected power generated at 620 W/m2 is 17.2 kW.

$$P_{\text{expected}} = (30\text{ kW})(620)/(1000)\ (0.925) = 17.205\text{ kW} \quad (1)$$

$$\eta_{\text{PV Plant}} = (15.3\text{ kW})/(17.205\text{ kW}) \simeq 90\ \% \quad (2)$$

The detected decrease in generation can be directly associated with shades formed by trees, lack of maintenance, uncertainty in the measurement process, and the aging of PV modules.

2.2. PV inverters

Through instantaneous power measurements in each cluster, it was determined that all the PV inverters were operating correctly, showing a regular conversion profile. Nevertheless, the display for PV inverter No. 2 was not working properly. To confirm the correct operation of PV inverters, power measurements were performed, obtaining the generation profile shown in Fig 3.

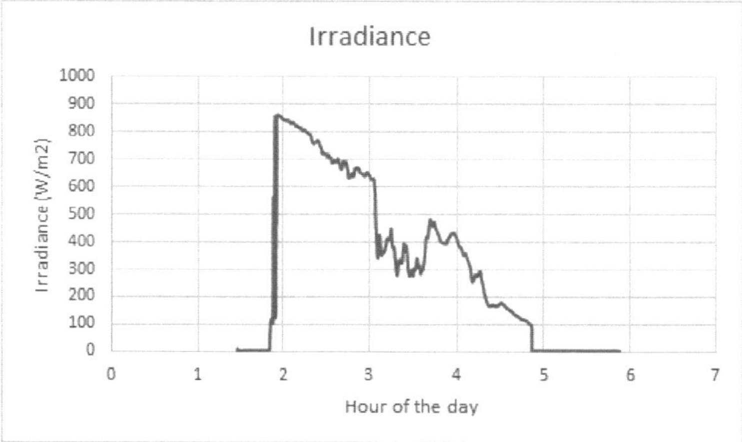

Figure 2. Irradiance profile over PV array for diagnosis purposes. This information was compared to DC power generated at 15:00 hrs. (3:00 PM), obtaining an efficiency of 90% from expected nominal power corrected by thermal decline.

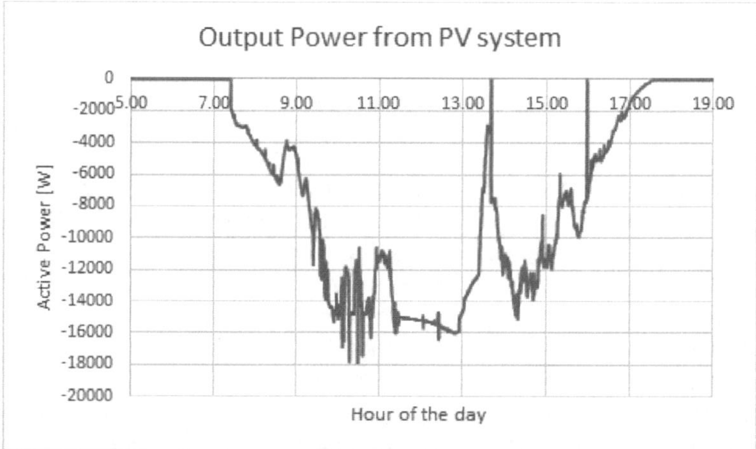

Figure 3. Instantaneous power of PV plant along the time. It is possible to see the generation profile, with two zero crosses due to activation and deactivation of the storage system around 13:00 hrs. and 16:00 hrs. The operation is performed manually due to inconveniences with automatic transference.

2.3. Storage inventers

According to voltage and current data obtained from the internal acquisition system of storage inverters and compared to external measurements, the corresponding management given to the storage system by the community was determined. Empirically and with no instructions or indications at all, the operator of the electrical system of the island connects the main load around 13:45 hrs. just before the storage group reaches the inverter's output voltage. Then, at 16:00 hrs. the island's load is disconnected to charge the batteries again. At 18:00 hrs. the main loads are connected again, and the diesel generator operates feeding the loads until 23:00 hrs. to finally use the remaining resource saved in batteries until the next day. It is possible to identify a higher charge voltage over storage group No. 1 when charging current is high, which can correspond to an improper configuration of the storage inverters. This behavior does not match with storage group No. 2, which shows constant charge voltages and smooth decrease during battery charge process.

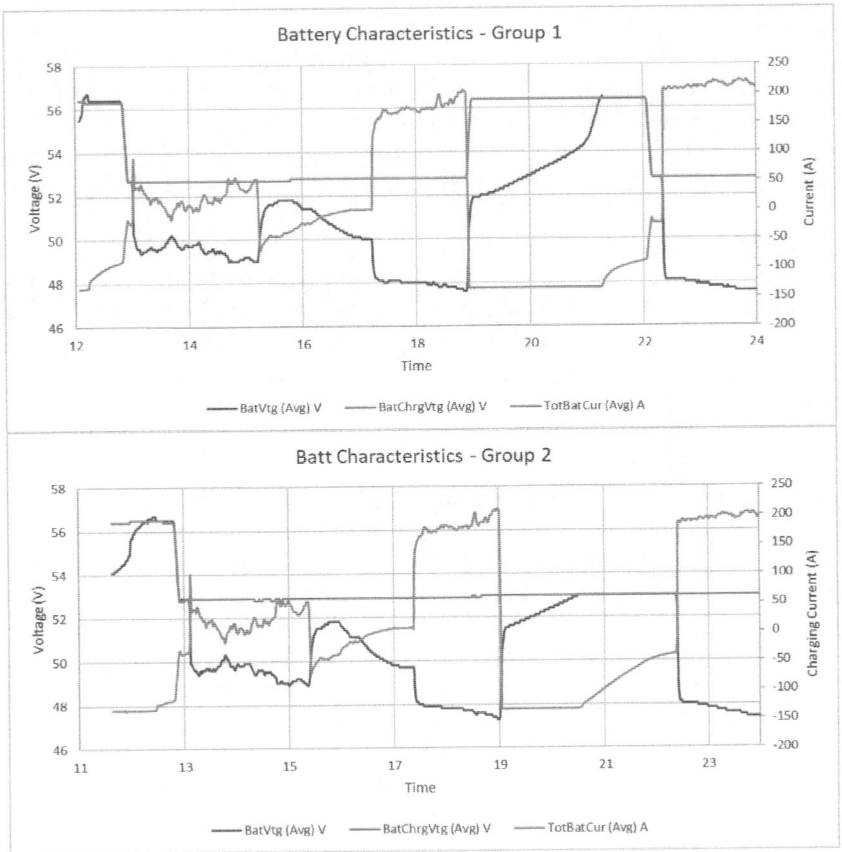

Figure 4. Stored energy profile in battery groups 1 (up) and 2 (down) taken from master inverters. The profiles show the management performed by the electrical system operator, over the storage groups. Charging voltage in group 1 is slightly higher than in group 2, which can be caused due to a wrong inverter configuration.

2.4. Storage system

As a preliminary diagnostic, it is evident that the energy storage system is not properly used due to lack of information and unknown real status of batteries in real time. The minimum average load percentage reached in batteries is around 60%, showing a positive index for battery cells, guaranteeing up to 3000 cycles according to manufacturer's information.

On the other hand, there are poor maintenance practices for battery cells. The Pb – acid batteries require frequent liquid change for the electrolyte, corresponding to a mixture of sulfuric acid diluted in water. To determine the current characteristics of electrolytes and charge of cells, specific gravity is measured. It was found that several cells were completely dry, identifying a critical scenario for batteries.

3. Battery recovery tests

From the current status of battery cells, it is concluded that to preserve and maximize the life-time cycle of the storage system, is necessary to recover the functionality of the cells diminishing the quantity of sediments formed and deposited on the electrodes, due to a lack in electrolytic solution caused by a poor preventive maintenance [3].

Several activation tests aimed to recover the nominal capacity of battery cells are now available in the literature. They are also performed on sulfated cells and are no longer working properly. Three different recovery methods and their results were studied and compared as follows:

3.1. High voltage method

It involves the application of a load voltage between terminals in the range of 1.3 and 1.5 times the nominal voltage of the cell during a short time, assuring that the cell's temperature does not exceed 45°C to avoid an important gasification of electrodes [4].

From the performed tests, it was concluded that this method is recommended for low sulphation levels, because it represents an important risk for the cells due to fast rising in cell's temperature.

3.2. Full charge and discharge cycles method

It is aimed to restore cells diagnosed with small damages, looking for the activation of active substances (transformation of Pb-sulphate to active components of the battery)[5]. This test consists in the full charge and deep discharge of the corresponding cells, and it was concluded from the performed tests that it is necessary to rigorously follow the methodology, as an erroneous performance of these cycles can easily lead to a fatal damage of the battery.

3.3. Pulsed loads or Phoenix method

This method consists in the continuous injection of voltage peaks between the electrodes of the battery, aimed to break the precipitations of Pb-sulphate and diminish their existence and growth. Its principal objective is to weaken the sulphate depositions without generate additional damages to the cell[6], [7].

The method is based on the resonant frequency of sulphate crystals. Due to the high harmonic components of pulsed peaks, it is possible to excite the depositions with their resonance frequency which principally depends on their size, physical form, and molecular structure.

It was found that big deposition crystals respond and disintegrate under the excitation of low frequencies, meanwhile small particles are sensible to high frequencies, allowing to decompose them and perform a sweep in the cases it could be necessary. The capacity of the battery can be recovered partially with outstanding results compared to the two previous methods.

4. Energy consumption management

Grid connection and management is performed manually because of a fatal failure of the automatic management set up, caused by absence of preventive and corrective maintenance. Then, a worker oversees the operation and monitoring of generation and storage systems, who connects and disconnects the different elements of the grid, according to the experience, the time and a display which shows the charge status of the batteries.

On the other hand, the access and usage of energy is ruled by demand, under common agreement. Every inhabitant reports the energy needs according to the type and quantity of electrical appliances planned to feed from the grid and then, monthly, the chief of the local action board collects the corresponding payment associated with the electrical active equipment in each house to buy fuel for the diesel generator. This methodology, despite is currently working, does not fully represent the real energy consumption for each customer, leading to misunderstandings about the monetary compensation.

Fig. 5 shows the daily management operations for the different stages of the system. Diesel generation is present for 4 hours a day to supply loads and battery charge, which leads to additional costs assumed by the community. The island's economy is based towards fishing and tourism, then it is identified a minimum consumption profile for devices that are connected 24 hours a day, as well as the increase of nocturnal use of appliances as TVs, sound equipment, among others.

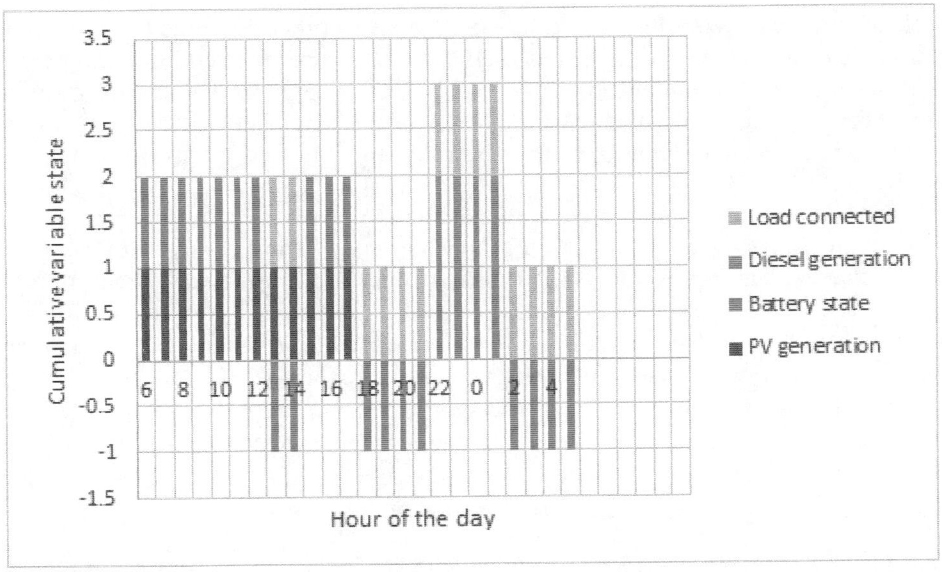

Figure 5. Daily management operations for the system's elements. Negative battery state means it is discharging while positive, charging. Positive states for PV and Diesel, mean they are generating energy, while a positive state in Load means it is connected.

Considering the consumption practices performed by locals and tourists, two methodologies were proposed to adapt the quantification and collection of energy rates, according to their real consumption:

- Include metered-based energy quantification for each user and establish a kWh flat rate to determine the apports from each consumer. This is the most optioned strategy because it represents the real energy consumption of each one of the users of the grid. In contrast, it requires an energy – meter for each consumer, increasing the implementation costs for the total solution.
- Include metered – based energy quantification for user groups and establish a base fare for all the users as well as a minimum energy coverage for each one of the consumers. This cost will be increased according to the difference in the energy user and will be divided in the whole group. This strategy strongly reduces costs of implementation, but it requires a careful selection of groups looking for equity in that conformation.

The following strategy was followed to implement a measurement system that helps to acquire data and establish the real consumption of energy per user: electronic meters were acquired, their software and hardware were adapted using current transformers, voltage dividers, and programming its digital processor unit capable of power calculations in the four quadrants to finally come up with an energy – meter with precision of seconds.
The technology used in these devices allows to monitor up to six independent single-phase loads and centralize the measurements in the same unit, the information can be accessed by local wi-fi connection by the inhabitants through their electronic devices. In this way, daily or weekly monitoring will help users to determine how much energy they are using, as well as indicate the wage they must pay for the energy service. Even more, if a stable internet signal is present, data can also be sent to the cloud, this feature would also help future operation if remote monitoring by agents outside the community is implemented, in this way, operation costs of the system will reduce since no operator needs to be physically present in the island. Proper implementation of this internet connection would also increase the amount of data available for generation/load analysis.

Even though energy-meters were tested prior its installation, put in service and their functionality was validated a couple of days, their operation failed almost after a month of service in the island, this was due to improper design of lightning protection system, which provoked overvoltage failures in the main digital board. However, implementation of these measurement scheme still preserves high potential for usage as a guidance tool to achieve a fairer charge for energy consumption.

5. Electrical system improvement and recommendations

From the collected data and the identification of generation and consumption profiles, the following elements were identified as improvements in the performance of the electrical system from final user's side:

- Usage of the PV array to feed batteries and essential loads (refrigeration) during the day. This to dynamize the generation – consumption system and reduce the operation cycles for the batteries.
- Identify the generation efficiency of a PV system when the Diesel generator is used, recognize energy losses that can be used to charge the batteries without diesel fuel, and then reduce the operation costs of the electrical system.

6. Social Impact

Energy consumption and land distribution in Isla Múcura, and more specifically, in Puerto Caracol, where the locals reside, is very inequitable. Touristic complexes take advantage of the main part of the available and habitable land, and in the same order, their energy consumption is the most representative in the whole electrical system. Nevertheless, in most cases their payments are not completely representative, compared to the total consumption. This phenomenon is related to the lack of available information about energy use profiles and the fares established in the island as a method to quantify the price of energy supply.

The development of this project gives to the whole community not only tools to clearly identify the total consumption for each user connected to the grid, but also strategies to take care of the PV generation plant, increasing its performance and its expected lifetime, but reducing at maximum the use of the diesel generator.

Training of locals about preventive and corrective maintenance of the electrical system, plays a fundamental role in two ways: First, the monitoring and maintenance of the PV plant and the storage system prevents the damage of their components, allowing the operator to take decisions based on technical and quantitative information and reducing intervention costs as well as the unavailability time. Second, the appropriation of the electrical system by the community, could achieve a closer monitoring of energy profiles and an optimization of not only the operation of the PV and diesel plants, but also of their consumption practices to achieve a higher efficiency during their daily tasks.

7. Conclusions

Colombia, as part of its strategy to increase access to electrical energy and looking for the reduction of the breaches reported by the UN worldwide, has been using renewable energies as a powerful tool to reach the most remote places of the country, as a low cost and highly effective method to provide energy to citizens.

This strategy has many advantages as well as several disadvantages that can be easily covered by giving the community decision tools and training them to give a closer look of the daily operation for the overall system. The immersion of these projects inside local communities, needs to be accompanied by socialization and appropriation of the projects, showing the advantages in terms of equity, risks, and the importance of their active participation in the durability and optimization of the new resource they have.

8. References

[1] Unidad de Planeación Minero Energética - UPME, "*Índice de Cobertura de Energía Eléctrica - ICEE 2018*" SIEL (Sistema de Información Eléctrico Colombiano), 2018. http://www.siel.gov.co/, (accessed Mar. 15, 2021)

[2] Instituto de Planificación y Promoción de Soluciones Eléctricas para las Zonas No Interconectadas - IPSE, "*Funciones principales del IPSE*" Instituto de Planificación y Promoción de Soluciones Eléctricas para las Zonas No Interconectadas - IPSE, 2021. https://ipse.gov.co/mapa-del-sito/ipse/historia/ (accessed Mar. 15, 2021).

[3] Karami H and Asadi R, "Recovery of discarded sulfated lead-acid batteries" (2009), *J. Power Sources*, **vol. 191**, no. 1, pp. 165–175, doi: 10.1016/j.jpowsour.2008.12.153.

[4] Ordoñez J, Gago E J, and Girard A, "Processes and technologies for the recycling and recovery of spent lithium-ion batteries", (2016), *Renew. Sustain. Energy Rev.*, **vol. 60**, pp. 195–205, doi: 10.1016/j.rser.2015.12.363.

[5] Jin J, Jin D, Shim J, and Shim W, "Enhancing Reversible Sulfation of PbO 2 Nanoparticles for Extended Lifetime in Lead-Acid Batteries", (2017), *J. Electrochem. Soc.*, **vol. 164**, no. 7, pp. A1628–A1634, doi: 10.1149/2.1291707jes.

[6] H. Ikeda, S. Minami, S. J. Hou, Y. Onishi, and A. Kozawa, "Nobel High Current Pulse Charging Method for Prolongation of Lead-acid Batteries", (2005), *J. Asian Electr. Veh.*, **vol. 3**, no. 1, pp. 681–687, doi: 10.4130/jaev.3.681.

[7] Zhang Y, Hou S, Minami S, and Kozawa A, "A high current pulse activator for the prolongation of Lead-acid batteries", (2008), *IEEE Veh. Power Propuls. Conf. VPPC 2008*, pp. 3–6, doi: 10.1109/VPPC.2008.4677701.

Acknowledgments
The authors would like to thank the Puerto Caracol community in Isla Múcura, for their kind disposition, support, and contributions to the social and technical studies performed by the students involved as well as Universidad Nacional de Colombia for funding this case study through the Fondo Nacional de Extensión Solidaria 2018.

IOP Conf. Series: Earth and Environmental Science **897** (2021) 012007 doi:10.1088/1755-1315/897/1/012007

Field experiment on flow stabilization of working fluid in a top-heat-type thermosyphon

K Maruyama[1], T Kawaguchi[2], T Fujisawa[2] and T Kawashima[3]

[1] Mechanical Engineering Course, Kanagawa Institute of Technology, Graduate School, 1030, Shimo-ogino, Atsugi, Kanagawa, Japan
[2] Department of Vehicle System Engineering, Kanagawa Institute of Technology, 1030, Shimo-ogino, Atsugi, Kanagawa, Japan
[3] Department of Mechanical Engineering, Kanagawa Institute of Technology, 1030, Shimo-ogino, Atsugi, Kanagawa, Japan

kawashima@eng.kanagawa-it.ac.jp

Abstract. To address the problem of global warming, increasing efforts are being made to use renewable sources of energy, such as solar energy, wind energy, and geothermal energy. However, the effective use remains a major challenge for its sustainable development. In this study, we used a top-heat-type thermosyphon to heat water using solar energy and transport the low-density hot water from the source to the sink (high to low elevation) without an external power source. The transported hot water can be used for cooking, bathing, underfloor heating, and heating homes and buildings, and warming cold springs. However, a disadvantage of top-heat-type thermosyphon is the intermittent flow of the circulating working fluid under low solar radiation. To address this issue, the authors proposed and developed a control system to stabilize the intermittent flow and prevent equipment damage and failure due to the sudden boiling of water. Field experiments were conducted to assess the practicability of the developed controller. The results showed that the controller efficiently converted the intermittent flow of working fluid to continuous flow by reducing the pressure in the buffer chamber and thus lowering the boiling point of the working fluid in the header of the solar collector.

1. Introduction

Increasing efforts are being made to reduce the emission of greenhouse gases to address the critical issue of global warming. These strategies include the use of renewable energy, such as solar energy, hydraulic power, wind power, geothermal power, and biomass energy, which are low-density energy sources with a wide global distribution. However, the effective use of renewable energy remains a key challenge for its sustainable development; thus, it is important to determine new ways to utilize renewable energy.

A mainstream application of renewable energy is photovoltaic power generation, which conveniently converts solar energy into electricity. A certain amount of the generated power could be used as low-temperature heat, such as water heating. Solar energy can be used more effectively to produce hot water using solar collectors with a conversion efficiency of 40–50%, compared to producing hot water using the electric power generated by photovoltaic panels with a conversion efficiency of 15–20%.

To date, several measures have been adopted to utilize solar energy for heating water, including the use of a thermosyphon. Morrison [1], Jamar et al. [2], and Jafari et al. [3] have reviewed the applications of solar energy. In addition, there have been extensive studies on top-heat-type thermosyphons, which can transfer heat from high to low elevations without any external electric power source. The transported hot water can be used for cooking, bathing, underfloor heating, and heating homes and buildings, and warming cold springs. Several types of top-heat-type thermosyphons have been proposed. Imura and Koito [4] reviewed the studies conducted on heat transfer using four types of top-heat type thermosyphons, namely loop heat pipe using a wick, vapor-lift pumping loop thermosyphon including two-phase flow top-heat thermosyphon, osmotic heat pipe, and vapor pressure anti-gravity thermosyphon including loop thermosyphon with a switching chamber. In addition, they discussed the improvements made in thermosyphons and the challenges faced. The loop heat pipe using a wick requires capillary tubes and the osmotic heat pipe requires semipermeable membranes, which hinder the increase in flow rate. The vapor pressure anti-gravity thermosyphon and the loop thermosyphon with a switching chamber have a complicated structure. Therefore, in this study, a vapor bubble pumping type thermosyphon, which circulates the working fluid using buoyancy forces of vapor bubbles generated by solar energy, was investigated for simplicity and scalability. This type thermosyphon is an improved version of two-phase flow top-heat thermosyphon.

Since the 1970s, there have been several studies on vapor bubble pumping type thermosyphons. Among these, Hirashima et al. [5] and Ippohshi et al. [6] have conducted early research on this type of thermosyphon. In these studies, the prototypes were manufactured and the heat transfer characteristics were experimentally investigated. Ito et al. [7] improved the condenser, and Yoshida et al. [8–11] continued the research of Ito et al., demonstrating the installation of a thermosyphon in a prefabricated full-scale model house to circulate working fluid with a height difference of 4 m and generate electricity using a small-scale power generator. These studies revealed some of the problems associated with this type of thermosyphon. One of the issues is the intermittent flow of the working fluid under low solar radiation in the morning and evening [10]. This intermittent flow causes sudden boiling of water in the header of the solar collector, resulting in equipment failure and shortened service life. In addition, studies on stabilizing the flow of circulating fluid in thermosyphons are limited [12–13]. Thus, a control system is needed to stabilize the flow rate of the working fluid.

Therefore, we focused on a top-heat-type thermosyphon to utilize solar energy to heat water and transport low-density hot water from the top to the bottom of the thermosyphon without an external power source. Further, to solve the issue of the intermittent flow of the circulating working fluid in the thermosyphon, the authors proposed a control system to stabilize intermittent flow by reducing the buffer chamber pressure to lower the boiling point of the working fluid in the header of the solar collector. The effectiveness of the controller was validated using indoor models [14]. In this study, field experiments were conducted to evaluate the effectiveness of the control system by assessing its ability to change the flow of the working fluid from intermittent to continuous. In addition, the feasibility of the thermosyphon system with the controller was confirmed.

2. Operating principle of the top-heat-type thermosyphon

Figure 1 shows a schematic of the top-heat-type thermosyphon used in this study.

The thermosyphon system consists of a solar collector, condenser, buffer chamber, heat exchanger, recuperator, and pipes. The vacuum pump decreases the pressure of the buffer chamber to lower the boiling point of the working fluid heated at the header of the solar collector. This results in buoyancy force-mediated upward flow of the boiling water with vapor bubbles along the pipes to the condenser, thereby driving the flow of the working fluid. The vapor bubbles in the two-phase flow disappear in the condenser upon cooling using the cold working fluid flowing through the recuperator, which is in contact with the condenser. The working fluid flows to the heat exchanger through the buffer chamber, and the thermal energy stored in the working fluid, that is, sensible heat, is transferred to water, cooled by the ice in the heat exchanger. The cooled working fluid is then preheated at the recuperator and returned to the header of the solar collector. Therefore, the collected thermal energy at the upper side of the system is transferred to the lower side without external power.

Under a theoretical steady-state condition, all thermal energy obtained at the solar collector is transferred to the cooled water in the heat exchanger with no heat dissipation. Meanwhile, the latent heat and some

IOP Conf. Series: Earth and Environmental Science **897** (2021) 012007 doi:10.1088/1755-1315/897/1/012007

amount of sensible heat in the working fluid circulate in the solar collector, condenser, and recuperator only.

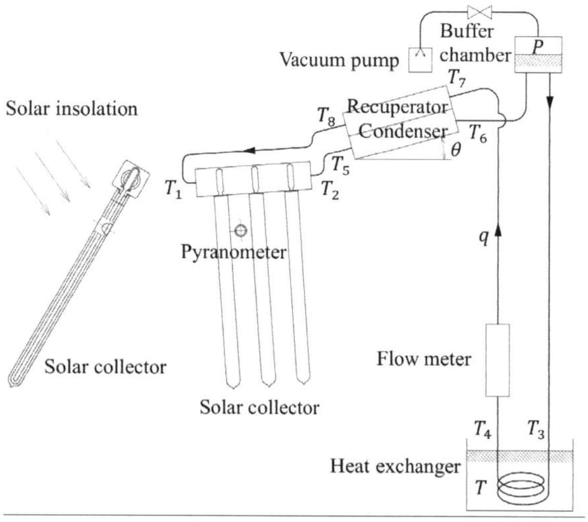

Figure 1. Schematic of the top-heat-type thermosyphon.

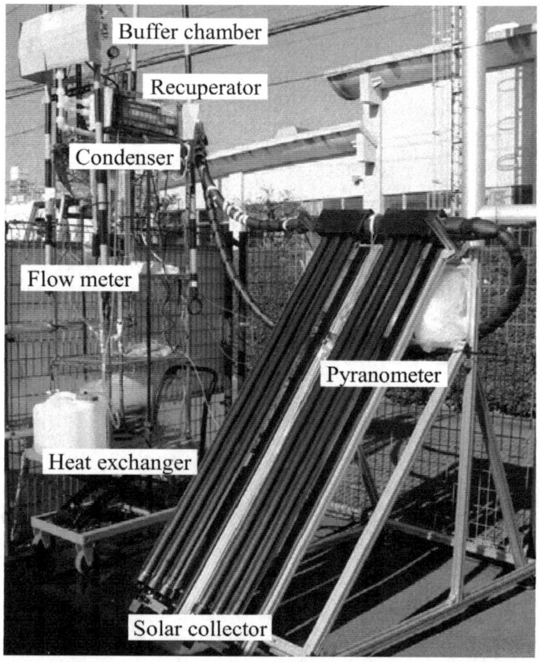

Figure 2. Photograph of the field experimental facility of the top-heat-type thermosyphon.

3. Field experimental facility of the top-heat-type thermosyphon

Figure 2 shows a photograph of the field experimental facility of the top-heat-type thermosyphon used in the study.

The flow rate of the working fluid was measured by a Coriolis flow meter FD-SS02A produced by KEYENCE Co. The pressure in the buffer chamber was measured by a pressure sensor AP10S (-100–100 kPa) produced by KEYENCE Co. The inlet and outlet temperatures of the solar collector, heat exchanger, condenser, and recuperator, and ambient air temperature were measured by a type K thermocouple. The amount of solar radiation perpendicular to the solar collector was measured with a pyranometer MS-602 produced by EKO Instruments Co. Ltd. The measured data were recorded using a multi-channel data logger system GL220 produced by Graphtec Co.

In this facility, one surface of the condenser and recuperator was made of a clear acrylic plate. Translucent flexible heat-resistant pipes were used from the condenser to the recuperator to observe the bubbles, especially their disappearance in the condenser. The working fluid used in the system was boiled tap water cooled to ambient air temperature to reduce the amount of dissolved air.

4. Intermittent flow of the working fluid

Figure 3 shows the measurement results of the thermosyphon without flow control, that is, atmospheric pressure exists in the buffer chamber on a sunny day.

From figure 3, it can be seen that the working fluid clearly exhibits a continuous flow because of sufficient thermal energy obtained from solar radiation on a sunny day. Under these conditions, the temperature of the working fluid in the header of the solar collector reaches its boiling point.

Figure 4 shows the results without flow control, that is, atmospheric pressure exists in the buffer chamber on a sunny day with occasional thin clouds.

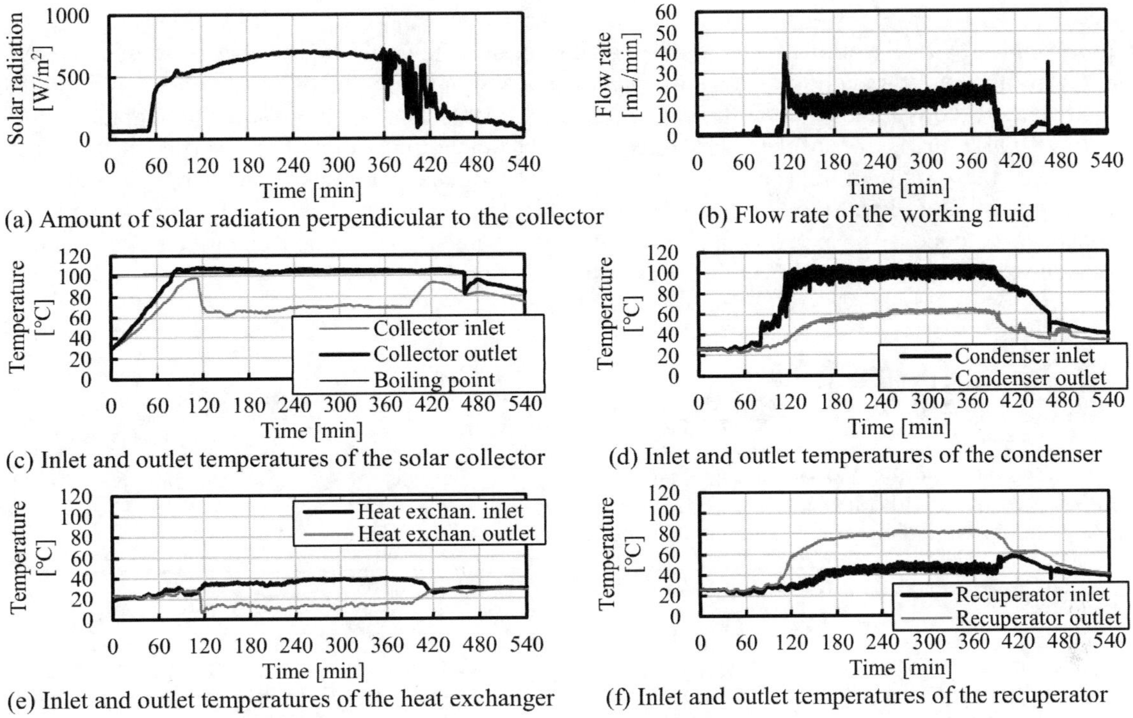

(a) Amount of solar radiation perpendicular to the collector
(b) Flow rate of the working fluid
(c) Inlet and outlet temperatures of the solar collector
(d) Inlet and outlet temperatures of the condenser
(e) Inlet and outlet temperatures of the heat exchanger
(f) Inlet and outlet temperatures of the recuperator

Figure 3. Measurement results of the thermosyphon without control on a sunny day, in which the working fluid exhibits a continuous flow.

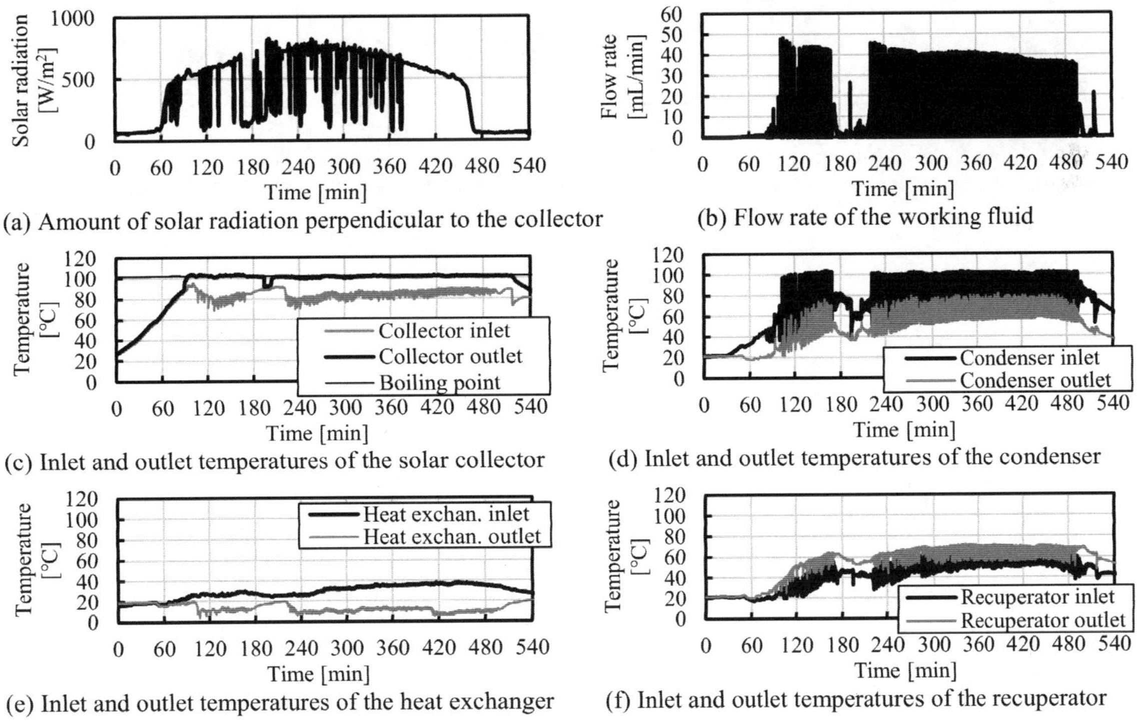

(a) Amount of solar radiation perpendicular to the collector
(b) Flow rate of the working fluid
(c) Inlet and outlet temperatures of the solar collector
(d) Inlet and outlet temperatures of the condenser
(e) Inlet and outlet temperatures of the heat exchanger
(f) Inlet and outlet temperatures of the recuperator

Figure 4. Measurement result of the thermosyphon without flow control on a sunny day with occasional thin clouds, in which the working fluid exhibits an intermittent flow.

Figure 4 shows that the working fluid exhibited an intermittent flow as it did not obtain sufficient thermal energy from the solar radiation on a sunny day with occasional thin clouds. Under this condition, the temperature of the working fluid at the condenser inlet fluctuates. Nonetheless, the temperature of the working fluid in the header of the solar collector reaches its boiling point. The insufficient latent heat absorbed by the working fluid results in discontinuous boiling under a limited amount of solar radiation. To convert the intermittent flow of the working fluid into a continuous flow, the pressure in the buffer chamber was reduced. The experimental results with the manual pressure control of the buffer chamber on a sunny day with occasional thin clouds are shown in figure 5.

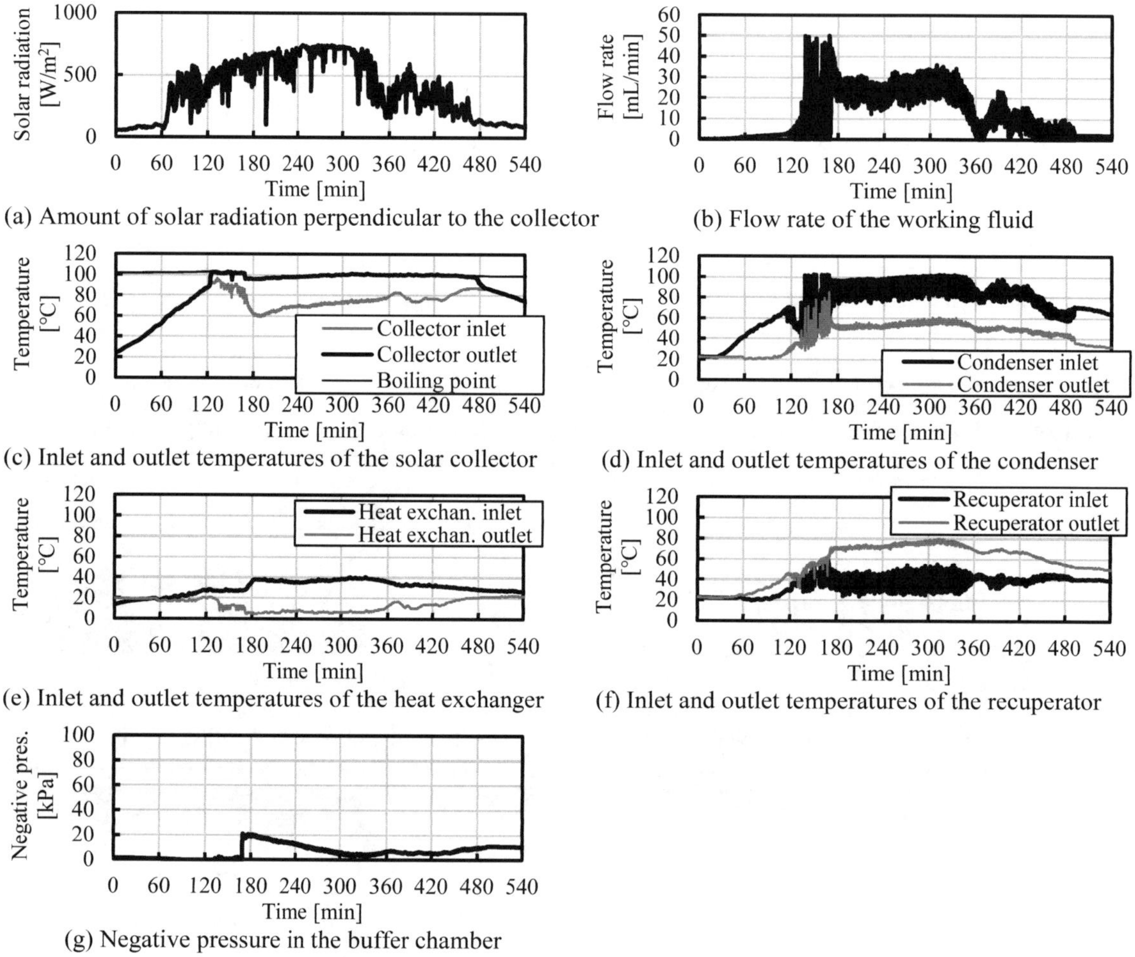

Figure 5. Experimental result of thermosyphon with manual pressure control in the buffer chamber to convert the intermittent flow of the working fluid to a continuous flow on a sunny day with occasional thin clouds.

As the amount of solar radiation increases, the working fluid exhibits an intermittent flow starting at approximately 120 min. The flow of the working fluid then changes to a continuous flow by slowly reducing the pressure in the buffer chamber at 168.9 min to prevent excess boiling. Therefore, these results confirm the effectiveness of the proposed method by changing the pressure of the buffer chamber to lower the boiling point of the working fluid.

5. Stabilizing the intermittent flow of the working fluid

A control system consisting of a solenoid valve, pressure transducer, vacuum pump with a tank, and embedded computer was subsequently employed to stabilize the intermittent flow of the working fluid. An algorithm encoded in a microcontroller board Arduino Uno was implemented to decrease the pressure in the buffer chamber to -80 kPa when the flow rate of the working fluid was above 30 mL/min and below 10 mL/min and this change was repeated three times within 18 min. The experimental results using the control system on a fairly sunny day are shown in figure 6. In this system, the pressure in the buffer chamber decreased from the beginning to promote a continuous flow of the working fluid.

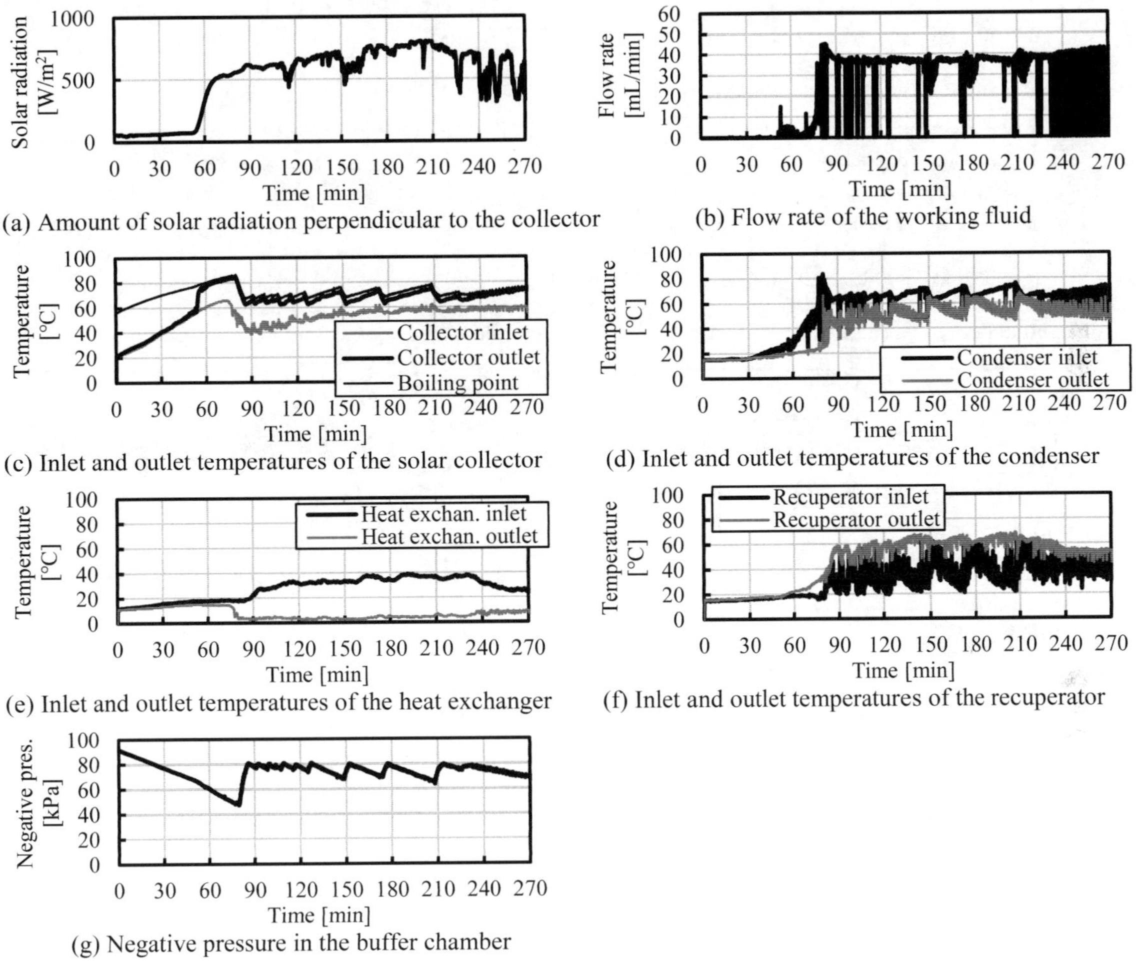

Figure 6. Experimental result of thermosyphon with a control system to control the pressure in the buffer chamber on a fairly sunny day, exhibiting the conversion of the intermittent flow of the working fluid to a continuous flow.

As the amount of solar radiation increased, the working fluid exhibited an intermittent flow at 77.15 min. The controller detected the intermittent flow of the working fluid and reduced the pressure in the buffer chamber. Subsequently, the intermittent flow changed to a continuous flow at 83.65 min. After a while, the continuous flow reverted to intermittent flow at 90.35 min because of the increase in the boiling point with a slow increase in the pressure in the buffer chamber due to a small air leak. Subsequently, the controller detected the intermittent flow and reduced the pressure in the buffer chamber. The interval of the changes in the intermittent flow depends on the amount of solar radiation received. The controller stopped working at 230 min and the intermittent flow of the working fluid continued thereafter. Therefore, these results confirm the feasibility of the thermosyphon with the controller.

Figure 7 shows the experimental result with the control system on a sunny day with the target pressure in the buffer chamber set to -75 kPa.

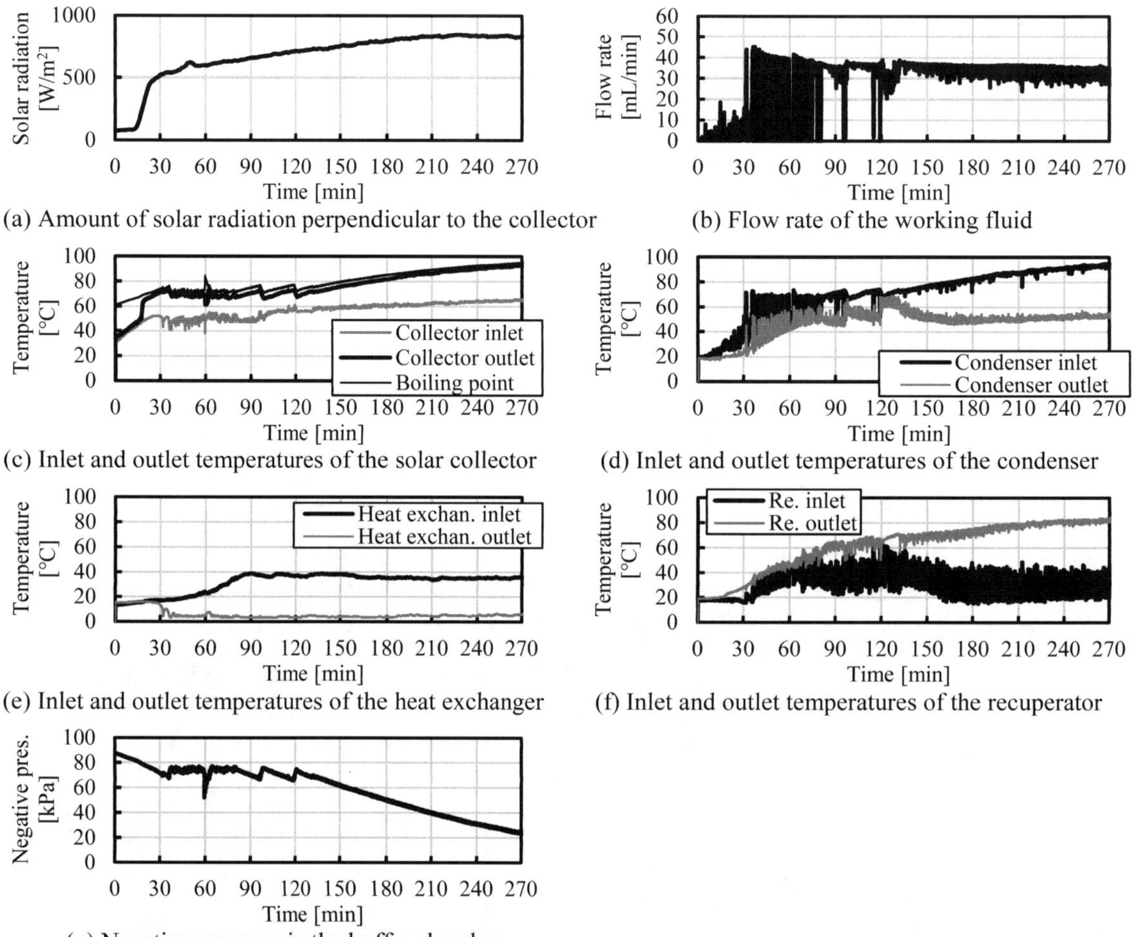

(a) Amount of solar radiation perpendicular to the collector
(b) Flow rate of the working fluid
(c) Inlet and outlet temperatures of the solar collector
(d) Inlet and outlet temperatures of the condenser
(e) Inlet and outlet temperatures of the heat exchanger
(f) Inlet and outlet temperatures of the recuperator
(g) Negative pressure in the buffer chamber

Figure 7. Experimental result of the thermosyphon with a pressure control system that set the buffer chamber pressure to -75 kPa on a sunny day, exhibiting the conversion of the intermittent flow of the working fluid to a continuous flow.

Even if the pressure reached its target value at 37.45 min, the working fluid continued to exhibit an intermittent flow until 80.45 min due to its boiling point, which was higher than that in the case with higher target pressure, as depicted in figure 6. Subsequently, the flow of the working fluid was converted to a continuous flow at 80.45 min as the amount of solar radiation increased until sufficient latent heat was obtained for continuous boiling. Although the pressure increased, the continuous flow was maintained due to a large amount of solar radiation received after 130 min. Therefore, the target pressure in the buffer chamber should be set according to the capacity of the thermosyphon system.

6. Considerations

The aim of this study was to assess the practicability of the proposed controller by conducting field experiments. The experimental equipment, i.e., thermosyphon, consisted of translucent flexible heat-resistant pipes to observe air bubbles instead of heat insulation. Although the temperature difference between the condenser outlet and the heat exchanger inlet is large, it does not affect the heat transfer performance and the control performance. In contrast, the temperatures of the condenser inlet and the header outlet of the solar collector are almost the same because of the insulated pipes between them, which affect the circulation of the working fluid.

The decrease in the temperature in the condenser is attributed to the transfer of latent heat and sensible heat to the recuperator. The transferred heat contributes to evaporation at the header. Therefore, the energy wastage is prevented and the effective transport of hot water from the top (source) to the bottom (sink) is ensured, when the pipes are insulated perfectly.

The flow of working fluid in the thermosyphon is greatly affected by the amount of solar radiation received. Small amounts of solar radiation do not cause a flow, whereas a large amount of solar radiation causes a continuous flow. Intermittent flow occurs when the amount of solar radiation received is insufficient to allow a continuous flow of the working fluid. It is not possible to conduct all field experiments under the same weather conditions and thus the amount of solar radiation received differs in each experimental result. In this paper, the experimental results showing intermittent flow were presented to assess the practicability of the proposed controller.

Even if the amount of solar radiation changes rapidly, the flow rate of working fluid does not change significantly unlike the amount of power generation of photovoltaic panels, because of the large time constant of the thermosyphon. The rapid changes are not cause of the intermittent flow.

The reason for intermittent flow is related to the solar input. When the amount of solar radiation received is large, the working fluid, which is cold water preheated in the recuperator moves to the header and is continuously boiled, resulting in its continuous flow. However, when the amount of solar radiation received is less, the inflowing cold water cannot be boiled continuously and the flow stops. Upon boiling, the water in the header flows out; the preheated cold water flows into the header and boiling stops. This process is repeated, resulting in the intermittent flow of the working fluid. Therefore, the occurrence of intermittent flow is related to the capacity of the solar collector and the volume of the header with respect to the amount of solar radiation.

7. Conclusions

In this study, we evaluated the ability of a top-heat-type thermosyphon to effectively use solar energy and its feasibility of the thermosyphon with a control system to stabilize the flow of the working fluid. Particularly, we studied the intermittent flow of the circulating working fluid in this type of thermosyphon under low solar radiation as this can damage the equipment due to the sudden boiling of water. The practicability of the proposed controller system was assessed by conducting field experiments, in which the intermittent flow of the working fluid was converted to a continuous flow by reducing the pressure in the buffer chamber due to lower the boiling point of the working fluid in the header of the solar collector after three pulsations of the flow rate of working fluid was detected. In addition, the results highlighted the significance of setting an appropriate target pressure of the controller. In this experimental facility, the target pressure was set to -80 kPa, and it should be set according to the capacity of the thermosyphon system. After all, the top-heat-type thermosyphon with the controller effectively uses solar energy to heat water that can be used for cooking, bathing, underfloor heating, and heating homes and buildings, and warming cold springs.

In the future, we will focus on harnessing solar energy using thermosyphons equipped with controllers without using an external power source.

8. References

[1] Morrison G L 2001 *Solar Energy -The state of the art-* ed J Gordon (ISES position papers) chapter 5 pp 223–289

[2] Jamar A, Majid Z A A, Azmi W H, Norhafana M and Razak A A 2016 *International Communications in Heat and Mass Transfer* **76** 178–187

[3] Jafari D, Franco A, Filippeschi S and Marco P D 2016 *Renewable and Sustainable Energy Reviews* **53** pp 575–593

[4] Imura H and Koito Y 2005 *Trans. of the JSRAE* **22**(1) pp 1–12 (In Japanese)
[5] Hirashima M, Kimura K, Utsumi Y, Kimura K, and Negishi K 1993 *Trans. of the JAR* **10**(2) pp 247–256 (In Japanese)
[6] Ippohshi S, Tabara S, Motomatu K, Mutoh A and Imura H 2003 *The 6th ASME-JSME Thermal Engineering Joint Conference* (Hawaii) TED-AJ03–578
[7] Ito S, Tateishi K and Miura N 2007 *ISES Solar World Congress 2007* (Beijing) **2** pp 930–934
[8] Hagino N and Yoshida H 2011 *Proc. ASME 2011 Power Conference* (Denver) 55358
[9] Imada H, Hagino N and Yoshida H 2013 *Trans. Japan. Society of Mechanical Engineers* B **79**(801) pp 809–813 (In Japanese)
[10] Yoshida H, Imada H, Hagino N and Yada N 2014 *EuroSun 2014 Conf. Proc.* (Aix-les-Bains, ISES) doi:10.18086/eurosun.2014.16.24
[11] Hagino N, Yoshida H and Imada H 2016 *Trans. Japan Society of Mechanical Engineers* **82**(837) DOI: 10.1299/transjsme.15-00596 (In Japanese)
[12] Bratsun D A, Zyuzgin A V, Polovinkin K V and Putin G F 2008 *Technical Physics Letters* **34**(8) pp 650–652
[13] Amirgaliyev Y, Kunelbayev M, Amirgaliyev B, Sundetov T, Yedilkhan D and Merembayev T 2019 *The 5th International Conference on Power Generation Systems and Renewable Energy Technologies* (Istanbul, IEEE) DOI: 10.1109/PGSRET.2019.8882650
[14] Fujisawa T, Kawaguchi T and Kawashima T 2018 *EuroSun 2018 Conf. Proc.* (Rapperswil, ISES) doi:10.18086/eurosun2018.10.15

IOP Conf. Series: Earth and Environmental Science **897** (2021) 012008 doi:10.1088/1755-1315/897/1/012008

Analytical models for adiabatic compressed air energy storage (A-CAES) systems in lined tunnels

Javier Menéndez[1,*], Jorge Loredo[2], Laura Álvarez de Prado[3], Jesús M. Fernández-Oro[4], Antonio Bernardo-Sánchez[3]

[1]SADIM, S.A. S.M.E, Mining and Civil Department, 33900 Ciaño, Spain
[2]UNIVERSITY OF OVIEDO, Mining Exploitation Department, 33005 Oviedo, Spain
[3]UNIVERSITY OF LEÓN, Department of Mining Technology, Topography and Structures, 24071 León, Spain
[4]UNIVERSITY OF OVIEDO, Energy Department, 33271 Gijón, Spain

*Corresponding author's e-mail: javiermenendezr@gmail.com

Abstract. Adiabatic compressed air energy storage (A-CAES) systems consist of an underground reservoir where compressed air is stored at high pressures. The ambient air is compressed by compressors located at the surface and the thermal energy is stored using thermal energy storage (TES) systems. The compressed air is stored in the subsurface reservoir (charge). Then, when the electricity is needed, the compressed air is released and expanded in gas turbines to produce electricity (discharge). In this paper, an analytical model has been developed to investigate the thermodynamic behaviour during air charge and discharge processes. Operating pressures from 4.5 to 7.5 MPa has been employed in lined tunnels in the compression and decompression stages. The model considers a 20 mm thick sealing layer, a 0.4 m thick concrete lining and a 1 m thick rock mass around the air. Air mass flow rates of 0.19 and 0.27 kg s^{-1} have been used in the charge processes for polymer material and steel, respectively. Finally, in the discharge processes the mass flow rate increases up to -0.38 and -0.45 kg s^{-1} for polymer and steel. The air temperature and pressure and the temperature and heat transfer in the sealing layer, concrete lining and rock mass have been analyzed for 100 cycles considering polymer material and steel as sealing layers. The heat transfer through the sealing layer reaches -150 and -95 W m^{-2} for steel and polymer, respectively. The results obtained show that the storage capacity increases when the heat transfer through the sealing layer increases.

1. Introduction

Adiabatic compressed air energy storage (A-CAES) and underground pumped-storage hydropower (UPSH) systems could be an alternative to store energy and facilitate the integration of variable renewable energies (VRE) in disused underground structures such as abandoned coal mining areas and salt deposits [1-6]. Figure 1 shows the location of coal mining areas and salt deposits where this systems may be installed in the European Union (EU) [7]. Álvarez et al. [8] developed a numerical and analytical model to analyze the thermodynamic performance of CAES reservoir in abandoned mines. An operation pressure range between 8 and 10 MPa was employed in the models. Fiber-reinforced plastic and steel were considered as sealing layers in the simulations. They concluded that no temperature fluctuation was obtained in the rock mass and the air temperature fluctuations decrease when the thermal conductivity of the sealing layer increases. In addition, good agreements were reached between

IOP Conf. Series: Earth and Environmental Science **897** (2021) 012008 doi:10.1088/1755-1315/897/1/012008

analytical and numerical models considering lined and unlined tunnels. Schmidt et al. [9] developed a numerical study to investigate the geomechanical performance of CAES systems in coal mines for 10,000 cycles of compression and decompression. They concluded that a volume increase less than 0.5% was obtained by applying the operating conditions (4.5-7.5 MPa) and the rock mass deformations reached moderated values. Zhou et al. [10] carried out a numerical analysis for the coupled thermo-mechanical performance in a lined rock cavern considering an operating air pressure of 4.5-6.9 MPa for 100 cycles. Different thickness of concrete lining, between 0 and 0.1 m were also considered. Kushnir et al. [11] carried out a study to investigate the temperature and pressure in CAES reservoirs.

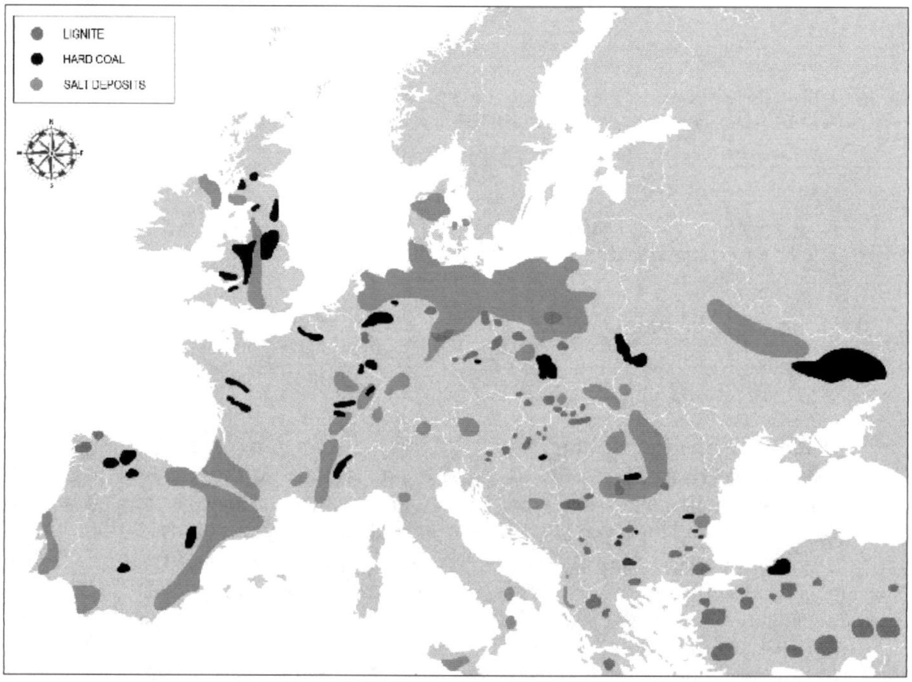

Figure 1. Coal mining areas and salt deposits in the EU.

In this paper, an analytical model has been developed in MATLAB to investigate the thermodynamic performance of A-CAES systems in lined tunnels. Polymer material and steel as sealing layers, concrete lining and rock mass are considered as model materials around the compressed air. In addition, operating air pressures from 4.5 to 7.5 MPa have been employed during air charge and discharge processes. Air mass flow rates of 0.19 and 0.27 kg s^{-1} have been used in the charge processes for polymer material and steel, respectively. The air temperature and pressure, and the temperature and heat flux in the sealing layer, concrete and rock mass were studied for 100 cycles of compression and decompression.

2. Methodology

2.1. Analytical model

The problem statement is shown in Figure 2. A lined tunnel with 4 m in diameter and 0.4 m thick concrete lining has been considered in the model. The values of the r1, r2, r3 and r4 are 1.58, 1.6, 2 and 3 m, respectively. In the analytical model developed in MATLAB, the energy equation has been applied as shown in Equation (1). In this equation, $\dot{Q}$ is the net heat transfer, in W, $\dot{W}$ is the net work, e is the specific energy, ρ is the air density, V is the reservoir volume, v is inlet velocity and S is the cross section.

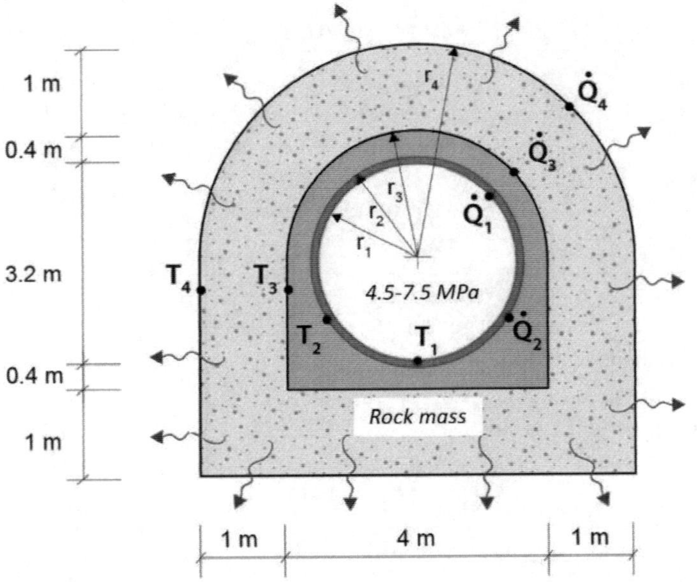

Figure 2. Cross-section of the underground lined tunnel. Compressed-air, sealing layer, concrete lining and rock mass.

The heat convection between the compressed air (fluid) and the sealing layer (solid) is represented in the Equation (2), where h is the heat transfer coefficient, in W $m^{-2}K^{-1}$ and m is the air mass flow rate, in kg s^{-1} and C_v is the specific heat at constant volume.

$$\dot{Q} - \dot{W} = \frac{\partial}{\partial t}\int \rho e\, dV + \oint \rho \hat{e}\,(\hat{v} dS) \tag{1}$$

$$-hA_1(T_a - T_1) = \frac{d}{dt}(mC_vT_a) - \dot{m}(C_vT_0 + RT_0 + \frac{v^2}{2}) \tag{2}$$

$$m = m_0 + \dot{m}\, t \tag{3}$$

The air temperature increases when the air pressure increases inside the reservoir. The air temperature is calculated by applying Equation (4). The temperature on the sealing layer (T_1), concrete lining (T_2) and rock mass wall (T_3) are obtained by applying non-stationary heat transfer Equations (5-7). An inlet temperature of 310 K is considered by the air.

$$\frac{dT_a}{dt} = \frac{\dot{m}}{m}\left[\frac{C_p}{C_v}T_0 + \frac{1}{2C_v}\left(\frac{\dot{m}}{m}L\right)^2 - T_a\right] - \frac{h2\pi r_1 L}{C_v m}(T_a - T_1) \tag{4}$$

The temperature on the sealing layer (T_1), concrete lining (T_2) and rock mass wall (T_3) are obtained by applying non-stationary heat transfer Equations (5-7). An inlet temperature of 310 K is considered by the air.

$$\frac{dT_1}{dt} = \frac{2U_h}{m_sC_{ps}}(T_a - T_1) - \frac{2U_s}{m_sC_{ps}}(T_1 - T_2) - \frac{dT_2}{dt} \tag{5}$$

$$\frac{dT_2}{dt} = \frac{2U_s}{m_cC_{pc}}(T_1 - T_2) - \frac{2U_c}{m_cC_{pc}}(T_2 - T_3) \tag{6}$$

$$\frac{dT_3}{dt} = \frac{2U_c}{m_{ss}C_{pss}}(T_2 - T_3) - \frac{2U_{ss}}{m_{ss}C_{pss}}(T_3 - T_4) \tag{7}$$

The boundary conditions for both polymer and steel sealing layers (air mass flow rates in charge and discharge processes, compression and decompression time and inlet temperature in the underground reservoir are presented in Table 1.

Table 1. Boundary conditions.

Boundary conditions	Polymer material	Steel
Mass flow rate - charge (kg/s)	0.19	0.27
Mass flow rate - discharge (kg/s)	-0.38	-0.45
Compression time (h)	8	8
Stored time (h)	6	6
Decompresion time (h)	4	4
Inlet temperature (K)	310	310

2.2. *Material properties*

The thermal properties of the materials used in the analytical model are shown in Table 2. The air is considered as compressible fluid and steel and polymer material (sealing layers), reinforced concrete and rock mass are considered as solid domains in the analytical model.

Table 2. Thermal properties of the materials used in the model.

Material	Specific Heat ($J\ kg^{-1}K^{-1}$)	Thermal conductivity ($W\ m^{-1}K^{-1}$)	Density ($kg\ m^{-3}$)
Air	1,006	0.0242	1.2
Concrete	1,000	1.6	2,500
Rock mass	700	5	2,600
Steel	500	45	7,800
Polymer material	1,950	0.095	920

3. Results and discussion

The analytical model has been conducted for 100 compression and decompression cycles using polymer material and steel as sealing layers. The temperature and pressure variations were observed during the simulations.

3.1. *Polymer material*

The temperature of air, polymer and concrete is shown in Figure 3a for the first cycle using polymer material as sealing layer. As shown in Figure 3b, the air temperature varies between 315 (charge) and 286 K (discharge) from the first cycle. In addition, the temperature of the polymer material and concrete is constant during the 100 cycles. The evolution of the air pressure is represented in Figure 4a for the first cycle (day 1). Due to the reduction of the air temperature, the pressure decreases from 7.5 to 7 MPa in the stored time (6 h). The heat transfer through the sealing layer and concrete lining is shown in Figure 5a for the first cycle. The heat flux reaches 140 and -85 W m^{-2} in the charge and discharge processes, respectively. As indicated in Figure 5b, from the first cycle the surface heat flux decreases down to 55 and -95 W m^{-2} in the charge and discharge processes through the polymer wall. The surface heat flux through the concrete wall reaches 52 and -75 W m^{-2} in the compression and decompression stages.

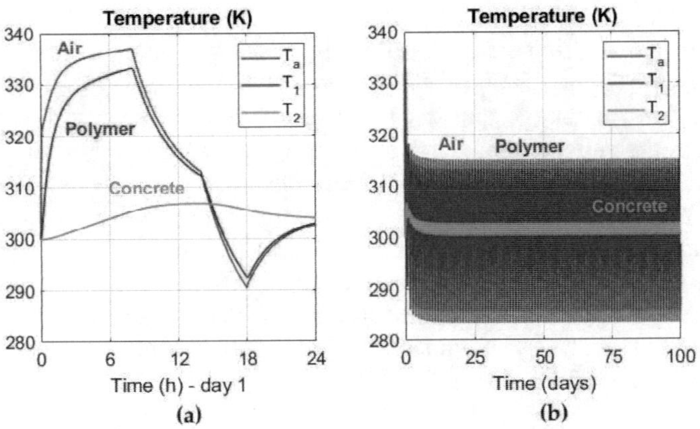

Figure 3. Temperature of air, polymer and concrete lining for 100 charging and discharging cycles. (a) First cycle; (b) 100 days of operation; (c) 100th cycle.

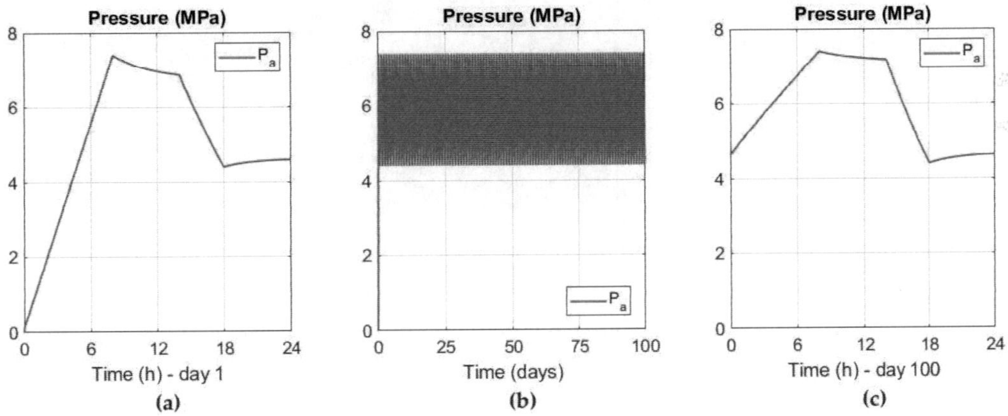

Figure 4. Air pressure for 100 charging and discharging cycles considering polymer as sealing layer. (a) First cycle; (b) 100 days of operation; (c) 100th cycle.

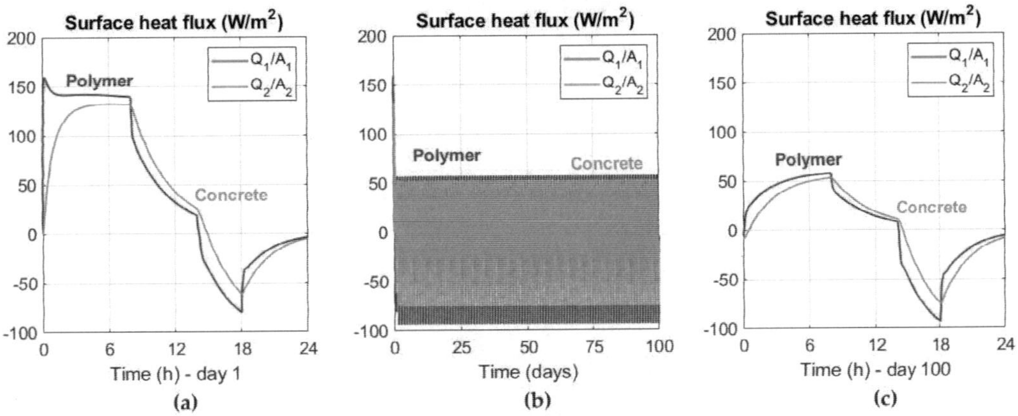

Figure 5. Surface heat flux of polymer and concrete lining for 100 charging and discharging cycles. (a) First cycle; (b) 100 days of operation; (c) 100th cycle.

3.2. Steel

The results obtained in the simulations considering steel as sealing layer are presented in this section. Note that the thermal conductivity increases up to 45 W m^{-1} K^{-1}. Figure 6b shows the temperature of air, sealing layer and concrete for 100 cycles of charge and discharge using steel as sealing layer. The variations in the air temperature within the subsurface reservoir are reduced with respect to the values obtained for the polymer material. The air temperature ranges between 305 and 297 K when steel is used as sealing layer. The air pressure for the first cycle using steel is shown in Figure 7a. The reduction of the air pressure during the stored time decreases when steel is used as sealing layer.

The surface heat flux through the sealing layer and concrete lining for the first cycle is presented in Figure 8a when steel is used as sealing layer. The heat transfer reaches 200 and -150 W m^{-2} in compression and decompression stages, respectively. Compared to polymer material, the heat transfer increases significantly when steel is used as sealing layer.

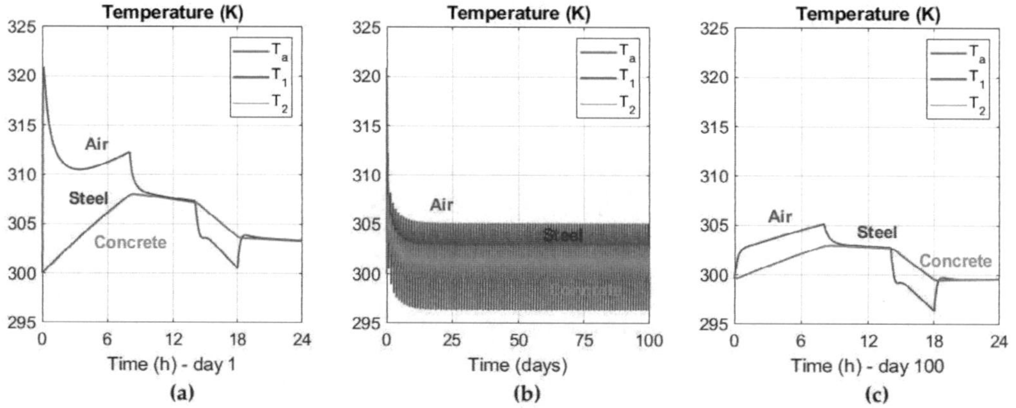

Figure 6. Temperature of air, steel and concrete lining for 100 charging and discharging cycles considering steel as sealing layer. (a) First cycle; (b) 100 days of operation; (c) 100th cycle.

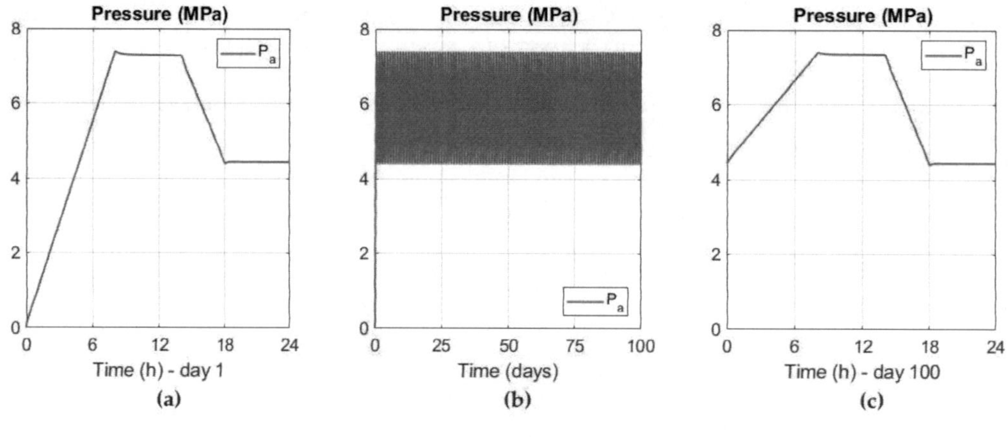

Figure 7. Air pressure for 100 charging and discharging cycles using steel as sealing layer. (a) First cycle; (b) 100 days of operation; (c) 100th cycle.

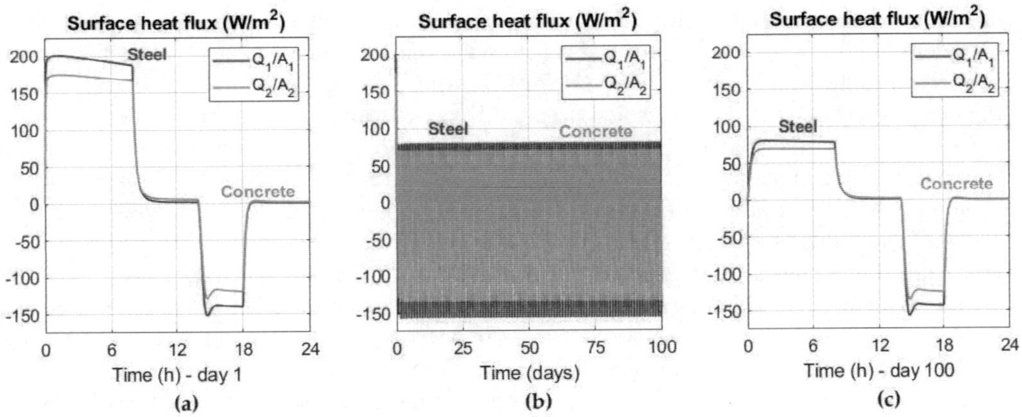

Figure 8. Surface heat flux of steel and concrete lining for 100 charging and discharging cycles using steel as sealing layer. (a) First cycle; (b) 100 days of operation; (c) 100th cycle.

Variations in air temperature decrease when the material used as sealing layer has a higher thermal conductivity. Finally, the storage capacity increases when the thermal conductivity of the materials around the fluid increases, and therefore the heat transfer through the sealing layer also increases.

4. Conclusions

An analytical model using MATLAB has been conducted to investigate the thermodynamic performance of A-CAES systems in lined tunnels. Two different sealing layers, polymer material and steel, have been considered as sealing layers. In addition, a 0.4 m thick concrete lining and a 1 m rock mass have been employed around the sealing layer. Air mass flow rates of 0.19 and 0.27 kg s^{-1} have also been considered in the charge processes for polymer material and steel, respectively. The air temperature and pressure within the reservoir, and the heat transfer and temperature in the materials considered around the fluid have been analyzed for 100 compression and decompression cycles between 4.5 and 7.5 MPa. The results obtained show that the air temperature fluctuations decrease when the thermal conductivity of the sealing layer increases. The heat transfer through the sealing layer reaches -150 and -95 W m^{-2} for steel and polymer material, respectively. Finally, the storage capacity increases when the air temperature variations decrease and the heat transfer around the fluid increases.

5. References

[1] Pujades E, Orban P, Archambeau P, Erpicum S, Dassargues A. Numerical study of the Martelange mine to be used as underground reservoir for constructing an Underground Pumped Storage Hydropower plant. Adv. Geosci. 45, 51–56 (2018).

[2] Pujades, E.; Orban, P.; Archambeau, P.; Kitsikoudis, V.; Erpicum, S.; Dassargues, A. Underground Pumped-Storage Hydropower (UPSH) at the Martelange Mine (Belgium): Interactions with Groundwater Flow. Energies, 13, 2353 (2020).

[3] Menendez J, Loredo J, Fernandez-Oro JM, Galdo M. Energy storage in underground coal mines in NW Spain: Assessment of an underground lower water reservoir and preliminary energy balance. Renew. Energy 134:1381-1391 (2019).

[4] Pujades E, Orban P, Bodeux S, Archambeau P, Erpicum S, Dassargues A. Underground pumped storage hydropower plants using open pit mines: how do groundwater exchanges influence the efficiency? Appl Energy 190:135–46 (2017).

[5] Menendez J, Loredo J. Use of closured open pit and underground coal mines for energy generation: Application to the Asturias Central Coal Basin (Spain). E3S Web of Conf ; 80: 01005 (2019).

IOP Conf. Series: Earth and Environmental Science **897** (2021) 012008 doi:10.1088/1755-1315/897/1/012008

[6] Menendez J, Fernandez-Oro JM, Galdo M, Loredo J. Pumped-storage hydropower plants with underground reservoir: Influence of air pressure on the efficiency of the Francis turbine and energy production. Renew. Energy 143:1427-1438 (2019).

[7] Menendez J, Ordóñez A, Álvarez R, Loredo J. Energy from closed mines: Underground energy storage and geothermal applications. Renew Sustainable Energy Rev 108:498–512 (2019).

[8] Alvarez A, Menendez J, Bernardo-Sanchez A, Galdo M, Loredo J, Fernandez-Oro J.M. Thermodynamic analysis of compressed air energy storage (CAES) reservoirs in abandoned mines using different sealing layers. App Sci 11, 2573 (2021).

[9] Schmidt F, Menendez J, Konietzky H, Pascual-Muñoz P, Castro J, Loredo J, Bernardo-Sanchez A. Converting closed mines into giant batteries: Effects of cyclic loading on the geomechanical performance of underground compressed air energy storage systems. J Energy Storage 32, 101882 (2020).

[10] Zhou S, Xia C, Zhao H, Mei S, Zhou Y. Numerical simulation for the coupled thermo-mechanical performance of a lined rock cavern for underground compressed air energy storage. Journal of Geophysics and Engineering, 14(6), 1382–1398 (2017).

[11] Kushnir R, Dayan A, Ullmann A. Temperature and pressure variations within compressed air energy storage caverns. Int. J. Heat Mass Transf. 55, 5616–5630 (2012).

IOP Conf. Series: Earth and Environmental Science **897** (2021) 012009 doi:10.1088/1755-1315/897/1/012009

Preliminary symptoms assessment of typical faults related to the fans and humidifiers of HVAC systems based on experimental data collected during Italian summer and winter

A Rosato[1,*], F Guarino[1,*], M El Youssef[1], S Sibilio[1] and L Maffei[1]

[1]Department of Architecture and Industrial Design, University of Campania Luigi Vanvitelli, Via San Lorenzo 4, 81031 Aversa, Italy

antonio.rosato@unicampania.it, francesco.guarino@unicampania.it

Abstract. The symptoms associated to the occurrence of typical faults in a heating, ventilation and air-conditioning (HVAC) system, including a single duct dual fan constant air volume air-handling unit, have been experimentally characterized. The operation of the HVAC unit with 3 artificially forced faults ((1) reduced velocity of the supply air fan, (2) reduced velocity of the return air fan, (3) the valve supplying the humidifier kept always closed) has been analysed and compared with that of healthy operation of the same plant under very similar boundary conditions (outside air temperature and initial indoor air temperature) during Italian summer and winter in order to preliminarily assess (i) the effects on the main operating parameters, and (ii) generate preliminary operation data to assist further research in fault detection and diagnosis of HVAC systems.

1. Introduction

The building sector plays a vital role in reducing greenhouse gases (GHG) emissions. In 2017, globally, the building sector accounted for 21% of the world's final energy use and it was responsible for 17% of the world's total energy-related GHG emissions [1]. Heating, ventilation and air-conditioning (HVAC) systems account for almost 50% of building energy consumption and about 10-20% of total energy demand in developed countries [2]. The integration of renewable energy sources into HVAC units is considered as one of the best options to reduce fossil fuel demand and, thus, related GHG emissions [3]. HVAC systems are often run under faulty conditions due to lack of proper maintenance, failure of components or incorrect installation. A study conducted on more than 55.000 Air Handling Units (AHUs) showed that a fraction of 90% runs with one or multiple faults [4]. Lin et al. [5] estimated that an effective identification and diagnosis of the faults in HVAC systems could save from 15% to 30% of the total energy consumed by buildings. Automated Fault detection and diagnosis (AFDD) is an automated process of detecting deviations from normal or expected operation (faults) and diagnosing the type of problem or its location [6-8]. AFDD technologies can offer several interrelated benefits, including energy savings and improved operational efficiency, utility cost savings, streamlining operations and maintenance processes as well as support for continuous energy management practices [5-7]. AFDD strategies represent one of the most active areas of research as well as a very fast-growing market segment in the context of building analytics technologies [5-7]. The methods used for performing AFDD analyses can be classified in (i) quantitative model-based, (ii) qualitative model-based and (iii) data driven-based [5,7]. The last category includes methodologies exploiting operational data collected from the system under investigation [5,7]. Data-driven AFDD approaches have gained encouraging results thanks to their flexibility in practical applications as well as the developments in data analytics [5,6]; nevertheless, the adoption of data-driven AFDD models is mainly limited by the lack of labelled data due to the fact that the architecture of sensor networks in HVAC units is not designed for AFDD purposes and, therefore, some important variables are generally not measured. Even if several scientific works focusing on data-driven AFDD methods are available in current literature [6,7], most of them refer to the ASHRAE RP-1312 data set dated 2011 [9]; relatively few studies give detailed information on how the faults are experimentally implemented in a real HVAC system or modelled [7,9]; even fewer studies describe how their simulated faulty operation data are validated [7,9]; nearly all the studies only consider one HVAC operation condition with changing weather conditions [9]; not many studies quantitatively examine how various faults and fault severities impact energy consumption, user comfort, maintenance cost and equipment life cycle [9]. According to [5], additional works to characterize the faults' prevalence based on field empirical data could prove valuable in guiding future AFDD development and implementation efforts. Lin et al [5] also highlighted that future AFDD studies should focus on expansion of test datasets as well as their provision for public use. In this paper, the symptoms associated to the occurrence of typical faults in the HVAC system serving the integrated test room of the SENS i-Lab of the Department of Architecture and Industrial Design of the University of Campania Luigi Vanvitelli (Aversa, south of Italy) have been experimentally characterized. The operation of the system with 3 artificially forced faults ((1) reduced velocity of supply air fan, (2) reduced velocity of return air fan, (3) humidifier valve kept always closed) has been analysed and compared with that of healthy operation of the same plant under similar boundary conditions (outside air temperature and initial indoor air temperature) during summer and winter. The analysis has been performed with the aim of (i) preliminarily assessing the effects on key operating parameters, and (ii) generating preliminary operation data to assist further research in fault detection and diagnosis of HVAC systems.

2. Description of the experimental set-up

The SENS i-Lab is an innovative, multi-sensorial and multi-purpose laboratory of the Department of Architecture and Industrial Design of the University of Campania Luigi Vanvitelli (Aversa, south of Italy, latitude: 40°58'21" N, longitude: 14°12'26" E). It mainly consists of an integrated test room served by a typical HVAC system, including a single duct dual fan constant air volume air-handling unit, able to control indoor air temperature, indoor air relative humidity, indoor air velocity and indoor air quality. The integrated test room is characterized by a floor area of 16.0 m^2, with a height of 3.6 m. Figure 1 reports the schematic of the AHU.

Table 1 describes the main characteristics of the AHU components. The heat carrier fluid is a mixture of water and ethylene glycol (90%/10% by volume). The AHU also includes the return air damper (D_{RA}), the outside air damper (D_{OA}), the exhaust air damper (D_{EA}) and the heat-recovery system damper (D_{HRS}). Three-way valves (V_{PreHC}, V_{PostHC}, V_{CC} and V_{HUM}) supply the pre-heating coil, the post-heating coil, the cooling coil and the humidifier, respectively.

The AHU is fully equipped in order to monitor, control and record the key parameters of the system. The main characteristics of the AHU sensors are reported in table 2. The AHU is operated according to a specific control logic. The following parameters are manually set (and eventually modified during the test) by end-users: (i) the desired targets of both indoor air temperature ($T_{SP,Room}$) and relative humidity ($RH_{SP,Room}$) to be achieved inside the test room; (ii) the deadbands DB_T and DB_{RH} for both $T_{SP,Room}$ and $RH_{SP,Room}$, respectively; (iii) the velocity of both the return air fan (OL_{RAF}) and the supply air fan (OL_{SAF}); (iv) the opening percentages of the return air damper (OP_{DRA}), the outside air damper (OP_{DOA}) and the exhaust air damper (OP_{DEA}); (v) the activation of the static heat recovery system damper (OP_{DHRS}). Once the previous parameters are manually set by end-users, the opening percentages (OP_{V_PreHC}, OP_{V_PostHC}, OP_{V_CC} and OP_{V_HUM}) of the valves supplying the pre-heating coil, the post-heating coil, the cooling coil and the humidifier are automatically managed in the range 0÷100% by PID (proportional-integral-derivative) controllers in order to achieve the desired targets inside the test room. However, alternatively the end-users are allowed to alter the components' operation based on specific research purposes by manually forcing (at the beginning or during the test) the opening percentages of the valves supplying the pre-heating coil, the post-heating coil, the cooling coil and the humidifier.

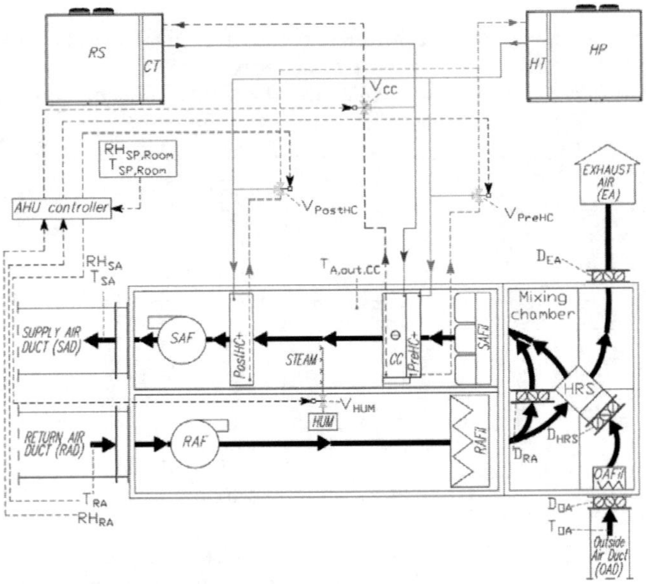

Figure 1. Schematic of the AHU.

Table 1. Characteristics of the main AHU components.

Supply air fan (SAF)/Return air fan (RAF)	Nominal supply/return air flow rate (m^3/h)	600/600
Cross flow heat recovery system (HRS)	Nominal efficiency (%)/recovery capacity (kW)	74.7/3.1
Pre-heating coil (PreHC)	Nominal heating capacity (kW) Nominal heat carrier fluid/air flow rate (m^3/h)	4.1 0.71/600
Cooling coil (CC)	Nominal cooling capacity (kW) Nominal heat carrier fluid/air flow rate (m^3/h)	5.0 0.86/600
Humidifier (HUM)	Nominal steam capacity (kg/h)/power (kW)	5.0/3.7
Post-heating coil (PostHC)	Nominal heating capacity (kW) Nominal heat carrier fluid/air flow rate (m^3/h)	5.0 0.86/600
Heat pump (HP)/Refrigerating system (RS)	Nominal capacity (kW) Nominal input power (kW) Nominal heat carrier fluid flow rate (m^3/h)	14.0/13.4 4.75/4.48 2.41/2.31

Table 2. Characteristics of the AHU sensors.

Monitored Parameter	Measuring range	Accuracy
Return air temperature T_{RA} / Supply air temperature T_{SA}	0 ÷ 50 °C	±0.8 K
Return air relative humidity RH_{RA} / Supply air relative humidity RH_{SA}	0 ÷ 100%	±3%
Outside air temperature T_{OA} / Cooling coil outlet air temperature $T_{A,out,CC}$	-50 ÷ 50 °C	±0.75 K

The refrigeration device operates to maintain a temperature T_{CT} of 7 °C inside the 75 liters cold tank (CT), while the heat pump is activated in order to achieve a temperature T_{HT} of 45 °C inside the 75 liters hot tank (HT). Air flow rate moved by the supply air fan can be varied between 0 (OL_{SAF} = 0%) and 1080 m^3/h (OL_{SAF} = 100%), while air flow rate of the return air fan is in the range from 0 (OL_{RAF} = 0%) up to 1460 m^3/h (OL_{RAF} = 100%). The parameters OP_{DRA}, OP_{DOA} and OP_{DEA} can be varied in the range 0–100%, where 100% means that the damper is fully open. The parameter OP_{DHRS} can be set to 100% (no heat recovery) or 0% (heat recovery takes place).

3. Experimental tests

Ten fault free and faulty experiments have been carried out to preliminarily investigate the HVAC behaviour under summer and winter conditions with reference to the occurrence of 3 specific faults. Table 3 describes the operating conditions of the fault free tests, while table 4 describes the operating conditions of the faulty tests.

Table 3. Operating conditions of the fault free tests.

	S1_FF	**W1_FF**	**W2_FF**	**W3_FF**
$\mathbf{T_{SP,Room}}$	26 °C	20 °C	20 °C	20 °C
$\mathbf{RH_{SP,Room}}$	50 %			
$\mathbf{DB_T}$	1 °C			
$\mathbf{DB_{RH}}$	5 %			
$\mathbf{OL_{SAF}}$	50 %			
$\mathbf{OL_{RAF}}$	50 %			
$\mathbf{OP_{V\_PostHC}}$, $\mathbf{OP_{V\_CC}}$, $\mathbf{OP_{V\_HUM}}$	0 ÷ 100 %			
$\mathbf{OP_{DRA}}$, $\mathbf{OP_{DHRS}}$	100 %			
$\mathbf{OP_{DOA}}$, $\mathbf{OP_{DEA}}$	20 %			
Date (dd/mm/yyyy)	21/07/2020	23/12/2020	29/12/2020	05/01/2021

Table 4. Operating conditions of the faulty tests.

	S2_F1	**S3_F2**	**S4_F3**	**W4_F1**	**W5_F2**	**W6_F3**
$\mathbf{T_{SP,Room}}$	20 °C	20 °C	20 °C	26 °C	26 °C	26 °C
$\mathbf{RH_{SP,Room}}$	50 %					
$\mathbf{DB_T}$	1 °C					
$\mathbf{DB_{RH}}$	5 %					
$\mathbf{OL_{SAF}}$	20 %	50 %	50 %	20 %	50 %	50 %
$\mathbf{OL_{RAF}}$	50 %	20 %	50 %	50 %	20 %	50 %
$\mathbf{OP_{V\_PostHC}}$, $\mathbf{OP_{V\_CC}}$	0 ÷ 100 %					
$\mathbf{OP_{V\_HUM}}$	0 ÷ 100 %	0 ÷ 100 %	0 %	0 ÷ 100 %	0 ÷ 100 %	0 %
$\mathbf{OP_{DRA}}$, $\mathbf{OP_{DHRS}}$	100 %					
$\mathbf{OP_{DOA}}$, $\mathbf{OP_{DEA}}$	20 %					
Date (dd/mm/yyyy)	31/07/2020	03/08/2020	18/09/2020	12/01/2021	14/01/2021	12/02/2021

Four tests (S1_FF, W1_FF, W2_FF, W3_FF) have been performed under fault free (FF) operating conditions; in more detail, the test S1_FF has been performed during the summer 2020 (S), while the other tests W1_FF, W2_FF and W3_FF have been carried out during the winter 2020/2021 (W). The remaining 6 tests (S2_F1, S3_F2, S4_F3, W4_F1, W5_F2, W6_F3) have been carried out while forcing the operation of specific AHU components (supply air fan, return air fan, humidifier valve) in order to artificially simulate 3 specific typical faults. The tests S2_F1, S3_F2 and S4_F3 have been performed during the summer 2020 (S), while the other tests W4_F1, W5_F2, W6_F3 have been carried out during the winter 2020/2021 (W). In greater detail, the fault 1 (F1), i.e. velocity of supply air fan forced at 20% (instead of the nominal value of 50%), has been artificially implemented during the tests S2_F1 and W4_F1; the fault 2 (F2), i.e. velocity of return air fan forced at 20% (instead of the nominal value of 50%), has been artificially implemented during the tests S3_F2 and W5_F2; the fault 3 (F3), i.e. opening percentage of the humidifier valve kept closed (instead of allowing its healthy operation with a value of OP_{V_HUM} in the range 0÷100%), has been artificially implemented during the tests S4_F3 and W6_F3. During all the experiments, the pre-heating coil has been always manually de-activated and the parameters indicated in table 2 have been measured every minute.

4. Comparison of boundary conditions between fault free and faulty tests

The faulty tests (S2_F1, S3_F2, S4_F3, W4_F1, W5_F2, W6_F3) have been carried out under very similar boundary conditions (i.e. outside air temperature as well as initial indoor return air temperature) with respect to the fault free tests (S1_FF, W1_FF, W2_FF, W3_FF). The fault free test S1_FF has very similar boundary conditions in comparison to the faulty tests S2_F1, S3_F2 and S4_F3 (in greater detail, the test S1_FF is longer and, therefore, different portions of the entire test S1_FF have been compared with the entire tests S2_F1, S3_F2 and S4_F3); the fault free tests W1_FF, W2_FF and W3_FF are characterized by very similar boundary conditions with respect to the faulty tests W6_F3, W4_F1 and W5_F2, respectively. Boundary conditions of the above-mentioned fault free and faulty experiments are compared in terms of outside air temperature and initial indoor return air temperature to assess their similarity. Figures 2a, 2b and 2c compare the test S1_FF with the tests S2_F1, S3_F2 and S4_F3, respectively. Figure 2d compares the tests W2_FF and W4_F1, figure 2e compares the tests W3_FF and W5_F2 and figure 2f compares the tests W1_FF and W6_F3.

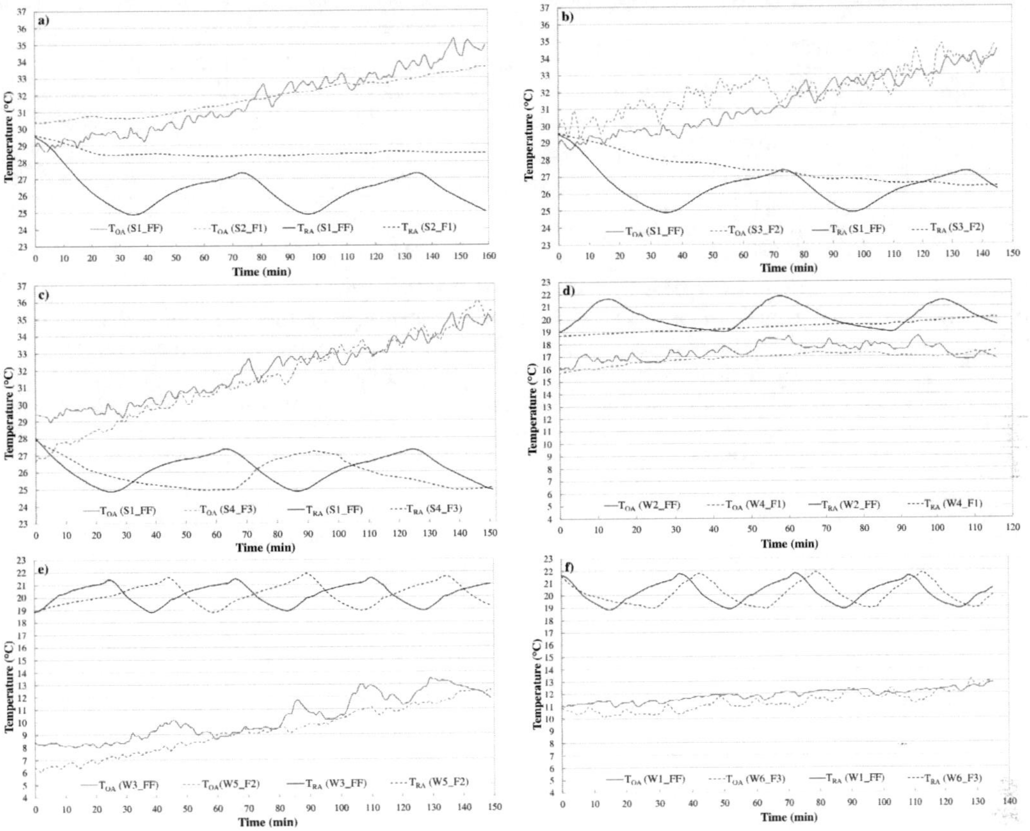

Fig. 2. Comparison of fault free and faulty tests in terms of T_{OA} and T_{RA}: comparison between S1_FF and S2_F1 (a), S1_FF and S3_F2 (b), S1_FF and S4_F3 (c), W2_FF and W4_F1 (d), W3_FF and W5_F2 (e), W1_FF and W6_F3 (f).

Figures 2a-f highlight very similar boundary conditions in terms of T_{OA} and initial T_{RA} between the corresponding fault free and faulty tests. The comparison in terms of outside air temperature has also been performed by means of the following metrics (the average difference $\bar{\varepsilon}$, the average absolute difference $|\bar{\varepsilon}|$, the root mean square difference RMSD):

$$\varepsilon_i = T_{OA,Fault\_free,i} - T_{OA,Faulty,i} \tag{1}$$

$$\bar{\varepsilon} = \sum_{i=1}^{N} \varepsilon_i / N \tag{2}$$

$$|\bar{\varepsilon}| = \sum_{i=1}^{N} |\varepsilon_i| / N \tag{3}$$

$$\text{RMSD} = \sqrt{\sum_{i=1}^{N} [(\varepsilon_i - \bar{\varepsilon})^2] / N} \tag{4}$$

where N is the number of experimental data points, while $T_{OA,Fault\_free,i}$ and $T_{OA,Faulty,i}$ are, respectively, the experimental values of outside air temperature at time step i under healthy and faulty operations. Table 5 summarizes the values of $\bar{\varepsilon}$, $|\bar{\varepsilon}|$ and RMSD as result of comparisons of the above-mentioned fault free and faulty tests. The results highlight how the values of $\bar{\varepsilon}$, $|\bar{\varepsilon}|$ and RMSD are at most 1.09 °C, 1.12 °C and 0.89

IOP Conf. Series: Earth and Environmental Science **897** (2021) 012009 doi:10.1088/1755-1315/897/1/012009

°C, respectively. Therefore, it can be stated that, according to [9], the boundary conditions of the corresponding fault free and faulty tests are very similar; as a consequence, comparing the above-mentioned healthy and faulty tests is possible for assessing the impact of each investigated fault on the AHU behavior/performance.

Table 5. Comparison of healthy and faulty tests in terms of outside air temperature T_{OA}.

Fault free against faulty tests	$\bar{\varepsilon}$ (°C)	$\overline{\lvert\varepsilon\rvert}$ (°C)	RMSD (°C)
S1_FF compared with S2_F1	-0.15	0.77	0.88
S1_FF compared with S3_F2	-0.75	0.96	0.89
S1_FF compared with S4_F3	0.44	0.67	0.80
W2_FF compared with W4_F1	0.53	0.60	0.74
W3_FF compared with W5_F2	1.09	1.12	0.70
W1_FF compared with W6_F3	0.54	0.58	0.40

5. Experimental symptoms of typical faults on HVAC performance

In Figure 3 the experimental performance of the HVAC system during the fault free and faulty tests have been compared in terms of (1) return air temperature (T_{RA}), (2) air temperature at the outlet of the cooling coil ($T_{A,out,CC}$), (3) supply air temperature (T_{SA}), (4) return air relative humidity (RH_{RA}) and (5) supply air relative humidity (RH_{SA}). In particular, figures 3a, 3b and 3c compare the test S1_FF with the tests S2_F1, S3_F2, S4_F3, respectively. Figure 3d compares the tests W2_FF and W4_F1. Figure 3e compares the tests W3_FF and W5_F2. Figure 3f compares the tests W1_FF and W6_F3.

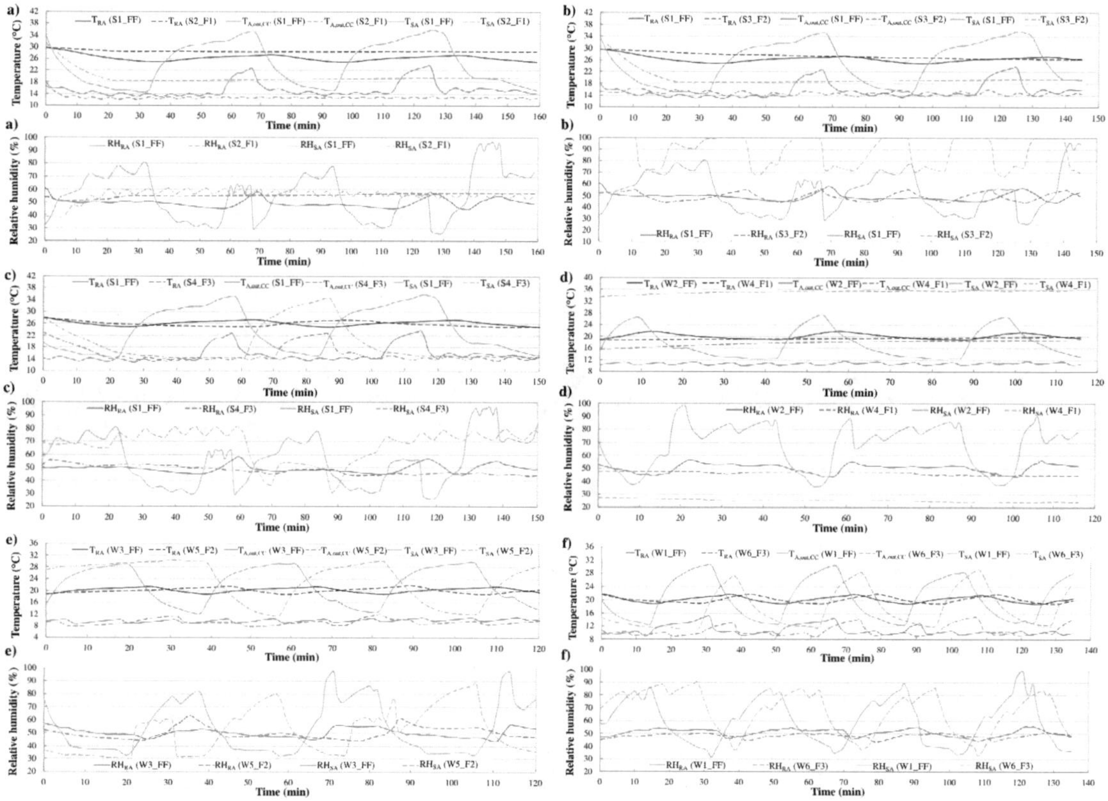

Fig. 3. Comparison of fault free and faulty tests in terms of T_{RA}, $T_{A,out,CC}$, T_{SA}, RH_{RA}, RH_{SA}: comparison between S1_FF and S2_F1 (a), S1_FF and S3_F2 (b), S1_FF and S4_F3 (c), W2_FF and W4_F1 (d), W3_FF and W5_F2 (e), W1_FF and W6_F3 (f).

The comparison between normal and faulty tests in terms of (1) $T_{A,out,CC}$, (2) T_{RA}, (3) RH_{RA}, (4) T_{SA} and (5) RH_{SA} has also been performed by contrasting the arithmetic mean μ as well as standard deviation σ of the above-mentioned 5 parameters calculated as follows:

$$\mu = \sum_{i=1}^{N} d_i / N \quad (5)$$

$$\sigma = \sqrt{\sum_{i=1}^{N} (d_i - \mu)^2 / N} \quad (6)$$

where N is the number of experimental points and d_i is the experimental value at time step i. The calculated values of μ and σ are reported in table 6; this table also shows the percentage difference %D between the arithmetic mean and the standard deviation of each above-mentioned parameters calculated during a faulty test with respect to those obtained during the similar fault free test.

Table 6. Differences between faulty and normal tests in terms of $T_{A,out,CC}$, T_{RA}, RH_{RA}, T_{SA} and RH_{SA}.

	ID Test	$T_{A,out,CC}$		T_{SA}		RH_{SA}		T_{RA}		RH_{RA}	
		μ (°C)	σ (°C)	μ (°C)	σ (°C)	μ (%)	σ (%)	μ (°C)	σ (°C)	μ (%)	σ (%)
Summer tests	S1_FF	15.52	2.57	24.39	3.23	55.17	18.25	26.36	1.03	49.43	3.23
	S2_F1	12.76	0.54	19.61	2.17	55.64	4.70	28.55	0.27	55.39	1.72
	%D	-17.78%	-79.11%	-19.59%	-33.04%	0.85%	-74.25%	8.31%	-73.69%	12.05%	-46.70%
	S1_FF	15.57	2.69	25.08	7.11	53.16	17.57	26.44	1.01	49.19	3.08
	S3_F2	14.82	0.61	18.07	1.48	88.25	12.39	27.42	0.89	49.50	3.55
	%D	-4.81%	-77.18%	-27.94%	-79.17%	66.02%	-29.44%	3.74%	-12.05%	0.63%	15.22%
	S1_FF	15.49	2.63	24.20	7.33	55.98	18.70	26.17	0.80	49.19	3.03
	S4_F3	15.68	2.70	19.21	6.26	67.22	11.87	25.89	0.83	48.07	3.18
	%D	1.22%	2.49%	-20.60%	-14.64%	20.08%	-36.52%	-1.06%	3.51%	-2.29%	5.10%
Winter tests	W2_FF	10.77	0.35	17.76	4.92	67.83	17.23	20.20	0.88	50.57	3.35
	W4_F1	17.81	0.97	35.18	0.59	25.27	0.92	19.36	0.38	46.50	1.16
	%D	65.37%	172.97%	98.07%	-87.95%	-62.75%	-94.64%	-4.13%	-56.45%	-8.06%	-65.34%
	W3_FF	9.90	0.60	22.38	5.96	53.16	19.34	20.22	0.78	50.09	3.33
	W5_F2	9.25	0.89	22.98	6.72	53.37	18.78	20.22	0.82	50.04	4.49
	%D	-6.58%	47.29%	2.69%	12.77%	0.39%	-2.93%	0.00%	5.55%	-0.10%	34.58%
	W1_FF	10.86	1.46	20.43	6.18	59.91	17.61	20.20	0.87	50.85	2.75
	W6_F3	10.65	1.57	17.22	5.65	67.35	15.82	20.11	0.91	48.27	1.71
	%D	-1.95%	7.93%	-15.74%	-8.50%	12.42%	-10.18%	-0.41%	3.78%	-5.07%	-37.77%

Table 6 indicates that, with respect to the cases without faults, in the case of the fault 1 (velocity of the supply air fan kept reduced at 20% instead of 50%):

- during summer, the maximum %D in terms of standard deviation is -79.1% for $T_{A,out,CC}$, while the minimum is -33.0% for T_{SA}; the maximum %D in terms of arithmetic mean is -19.6% for T_{SA}, while the minimum is +0.9% for RH_{SA}; during summer, the percentage difference in terms of standard deviation is significant for all the parameters ($T_{A,out,CC}$, T_{RA}, RH_{RA}, T_{SA}, RH_{SA}); the values of %D in terms of arithmetic mean can be assumed as not really relevant for all the parameters ($T_{A,out,CC}$, T_{RA}, RH_{RA}, T_{SA}, RH_{SA});
- during winter, the maximum %D in terms of standard deviation is 173.0% for $T_{A,out,CC}$, while the minimum is -56.5% for T_{RA}; the maximum %D in terms of arithmetic mean is 98.1% for T_{SA}, while the minimum is -4.1% for T_{RA}; during winter, the percentage difference in terms of standard deviation is relevant for all the parameters ($T_{A,out,CC}$, T_{RA}, RH_{RA}, T_{SA}, RH_{SA}); the %D in terms of arithmetic mean is significant only for $T_{A,out,CC}$, T_{SA} and RH_{SA}.

Table 6 also shows that, in comparison to the cases without faults, in the case of the fault 2 (velocity of the return air fan kept reduced at 20% instead of 50%):

- during summer, the maximum %D in terms of standard deviation is -79.2% for T_{SA}, while the minimum is -12.1% for T_{RA}; the maximum %D in terms of arithmetic mean is 66.0% for RH_{SA}, while the minimum is -0.6% for RH_{RA}; during summer, the values of %D in terms of standard deviation are significant for $T_{A,out,CC}$, T_{SA} and RH_{SA} only; the values of %D in terms of arithmetic mean can be assumed relevant only for T_{SA} and RH_{SA};
- during winter, the maximum %D in terms of standard deviation is 47.3% for $T_{A,out,CC}$, while the minimum is -2.9% for RH_{SA}; the maximum %D in terms of arithmetic mean is -6.6% for $T_{A,out,CC}$,

while the minimum is 0.0% for T_{RA}; during winter, the values of %D in terms of standard deviation are significant only for $T_{A,out,CC}$ and RH_{RA}; the values of %D in terms of arithmetic mean can be assumed as negligible for all the parameters ($T_{A,out,CC}$, T_{RA}, RH_{RA}, T_{SA}, RH_{SA}).

Finally, table 6 demonstrates that, with reference to the cases without faults, in the case of fault 3 (the opening percentage of the valve regulating the flow to the humidifier has been kept closed):

- during summer, the maximum %D in terms of standard deviation is -36.5% for RH_{SA}, while the minimum is 2.5% for $T_{A,out,CC}$; the maximum %D in terms of arithmetic mean is -20.6% for T_{SA}, while the minimum is -1.1% for T_{RA}; during summer, the values of %D in terms of standard deviation are significant only for RH_{SA}; the values of %D in terms of arithmetic mean are relevant for T_{SA} and RH_{SA} only;
- during winter, the maximum %D in terms of standard deviation is -37.8% for RH_{RA}, while the minimum is 3.8% for T_{RA}; the maximum %D in terms of arithmetic mean is -15.7% for T_{SA}, while the minimum is -0.4% for T_{RA}; during winter, the value of %D in terms of standard deviation is significant only for RH_{RA}; the values of %D in terms of arithmetic mean can be assumed as not really relevant for all the parameters ($T_{A,out,CC}$, T_{RA}, RH_{RA}, T_{SA}, RH_{SA}).

For each type of fault and parameter, a performance index has been assigned in table 7 with the following sings: "+" indicates that the fault causes substantial positive changes (greater than 20%) of %D; "-" indicates that the fault causes substantial negative changes (greater than -20%) of %D; "0" indicates that the fault causes not substantial changes (between -20% and 20%) of %D.

Table 7. Symptoms' relevance of the faults.

	Fault	$T_{A,out,CC}$		T_{SA}		RH_{SA}		T_{RA}		RH_{RA}	
	name	μ	σ	μ	σ	μ	σ	μ	σ	μ	σ
	F1	0	-	0	-	0	-	0	-	0	-
Summer tests	F2	0	-	-	-	+	-	0	0	0	0
	F3	0	0	-	0	+	-	0	0	0	0
	F1	+	+	+	-	-	-	0	-	0	-
Winter tests	F2	0	+	0	0	0	0	0	0	0	+
	F3	0	0	0	0	0	0	0	0	0	-

6. Conclusions

The symptoms of 3 typical faults related to the supply/return air fans and the humidifier valve of a typical HVAC system have been preliminarily assessed based on measured data. The preliminary analysis highlighted that: a velocity of the supply air fan reduced at 20% (instead of 50%) mainly affects the values of $T_{A,out,CC}$, T_{SA} and RH_{SA} during winter; a velocity of the return air fan reduced at 20% (instead of 50%) significantly influences the values of T_{SA} and RH_{SA} during summer; the impact is also relevant with reference to the values of RH_{SA} during summer in the case of the humidifier valve is kept fully closed. In the future the authors would like to perform additional fault free and faulty tests with the aims of (i) extending the analysis of the faults investigated in this work under different operating scenarios, as well as (ii) assessing the impact of new faults regarding sensors, controlled devices, equipment and controllers. In addition, it would be desirable to develop and experimentally validate a simulation tool able to predict the behaviour of the HVAC system in order to assist future researches on AFDD methods.

7. References

[1] Li B, Rowe A, Wild P 2021 Energy code effectiveness on GHG emission mitigation for single-family houses in Canada *J. Clean. Prod.* **299**

[2] Cao X, Dai X, Liu J 2016 Building energy-consumption status worldwide and the state-of-the-art technologies for zero-energy buildings during the past decade *Energy Build.* **128**

[3] Roselli C, Sasso M, Tariello F 2019 Assessment of a solar PV-driven desiccant-based air handling unit with different tracking systems *Sustain. Energy Technol. Assess.* **34**

[4] Yan K, Zhong C, Ji Z, Huang J 2018 Semi-supervised learning for early detection and diagnosis of air handling unit faults *Energy Build.* **181**

[5] Lin G, Kramer H, Granderson J 2020 Building fault detection and diagnostics: Achieved savings, and methods to evaluate algorithm performance *Build. Environ.* **168**

[6] Rosato A, Guarino F, Filomena V, Sibilio S, Maffei L 2020 Experimental calibration and validation of

IOP Conf. Series: Earth and Environmental Science **897** (2021) 012009 doi:10.1088/1755-1315/897/1/012009

a simulation model for fault detection of HVAC systems and application to a case study *Energies* **13**

[7] Piscitelli M S, Mazzarelli D M, Capozzoli A 2020 Enhancing operational performance of AHUs through an advanced fault detection and diagnosis process based on temporal association and decision rules *Energy Build.* **226**

[8] Guarino F, Filomena V, Maffei L, Sibilio S, Rosato A 2019 A Review of Fault Detection and Methodologies for Air-Handling Units *Glob. J. Energ. Technol.* **6**

[9] Wen J, Li S 2012 ASHRAE 1312-RP - Tools for evaluating fault detection and diagnostic methods for Air-Handling Units, American Society of Heating, Refrigeration and Air-Conditioning Engineers: Atlanta, GA, USA.

Acknowledgments

This work was undertaken as part of the program “PON FSE-FESR Ricerca e Innovazione 2014–2020” of the Italian Ministry of Education, University and Research, Action I.1 “Dottorati Innovativi con caratterizzazione industriale”.

IOP Conf. Series: Earth and Environmental Science **897** (2021) 012010 doi:10.1088/1755-1315/897/1/012010

Influence of the volume of ascorbic acid in the synthesis of copper nanoparticles mediated by chemical pathway and its stability over time

D. Asmat-Campos[1,2]*, D. Delfin-Narciso[2], L. Juárez-Cortijo[2] and R. Nazario-Naveda[2]

[1]Universidad Privada del Norte, Dirección de Investigación e Innovación, Trujillo, Perú
[2]Universidad Privada del Norte, Grupo de Investigación en Ciencias Aplicadas y Nuevas Tecnologías, Trujillo, Perú

*Corresponding author's e-mail: david.asmat@upn.edu.pe

Abstract. In the present investigation, the effect of ascorbic acid volume in the synthesis of copper nanoparticles (Cu NPs) mediated by chemical route and their stability over time was evaluated. For the synthesis, copper sulfate pentahydrate $CuSO_4$ ($5H_2O$) was used as a precursor agent and ascorbic acid (AA) as a reducing agent. Cu NPs was characterized by the following techniques: UV-Visible spectrophotometry to evaluate structural changes that are evidenced in the absorbance peak and atomic absorption spectrophotometry to define nanoparticulate concentrations material in the precipitated and supernatant phases generated. On the methodology it was possible to observe a controlled formation based on the increase in the volume of ascorbic acid in the presence of sodium hydroxide, noticing a production of Cu nanostructures with a tendency to oxidation over time. The UV-visible results showed characteristic surface plasmon resonance peaks of metallic copper for the colloid containing 1.2 mL of A.A; as well as a specific copper concentration of 0.14 ppm in the supernatant and 1519.1 ppm in the precipitate. It is also evidenced that the solution exhibits a rapid reaction on exposure to air by shifting the absorbance peak to 386 nm. In addition, it does not present notable photosensitivity with respect to exposure to sunlight.

1. Introduction

There is currently a growing demand for nanomaterials due to the novel properties they exhibit and the wide field of application, among which optoelectronics, electronics, food industry, textile industry, biosensors, catalysts, and especially medical and biological applications [1–5]. Among these materials there is special interest in the metallic nanostructures of copper, silver and gold because they show a wide spectrum of antimicrobial activity against different species of microorganisms such as fungi, Gram positive and Gram negative bacteria. However, among these, copper turns out to be more attractive for some researchers since, in addition to its excellent physical and chemical properties, it has low cost and abundant availability [6–9].

Previous studies have confirmed that Cu NPs have antimicrobial activity against E. coli and Staphylococcus among other species, as well as antifungal properties. The ability of Cu NPs to interact with and neutralize microorganisms lies in their morphological characteristics of shape, size, and stability to form other compounds; however, its synthesis in an uncontrollable atmosphere continues to be a challenge due to its electrical properties, which generate a rapid reaction on exposure to air [10–12]. It has been reported that copper oxidizes forming cuprous oxide (CuO) and cupric oxide (Cu2O) over time and its properties depend on a series of factors such as the type of reactants, concentrations, temperature, pH, among others. Nanostructures developed in aqueous medium or other fluids, for example, usually present a deposition phase which implies the existence of manometric material both in the precipitate and in the suspension. Under these conditions, the atomic absorption spectrophotometry test performed on the sediment and the supernatant can quantify the concentration of the analyte in a sample [13].

Copper oxide (CuO) is also an accessible material compared to noble metals and polymeric compounds can be made due to its stability [14]. A series of methods have been tested, the best known being the wet chemistry method, electrochemical synthesis, sonochemical synthesis, laser ablation and heat treatment, among others [15–17]. There are reports of copper nanoparticles obtained from a copper salt as a precursor and AA as a reducing agent, revealing an XRD pattern of a mixture of metallic Cu and copper (I) oxide of cubic shape and average size of 28.73 and 25.19 nm respectively, results that were corroborated with EDX spectroscopy and SEM analysis, the process was carried out at a temperature of 80 ° C and in aqueous solution, using a starch solution as dispersant without being able to avoid a later and slow oxidation process in which the role of AA plays an important role in slowing down oxidation and agglomeration, helping nanoparticles to obtain better stability. Different states of copper matrices have been identified by UV-vis spectroscopy: absorption bands at 250nm (Cu +), 320-370nm (cuprous peroxide), 400-440 (charge transfer bands of cuprous oxide), 510-580 (neutral Cu plasmon resonance) and 620-850 nm (dd transitions in Cu2 ions) [18–21].

One of the new areas that takes advantage of the antipathogenic properties of nanomaterials is agriculture. Copper and copper oxide ions have been reported as pesticides, fungicides, and fertilizers, being considered friendly control strategies because they require a low concentration of the metal [19]. This was proven in the effective inhibition of *Xanthomonas axonopodis pv. Punicae* (Xap), a bacterium that affects the cultivation of *Punica granatuma* in India, inducing great economic losses for the producers of the sector. Nanomaterials with a particle size less than 100 nm not only influence important events in plant life but could also reduce the dangerous effects of pesticides by improving the quality and yield of crops by delivering the macronutrients and micronutrients required with a reduction in the use of conventional chemical fertilizers and pesticides [22,23]. Alternative pesticides were also explored in the control of S. littoralis based on oxides derived from essential nutritive elements of the soil, for their evaluation, nanostructures of CuO and CaO synthesized by means of wet chemistry methods were elaborated, both showed very efficient pesticidal activity against S. littoralis, being those of CuO those of fast effect, compared to those of CaO. The efficiency shown by metal oxides poses a new role for nanomaterials in the generation and formulation of pesticides [24]. Copper and copper oxide also have optical properties, in a report made on CuO, which was exposed to ultraviolet radiation, it was concluded that there was no change in the structure, this is important if it is to perform optical testing or use these nanomaterials in capturing light for solar cells [25–27].

The objective of the present investigation is to determine the influence of the volume of A.A. in obtaining copper nanoparticles by chemical route and their stability over time, determining if there is an influence of exposure to air and light on the stability of the nanostructure under uncontrolled atmospheric conditions, evaluated by the changes that can be generated in the absorbance peak by UV-vis spectrophotometry. For future application purposes in pest control on agricultural crops.

2. Materials and methods

2.1. Materials

The chemicals used in the experiment are described below: Copper Sulfate II pentahydrate ($CuSO_4$ * 5 H_2O) p.a. EMSURE® ACS, ISO, Reag. PH Eur (CAS number 7758-99-8), ultrapure water with 0.2 μm final filtration (reverse osmosis water purification system / UV light / activated carbon / Thermo

Scientific brand), Sodium hydroxide (NaOH) supplied by Merck Millipore (CAS number 1310-73-2), and Ascorbic Acid (CAS number 50-81-7).

2.2. Synthesis of Copper Nanoparticles
The experimental scheme for the synthesis of copper nanoparticles by chemical route consisted of adding, dropwise, to a solution of 10 mL of copper sulfate [C] 0.05 M, 0.5 mL of sodium hydroxide [C] 7.5 M, by means of vigorous stirring until convert it into a mixture of intense blue color and viscous consistency and then add a series of variant volumes of AA [C] 1.13 M as in Figure 1, which are 0.3 mL, 0.6 mL, 1.2 mL, 1.8 mL and 2.4 mL. This was done at room temperature and under conditions of uncontrolled atmosphere and lighting. The stability of the synthesis product was then analyzed after 4 days by recording the absorbance by UV-Vis spectrophotometry during days 0 and 4.

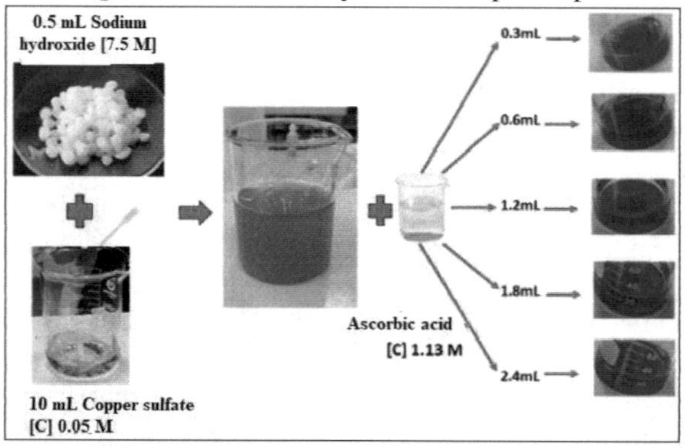

Figure 1. Process that shows color changes in the sample as an effect of varying the volume of A.A with values of 0.3 mL, 0.6 mL, 1.2 mL, 1.8 mL, and 2.4 mL from top to bottom in the extreme left of the image

2.3. Effect of light and air exposure on the absorbance peak
Due to the results reported on the influence of the atmosphere in the oxidation processes of Cu NPs, a follow-up was carried out to possible changes that the absorbance peak of a sample exposed to different light and air conditions may present, it was divided and stored in test tubes under the following conditions: one sample was left exposed to natural light and without a lid, a second sample remained with a lid and exposed to light, and the third sample was kept without exposure to light and with a lid; recording the ultraviolet absorption spectrum on days 0, 7 and 11 from the synthesis.
Also due to the presence of two phases of the mixture formed over time, precipitate and supernatant, an atomic absorption test was carried out to determine the concentrations of copper in the respective phases of the colloid.

2.4. Characterization of copper nanoparticles
The synthesis and experimental tests were carried out in the Physics Laboratory of the Department of Sciences of the Universidad Privada del Norte (UPN), Trujillo. Of the characterizations carried out by UV-vis spectrophotometry, the absorbance spectrum was monitored by means of a UV-Vis spectrophotometer (Shimadzu, UV 1900), in the range of 200 to 900 nm for all reactions. Thus, an analysis of the copper concentration in the precipitate and supernatant was also carried out with an Atomic Absorption Spectrophotometry test.

3. Results and Discussion
The obtained colloid has two defined absorption bands, one in the visible spectrum region and the other close to the ultraviolet. In the UV-Vis absorption spectrum that characterizes each sample according to the variation in the volume of A.A added to the base solution (Figure 2). The results are an indication that the colloid contains copper nanoparticles formed for the volumes of 0.6 and 1.2 mL

of A.A, since the peak close to the characteristic of metallic copper is observed in the visible range around 501nm. Thus, it can also be concluded that there is the presence of cuprous oxide nanoparticles, which are characterized by having an absorbance peak around 400 nm.

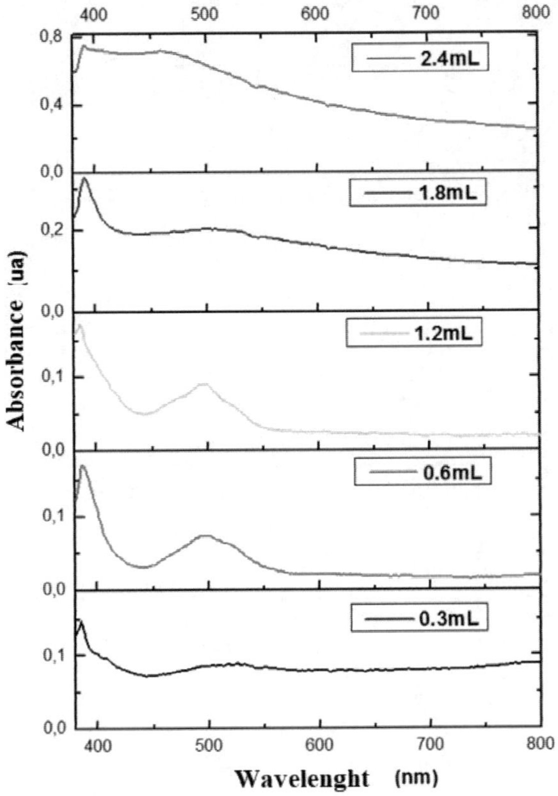

Figure 2. UV-vis spectrum of samples containing 0.3 mL, 0.6 mL, 1.2 mL, 1.8 mL and 2.4 mL of A.A

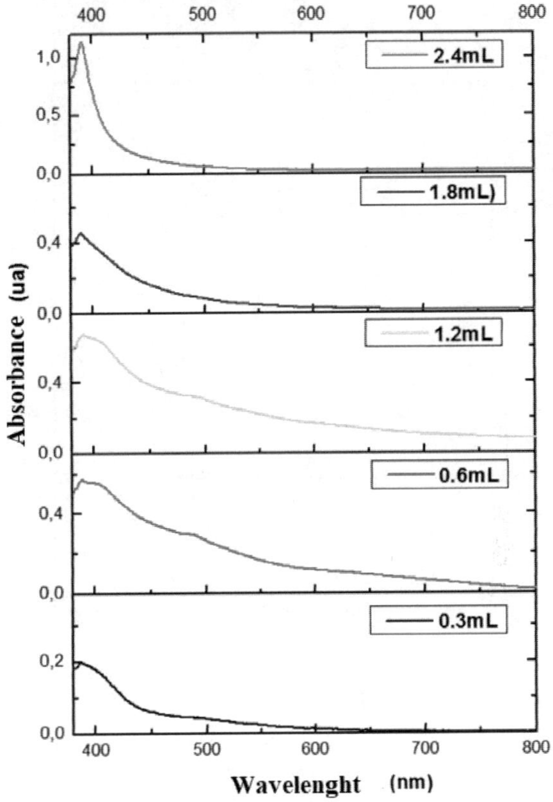

Figure 3. UV-vis spectrum of colloids containing 0.3 mL, 0.6 mL, 1.2 mL, 1.8 m and 2.4 mL of A.A after 4 days of synthesis.

However, volume additions above this value generate a shift towards the ultraviolet, evidence of a possible oxidation process in addition to a broadening of the peak indicating polydispersity in the size of the particles formed or agglomerations of them.

After four days, the spectrophotometry results obtained from the colloids stored in closed flasks, without exposure to air, show instability over time (Fig. 3), due to the shift of the 501 nm peak to the right for the samples with 0.3, 0.6 and 1.2 mL of AA, which register a shift to the left close to the ultraviolet range, which is an index of an oxidation process. A complete shift towards the 392 nm peak is also observed for the 1.8 and 2.4 mL volumes, indicating the possibility of the generation of cuprous oxide nanoparticles [15].

The spectrophotometric tests carried out to evaluate the stability to the exposure of air (Fig. 4) and light (Fig. 5) yielded the following results: After 7 days, the sample exposed to air (D-7-A) has a shift to the right at 385 nm unlike the unexposed sample (D-7-C) whose absorbance shows the existence of nanoparticles with a peak at 490 nm showing an incomplete oxidation process. In a similar way, the behavior is observed for the solution after 11 days (D-11-A and D-11-C) in which the sample open to air shows a greater tendency to shift the peak to the right side.

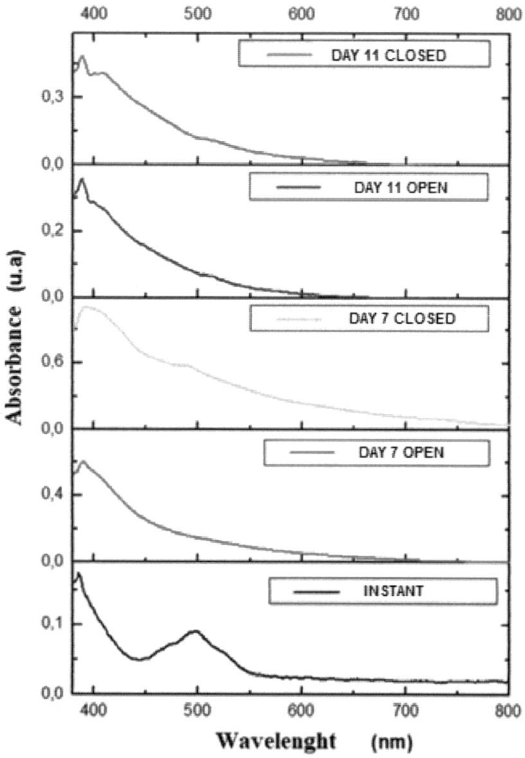

Figure 4. UV-vis spectrum of 1.2 mL AA samples at: 0 days (D-0), 7 days exposed to air (D-7-A), 7 days closed without air (D-7-C), 11 days exposed to air (D-11-A), 11 days closed without air (D-11-C).

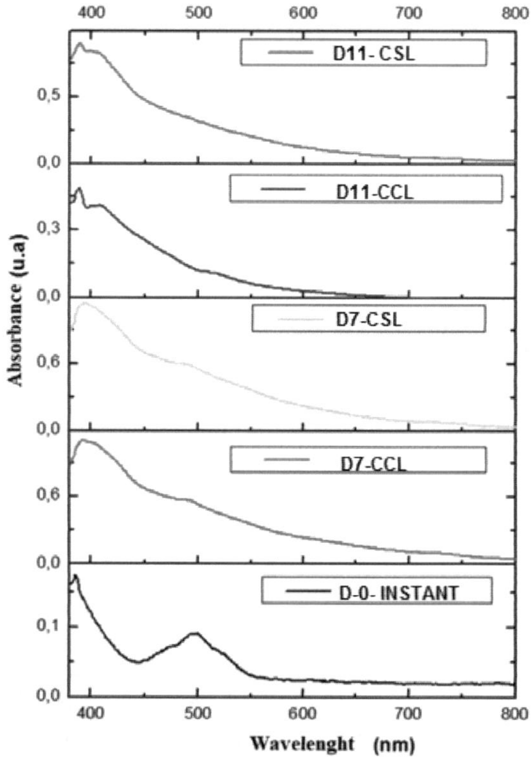

Figure 5. UV-vis spectrum of 1.2 mL AA samples at: 0 days (D-0), 7 days with light exposure (D7-CCL), 7 days without light exposure (D7-CSL), 11 days with light exposure (D11-CCL), 11 days without light exposure (D11-CSL).

Regarding the effect of exposure to light, there are no notable differences in the wavelength that defines the absorbance peak, at seven days both samples (D7-CCL and D7-CSL) have a peak shifted to the right in the visible range of 490 nm, indicative of the oxidation process in progress; similar behavior is observed in the case of the samples with 11 days of aging (D11-CCL and D11-CSL), with a predominance in the order of 386 nm, close to the characteristic peak of cuprous peroxide.
Due to the presentation of two phases that are formed when the sample is left to rest for 14 hours, an atomic absorption spectrophotometry was performed on both the precipitate and the supernatant, this in order to define in which phase we found the highest concentration of the desired metal.

Table 1. Precipitate and supernatant atomic absorption spectrophotometry of the NPS Cu Solution

Samples	[C]Cu (ppm) Samples NPs	[Cu]Real ppm Standard
Supernatant	0.14	295
Sediment	1519.1	30382

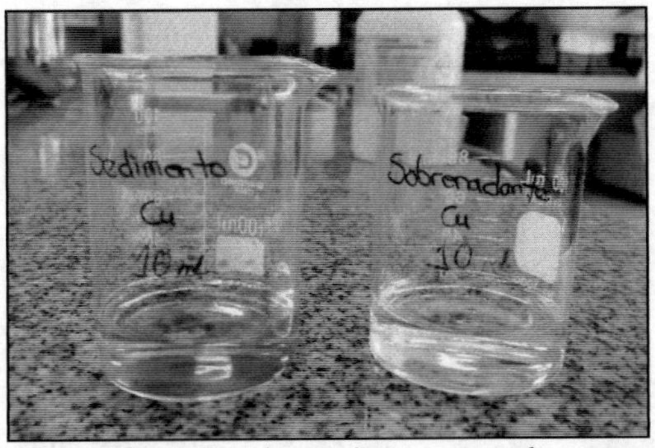

Figure 6. Atomic absorption spectrophotometry NP Cu chemical synthesis method, sediment, and supernatant analysis. (a) Digestated sample of NP Cu sediment cyan color. (b) Colorless supernatant NP Cu digested sample

4. Conclusions

The results obtained in this article suggest that copper nanoparticles are formed by using a volume of 1.2 mL of AA as a reducing agent because it presents a well-defined absorbance peak, which resulted in the order of 501 nm. However, this solution is not stable over time as it shows a shift of peak to the right, evidencing a possible oxidation process that converts copper into cuprous oxide with the passing of days and can even form cuprous peroxide. If the sample is stored exposed to air, the compound formed is not stable and is evidenced by a spectrum whose absorbance peak undergoes a more accelerated shift to the left, stabilizing in the order of 380 nm characteristic of copper peroxide.

It was also determined that the nanoparticles obtained are not photosensitive, as they do not present differences in absorbance in samples exposed to and without light, but they do react to prolonged exposure to air.

Thus, it was also concluded that in the two phases that occur of the colloid, precipitate and supernatant, the highest concentration of the metal is found in the precipitate with 1519.1 ppm compared to 0.14 ppm of the supernatant (Figure 6).

5. References

[1] Padilla-Vaca F, Mendoza-Macías C L, Franco B, Anaya-Velázquez F, Ponce-Noyola P and Flores-Martínez A 2018 El mundo micro en el mundo nano: importancia y desarrollo de nanomateriales para el combate de las enfermedades causadas por bacterias, protozoarios y hongos *Mundo Nano. Rev. Interdiscip. en Nanociencias y Nanotecnología* **11** 15–27

[2] Bordbar M M, Tashkhourian J, Tavassoli A, Bahramali E and Hemmateenejad B 2020 Ultrafast detection of infectious bacteria using optoelectronic nose based on metallic nanoparticles *Sensors Actuators B Chem.* **319** 128262

[3] Elamathi R, Ramesh R, Aravinthraj M, Manivannan M, Liakath Ali Khan F, Mphale K, Letsholathebe D, Kaviyarasu K, Kennedy J and Maaza M 2020 Investigation of structural and electrical properties of lithium cobalt oxide nanoparticles for optoelectronic applications *Surfaces and Interfaces* **20** 100582

[4] Kobayashi Y, Yasuda Y and Morita T 2016 Recent advances in the synthesis of copper-based nanoparticles for metal–metal bonding processes *J. Sci. Adv. Mater. Devices* **1** 413–30

[5] Oliveira F R, Galvão F M F, Do Nascimento J H O, Silva K K O S, Medeiros J I and Zille A 2015 Photocatalytic Properties of Sisal Fiber Coated with Nano Titanium Dioxide *Mater. Today Proc.* **2** 41–8

[6] Kiriyanthan R M, Sharmili S A, Balaji R, Jayashree S, Mahboob S, Al-Ghanim K A, Al-Misned F, Ahmed Z, Govindarajan M and Vaseeharan B 2020 Photocatalytic, antiproliferative and

antimicrobial properties of copper nanoparticles synthesized using Manilkara zapota leaf extract: A photodynamic approach *Photodiagnosis Photodyn. Ther.* **32** 102058

[7] Bashir F, Irfan M, Ahmad T, Iqbal J, Butt M T, Sadef Y, Umbreen M, Shaikh I A and Moniruzzaman M 2021 Efficient utilization of low cost agro materials for incorporation of copper nanoparticles to scrutinize their antibacterial properties in drinking water *Environ. Technol. Innov.* **21** 101228

[8] Rojas-Michea C, Morel M, Gracia F, Morell G and Mosquera E 2020 Influence of copper doping on structural, morphological, optical, and vibrational properties of ZnO nanoparticles synthesized by sol gel method *Surfaces and Interfaces* **21** 100700

[9] Yousefvand P, Mohammadi E, Zhuang Y, Bloukh S H, Edis Z, Gamasaee N A, Zanganeh H, Mansour F N, Heidarzadeh M, Attar F, Babadaei M M N, Keshtali A B, Shahpasand K, Sharifi M, Falahati M and Cai Y 2021 Biothermodynamic, antiproliferative and antimicrobial properties of synthesized copper oxide nanoparticles *J. Mol. Liq.* **324** 114693

[10] Cho K H, Park J E, Osaka T and Park S G 2005 The study of antimicrobial activity and preservative effects of nanosilver ingredient *Electrochim. Acta* **51** 956–60

[11] Argueta-Figueroa L, Mesta-Pichardo D, Torres-Gómez N, Martínez-Alvarez O, Paulino-González Á D and Arenas-Arrocena M C 2018 Acrílico termopolimerizable enriquecido con nanopartículas de cobre: evaluación antibacteriana y citotóxica *Mundo Nano. Rev. Interdiscip. en Nanociencias y Nanotecnología* **11** 45–60

[12] Cohen D, Soroka Y, Ma'or Z, Oron M, Portugal-Cohen M, Brégégère F M, Berhanu D, Valsami-Jones E, Hai N and Milner Y 2013 Evaluation of topically applied copper(II) oxide nanoparticle cytotoxicity in human skin organ culture *Toxicol. Vitr.* **27** 292–8

[13] Alsawafta M, Badilescu S, Packirisamy M and Truong V-V 2011 Kinetics at the nanoscale: formation and aqueous oxidation of copper nanoparticles *React. Kinet. Mech. Catal. 2011 1042* **104** 437–50

[14] Ren G, Hu D, Cheng E W C, Vargas-Reus M A, Reip P and Allaker R P 2009 Characterisation of copper oxide nanoparticles for antimicrobial applications *Int. J. Antimicrob. Agents* **33** 587–90

[15] Din M I and Rehan R 2016 Synthesis, Characterization, and Applications of Copper Nanoparticles *http://dx.doi.org/10.1080/00032719.2016.1172081* **50** 50–62

[16] Lopez R. G A and Ataucuri S. J C 2018 Estudio sobre metodologías aplicadas para la síntesis de nanopartículas de cobre (NPsCU) y plata (NPsAG) en diferentes formas y tamaños *Rev. Cient. Tec.* **27** 45

[17] Granata G, Yamaoka T, Pagnanelli F and Fuwa A 2016 Study of the synthesis of copper nanoparticles: the role of capping and kinetic towards control of particle size and stability *J. Nanoparticle Res. 2016 185* **18** 1–12

[18] Gawande M B, Goswami A, Felpin F X, Asefa T, Huang X, Silva R, Zou X, Zboril R and Varma R S 2016 Cu and Cu-Based Nanoparticles: Synthesis and Applications in Catalysis *Chem. Rev.* **116** 3722–811

[19] Jain S, Jain A, Kachhawah P and Devra V 2015 Synthesis and size control of copper nanoparticles and their catalytic application *Trans. Nonferrous Met. Soc. China* **25** 3995–4000

[20] Khan A, Rashid A, Younas R and Chong R 2015 A chemical reduction approach to the synthesis of copper nanoparticles *Int. Nano Lett. 2015 61* **6** 21–6

[21] Pestryakov A N, Petranovskii V P, Kryazhov A, Ozhereliev O, Pfänder N and Knop-Gericke A 2004 Study of copper nanoparticles formation on supports of different nature by UV–Vis diffuse reflectance spectroscopy *Chem. Phys. Lett.* **385** 173–6

[22] XG Z, CN X, Y T, E T, H Y and Y S 2000 Observation of charge stripes in cupric oxide *Phys. Rev. Lett.* **85** 5170–3

[23] Mondal K K and Mani C 2011 Investigation of the antibacterial properties of nanocopper against Xanthomonas axonopodis pv. punicae, the incitant of pomegranate bacterial blight *Ann. Microbiol. 2011 622* **62** 889–93

IOP Conf. Series: Earth and Environmental Science **897** (2021) 012010 doi:10.1088/1755-1315/897/1/012010

[24] Padhi S and Behera A 2021 Nano-enabled Approaches for the Suitable Delivery of Fertilizer and Pesticide for Plant Growth *Plant Growth-Promoting Microbes Sustain. Biot. Abiotic Stress Manag.* 355–94

[25] Ayoub H A, Khairy M, Elsaid S, Rashwan F A and Abdel-Hafez H F 2018 Pesticidal Activity of Nanostructured Metal Oxides for Generation of Alternative Pesticide Formulations *J. Agric. Food Chem.* **66** 5491–8

[26] Parveen F, Sannakki B, Mandke M V. and Pathan H M 2016 Copper nanoparticles: Synthesis methods and its light harvesting performance *Sol. Energy Mater. Sol. Cells* **144** 371–82

[27] Shi J, Abid A D, Kennedy I M, Hristova K R and Silk W K 2011 To duckweeds (Landoltia punctata), nanoparticulate copper oxide is more inhibitory than the soluble copper in the bulk solution *Environ. Pollut.* **159** 1277–82

IOP Conf. Series: Earth and Environmental Science **897** (2021) 012011 doi:10.1088/1755-1315/897/1/012011

A comparative approach in the utilisation of rice husk and empty palm bunch in the biomass boiler

Sivabalan Kaniapan[1], Kartikeyan Patma Nesan[2], Hamdan Ya[1], Suhaimi Hassan[1], Azizul Buang[2] and Mohd Safuan Zakaria[1]

[1] Mechanical Engineering Department, Universiti Teknologi Petronas, Seri Iskandar Perak, Malaysia
[2] Chemical Engineering Department, Universiti Teknologi Petronas, Seri Iskandar Perak, Malaysia

sivabalan_19001036@utp.edu.my, kartik_91@hotmail.com

Abstract. Growing world's population has immense contribution towards world economy and energy utilisation. The enormous usage of conventional fuel has contributed many environmental problems such as greenhouse gas emission (GHG), world climate change, and deterioration of human health. Recent study focuses on the generated power from EPB compared with methane in a typical biomass boiler. Also, there are very limited studies on the Air to Fuel (ATF) ratios value in boiler operation. In this paper, empty palm bunch (EPB) and rice husk (RH) have been selected as biomass fuel in the biomass boiler. The same recommended parameters of boiler and turbine was chosen for both EPB and RH feedstocks from previous study. Overall, the study proven to produce about 33% and 25% of energy from EPB and RH of what a methane (CH_4) can produce from the same amount of feeding rate, respectively, with EPB producing 13.31% of higher turbine power than RH. This directly contributes to the technical feasibility and adaptability of environmentally friendly elements by seizing the opportunity of carbon emission of conventional fuel and replacing it with natural resources such as EPB and RH which are part of the biomass fuel replacement regime. However, ATF ratio of RH is significantly minimal of what a CH_4 and EPB utilised to burn 1 kg of fuel. Therefore, EPB and RH would be suitable for future renewable biomass feedstock in comparison with conventional fuel for power generation purposes.

1. Introduction

Energy needs are essential for both social and economic growth of every nation in the world. The development of energy usage is expected to grow simultaneously with the world population index. According to the U.S Energy Information Administration (EIA) statistic [1], both non-OECD countries involving India, Indonesia, China, etc. and OECD countries such as Japan, Germany, U.K, etc. have shown a continuous increase in both energy consumption and population growth from 2010 to 2015 as shown in Figure 1 below. Today's world population reaches almost 7.7 billion people in 2019 and will continue to expand to 8.5 billion in 2030 [2]. In addition, according to the Department of Statistics Malaysia (DOSM), Malaysia population growth has increased from 28.6 million to approximately 32.4 million [3] in 2010 and 2018, respectively, which also causes an increment of 47.3% [4] from the overall expected power consumption for both domestic use and non-domestic use.

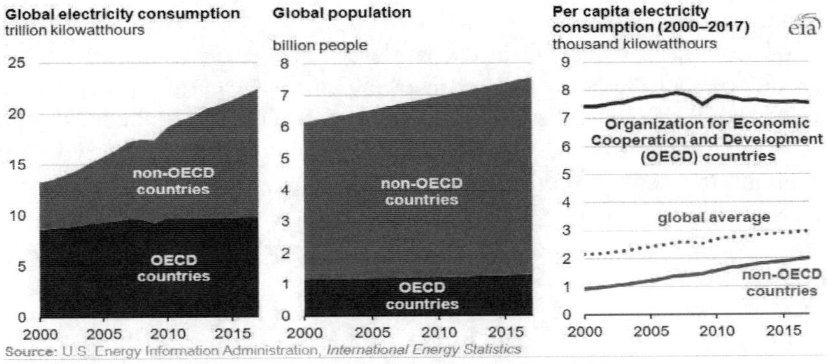

Figure 1. Shows global electricity consumption and global population growth from the year 2000 to 2015.

Moreover, power generation is also inter-related with the economic growth and living standards among individuals in the particular region. Energy is consumed predominantly in terms of electricity for cooking, washing, internet browsing, and others. However, nearly 86% [5] of generated electricity is harnessed from conventional fuel in Malaysia which is expected to extinct in another few decades. Also, the immense usage of fossil fuel for long term could potentially accumulate greenhouse gas emissions (GHG), world climate change, and others. Therefore, switching to biomass elements such as rice mill residues and oil palm residues is likely to be utilised as a power solution and indirectly able to reduce the conventional fuel usage due to the abundant availability of the feedstocks. Moreover, biomass is an exceptional and efficient option to be replaced conventional fuel as it is part of the renewable energy sources and will not extinct sooner. Biomass does not contribute to the release of carbon emission even after combustion as it uses carbon during the growing stage through the process of photosynthesis. Therefore, empty palm bunch (EPB) and rice husk (RH) is chosen as biomass fuels for this study as potential fuel replacement.

2. Rice mill residues

Rice has always been a staple food for many years among individuals worldwide, particularly in Southeast Asia, where nearly 80% of the world's rice is consumed by Asians [6]. The world rice production in 2019 is nearly 755 million tonnes, with an increase of 8.79% from the year 2010 [6]. In Malaysia, rice production has increased from 2.5 billion tonnes in 2018 to 2.9 billion tonnes in 2019 [6]. In line with rice's spur production rate, the accumulation of rice residues or waste has been a multi-faceted problem faced by the growers. Rice waste comes in a various forms and shapes, such as rice husk (RH) and rice straw (RS). Almost 40 % of RS and 20 % of RH are produced for every kilogram of harvested paddy depending on the cultivation method [7].

According to Kadam et al. [8], around 1.35 tons of rice waste will remain in the field after producing 1 ton of rice grain. Consequently, this rice waste was not properly discarded which causes environmental-related issues through the emission of greenhouse gases (GHG) from traditional open burning [9], short-cut to eradicate pest problem which eventually causes respiratory-related problem like asthma [10] due to the inhalation of suspended particles from this activity.

These rice residues (RH and RS) can be used in various thermochemical and biochemical processes to produce char, liquid biofuel, or steam for power generation. In 2017, it has been anticipated that around 638.03PJ of energy could be potentially harnessed from almost 0.77 billion tons of RS and 0.15 billion tons of RH [11]. Mohiuddin et al. [12] have estimated that about 70% rice husk is needed to produce 1328 GWh/ year of electricity with a cost of 47.36 cents/kWh in comparison with coal which needs 55.22cents to produce a kWh of power. Nevertheless, in Malaysia rice husk is still not being exploited for optimal utilisation in terms of power generation, S. M. Shafie [13] estimated that about 5652 GWh power could be produce from rice residue which accounted for 5.4% of the overall power demand in Malaysia.

3. Palm mill residues

Palm oil is one of the largest edible oil in the world accounted for nearly 71.48 million tonnes [14] surpassing other oil commodities such as sunflower oil and soybean oil. Palm oil is very distinctive in its own way, as it is edible oil which is also common in biofuel and bio-chemical production. Palm oil production is a very prominent business commodity in Malaysia where nearly 425 palm oil mills in operation throughout the region produces roughly 19.8 million tonnes of oil in 2019, carrying approximately 38.23 billion ringgit business [3], [15]. Apart from being the secondary palm oil producers globally, Malaysia carries a lot of research and development to yield more oil for every ton of fresh fruit bunch.

In line with the huge production of palm oil, Malaysia is producing large quantity of waste where 22-23 million tons of EPB is produced annually [16]. Empty Palm bunch (EPB) is the main source of palm biomass with huge market potential that can be used in enormous ways such as in power production and other waste to value-added processes. Currently, EPB is the least used element in power fuel production due to its high moisture absorption rate, lower bulk density, and lower calorific value in comparison with the conventional fuel properties [17]. However, EPB can be altered in a way that it fits with the current biomass boiler application to produce maximum power output.

Recent study of Nnaemeka Sp et al. [18] has only focused on the power yield from EPB in contrast with methane in biomass boiler. Also, there are very limited studies on the air to fuel ratios computation in biomass boiler operation. Thus, this research paper computes by comparing the usage of RH and EPB in biomass boiler through air to fuel stoichiometric derivations and the turbine output. Both EPB and RH have been focused as an important biomass material in this research article due to the abundant availability, lower emission rate and produces new employment opportunity.

4. Materials and method

Figure 2. Sample materials of EPB and RH.

The material of EPB and RH for this research article was procured from Palm Oil Mill, Teluk Intan, Perak, Malaysia and Sungai Manik Rice Mill, Perak, Malaysia, respectively. The EPB is received in a fibre form which has been pressed and shredded into small pieces in the palm oil mills, whereas the RH is received as it is from the production mill. Both EPB and RH were then left to dry in the oven under 105°C for about 24 hours to remove any excessive moisture. Both the dehydrated EPB and RH were then sent to the grinding section to produce fine powder and sieved to make even sizing of $0.30 < dp1 < 0.50$ mm. Finally, powdered EPB and RH were kept in the container as shown in Figure 2.

4.1. Tools and equipment

In this research paper, ultimate analysis (UA), proximate analysis (PA) and high heating value (HHV) of the feedstock are identified by using lab equipment such as CHNS analyser, thermogravimetric analyser, and bomb calorimetric analyser respectively.

4.2. Power generation method

In this research article, steam turbine has been used which involves three stages of conversion such as thermal energy dissipation through combustion of the biomass feedstock in the boiler (converting water into high pressure steam), followed by generation of kinetic energy via ejected steam into turbine (injection of high-pressure steam to rotate steam turbine), and finally converting the mechanical energy of turbine rotation into electrical power. Figure 3 below shows the overall representation of biomass power plant process conversion into electricity.

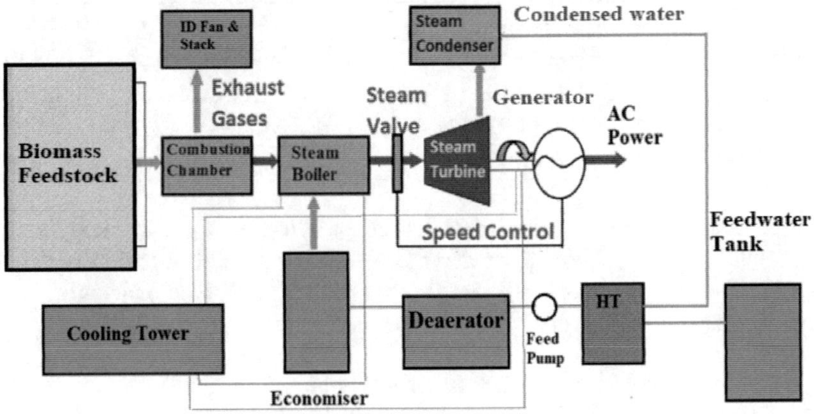

Figure 3. Operation of biomass power plant.

5. Results and discussion

Table 1 illustrates the chemical composition of EPB and RH from the experimental analysis of (PA, UA, and HHV).

Table 1. Chemical composition, HHV and bulk density of EPB and RH.

Parameter		
Proximate analysis		
Material	EPB	RH
MS, (%wt)	8.15	4.92
VM, (%wt)	74.20	63.20
AC, (%wt)	12.76	12.35
FC, (%wt)	4.89	18.19
Ultimate analysis (dry basis)		
C, (%wt)	46.00	42.50
H, (%wt)	6.45	5.42
N, (%wt)	0.48	0.43
O, (%wt) [a]	31.80	34.61
S, (%wt)	0.10	0.01
Components		
Cellulose, (%wt) [b]	37.82±0.57	32.67
Hemicellulose, (%wt) [b]	21.85±0.54	31.68
Lignin, (%wt) [b]	12.16±0.17	18.81
Related properties		
HHV, (MJ/kg)	16.96	14.70
Bulk Density, (kg/m^3) [b]	150	121

[a] Calculated by difference
[b] [19],[20]

IOP Conf. Series: Earth and Environmental Science **897** (2021) 012011 doi:10.1088/1755-1315/897/1/012011

5.1. Chemical formula derivation

Presumption has been made for both EPB and RH molecular formula as $C_a H_b O_c N_d S_e$ which consist of carbon, hydrogen, oxygen, nitrogen, and sulphur derived from the experimental result (ultimate analysis) from Table 1. Thus, the molecular formula derived for 100g of the provided sample (EPB and RH) is as shown in Table 2.

Table 2. Molecular formula derivation of EPB and RH.

Properties	EPB	RH
C	a = 0.460 / 12; a = 0.03730	a = 0.425/12; a = 0.03542
H	b = 0.0645 / 1; b = 0.06450	b = 0.0542 / 1; b = 0.05420
O	c = 0.318 / 16; c= 0.01990	c = 0.3461 / 16; c= 0.02163
N	d = 0.0048 / 14; d = 0.00034	d = 0.0001 / 32; d = 0.000003
S	e = 0.0010 / 32; e = 0.00003	e = 0.0001 / 32; e = 0.000003

Hence, the molecular formula of EPB is $C_{0.0373}H_{0.0645}O_{0.0199}N_{0.00034}S_{0.00003}$ and RH is $C_{0.03542}H_{0.0542}O_{0.02163}N_{0.00031}S_{0.000003}$ which is a dry molecular formula eliminating both ash and moisture presence in the molecular structure of both feedstocks. Also, the molecular weight (MW) of EPB and RH is 0.83622 and 0.82976 respectively.

5.2. Stoichiometric balance combustion equation of EFB and RH

The biomass calorific values, ash content, bulk density and moisture content play a vital role in determining the combustion efficiency in the boiler. Theoretically, in complete combustion system by-products such as carbon dioxide (CO_2), water vapour (H_2O), sulphur oxides (SO_X), nitrogen oxides (NO_X) and other by-products will be produced depending on the number of moles presence in each constituent in biomass feedstock and the reaction capabilities of the feedstock with the provided air supply and heat. In this study, further presumption has been made where both RH and EPB reacted with air in the boiler with the ratio of (79% nitrogen:21% oxygen) and the produced by-products is only composed of (CO_2), (H_2O), sulphur dioxide (S_2O) and nitrogen (N_2). Therefore, the remaining stoichiometric equation of complete combustion process for both EPB and RH can be derived with respect to equation (1) and (2) depicted from Nnaemeka Sp et al. [18] :

a) (EPB)

$$C_{0.0373}H_{0.0645}O_{0.0199}N_{0.00034}S_{0.00003} + \gamma\,(O_2 + 3.76N_2) \rightarrow \beta CO_2 + \alpha H_2O + \delta SO_2 + \rho N_2 \quad (1)$$

C: $\beta = 0.0373$

H : $\alpha = 0.0645 / 2,\ \alpha = 0.03225$

S : $\delta = 0.00003$

O : $0.0199 + 2\gamma = 2\beta + \alpha + 2\delta,\ \gamma = 0.043505$

N : $0.00034 + 2(\gamma)\,(3.76) = 2\rho,\ \rho = 0.1638$

Thus, the overall stoichiometric equation of EPB combustion would be as follows:

$$C_{0.0373}H_{0.0645}O_{0.0199}N_{0.00034}S_{0.00003} + 0.043505\,(O_2 + 3.79N_2) \rightarrow 0.0373CO_2 + 0.03225H_2O + 0.00003SO_2 + 0.1638N_2$$

b) (RH)

$$C_{0.03542}H_{0.0542}O_{0.02163}N_{0.00031}S_{0.000003} + \gamma\,(O_2 + 3.79N_2) \rightarrow \beta CO_2 + \alpha H_2O + \delta SO_2 + \rho N_2 \quad (2)$$

C: $\beta = 0.03542$

H : $\alpha = 0.0542 / 2,\ \alpha = 0.0271$

S : $\delta = 0.000003$

O : $0.02163 + 2\gamma = 2\beta + \alpha + 2\delta,\ \gamma = 0.037258$

N : $0.00031 + 2(\gamma)\,(3.76) = 2\rho,\ \rho = 0.1403$

Thus, the overall stoichiometric equation of RH combustion would be as follows:

$C_{0.03542}H_{0.0542}O_{0.02163}N_{0.00031}S_{0.000003}$ + 0.037258 (O_2 +3.79N_2) → 0.03542CO_2 + 0.0271H_2O + 0.000003SO_2 +0.1403N_2

Therefore, the stoichiometric of air to fuel ratios to burn one kg of EPB and RH would be as follows:

a) EPB
Mass of Air / Mass of fuel = (0.043505) (MW_{air} / MW_{EPB})
= [0.043505 ((16 x 2) + (3.76 x 14 x 2))] / 0.83622
= 7.14 kg

b) RH
Mass of Air / Mass of fuel = (0.037258) (MW_{air} / MW_{EPB})
= [0.037258 ((16 x 2) + (3.76 x 14 x 2))] / 0.829756
= 6.16 kg

5.3. Boiler feeding and heat dissipation of RH and EPB
Specific size and consistency of biomass materials should be achieved through size measurement protocols where larger size of EPB is shredded into smaller pieces to improve bulk density and further enhances the fuel thermo-chemical characteristic and vice-versa for smaller biomass feedstock like RH. Table 3 shows the calculated combustion rate and heat released for both EPB and RH.

Table 3. Combustion rate and heat released of EPB and RH.

Components	EPB	RH
Combustion rate, C_R (kg/hr) = Total mass of burnt fuel (kg) / burning time (hr)	900	900
Heat released, Q (GJ) = (calorific value x combustion rate)	15.20	13.17

5.4. Turbine output power
Rajput et al. [21] have identified equation (3) through the combustion of 10.35 tonnes of EPB.

$$\text{Turbine shaft power} = \frac{(m \times Cv \times \eta\text{circle} \times \eta\text{turbine} \times \eta\text{heat} \times \eta\text{combustion})}{3600} \quad (3)$$

The same equation used by Nnaemeka Sp et al. [18] to find the turbine power with the following recommended parameters except feedstock mass, calorific values, and cycle efficiency as shown in Table 4 below.

Table 4. Turbine parameter with EPB and RH.

Turbine parameters	Current feedstocks		Nnaemeka Sp et al. [20]	
	EPB	RH	EPB	CH_4
Cycle efficiency, η_{circle} (%)	30	30	45	45
Turbine efficiency, $\eta_{turbine}$ (%)	90	90	90	90
Heat transfer efficiency of boiler, η_{boiler} (%)	85	85	85	85
Combustion efficiency, $\eta_{combustion}$ (%)	95	95	95	95
Calorific value, *Cv* (kg/kJ)	16960.0	14700.0	19500.0	55178.2
Mass of feedstock, *m* (kg)	900	900	896	896

In this study, the same equation is utilised to calculate the power produced from the turbine for both EPB and RH with a little alteration of feeding mass value to 900 kg/hr.

a) EPB
Turbine shaft power = 0.924MW

b) RH
Turbine shaft power = 0.801 MW

Overall, the result obtained from this study shows a significant variance in the turbine output for both EPB and RH in comparison with the previous study's result. It is proven that the turbine output power was basically inter-related with the HHV of the biomass feedstock, existing studies have used EPB with the HHV of 19500 kg/kJ [18] which is approximately 13.03% and 24.62% higher than the current HHV values of both EPB and RH used in this research paper. In addition, EPB exhibits more turbine power of 13.31% than the RH with the constant parameters as depicted in Table 4. Figure 4 and Table 5 illustrate the comparison of turbine power with the current fuel of EPB and RH referring to Nnaemeka Sp et al. [18] output for different feeding rate. The study has revealed that EPB and RH produce almost one-third and one-fourth power of what a CH_4 gas could produce from previous study with the same amount of fuel feeding. On the contrary, the air to fuel ratios of RH is significantly lesser than the EPB or CH_4 needed for full combustion to take place. In RH combustion, only 6.16kg air is needed to burn 1 kg of fuel as due to the higher oxygen in the fuel molecular structure which contributed in the completion of the burning process.

Table 5. Power generated based on the feeding rate EPB, RH and CH4.

Feeding rate, (kg/hr)	Power generator, kW			
	Current Feedstock		Nnaemeka Sp et al. [18]	
	EPB (1)	RH	EPB (2)	CH_4
100	154.07	133.54	187.57	511.51
200	308.14	267.08	375.13	1023.02
300	462.21	400.62	562.70	1514.52
400	616.28	534.16	750.26	2046.03
500	770.36	667.70	937.83	2557.54
600	924.43	801.24	1125.40	3069.05
700	1078.50	934.78	1312.90	3580.60
800	1232.57	1068.32	1500.52	4092.06
900	1386.64	1201.86	1688.10	4603.60

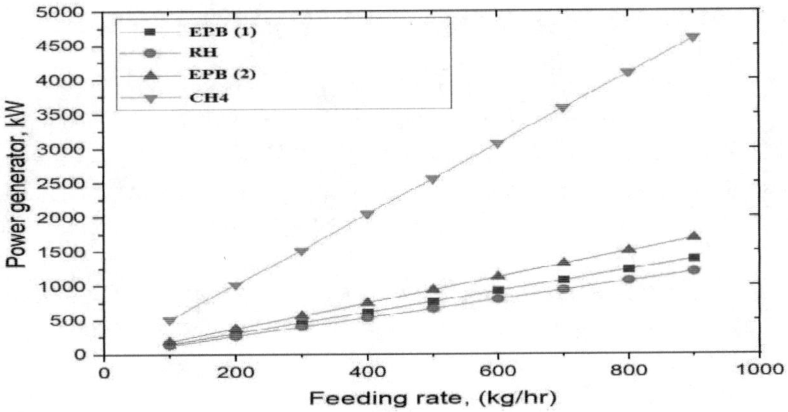

Figure 4. Power projected based on the given fuel feeding.

6. Conclusions

The research study has reached its objective. The usage of EPB and RH has been an ideal and sustainable replacement as biofuel in a biomass boiler. Overall, the study has proven to produce about 33% and 25% of energy from EPB and RH of what a methane (CH_4) can produce from the same amount of feeding rate, respectively, with EPB producing 13.31% of higher turbine power than RH. This directly contributes to the technical feasibility and adaptability of environmentally friendly elements by seizing the opportunity of carbon emission of conventional fuel and replacing it with natural resources such as EPB and RH which are part of the biomass fuel replacement regime. EPB and RH have abundant availability, lower emission rate and produces new employment opportunity in the bioenergy field. Malaysia has abundant natural resources such as palm oil waste, rice mill waste, coffee waste and others which can be utilised in many ways to produce valuable energy for electricity generation. Most of these biomass materials were left unattended on the mills or even on the production land which cause a lot of environmental defects in the form of landfill gaseous which are twenty times more dangerous than the conventional fuel emission. However, biomass fuels are always abandoned due to several reasons behind its inefficiency of being a boiler feedstock such as low bulk density, higher water absorption, low heating value and high transportation cost. Technically, these above-mentioned issues can be catered through several pre-processing methods such as pyrolysis, hydrothermal carbonization, torrefaction, and densification.

7. References

[1] Ari Kahan. Global electricity consumption continues to rise faster than population - Today in Energy - U.S.Energy Information Administration(EIA) [updated 2020 June 15; cited 2021 Aug 15]. Available from: https://www.eia.gov.todayinenergy/detail.php?id=44095

[2] Department of Economic and Social Affairs Population Division World Population Prospects 2019: Highlights(ST/ESA/SER./423)2019 [updated 2019; cited 2021 Aug 15]. Available from: https://population.un.org/wpp/ Publications/Files/WPP2019_Highlights.pdf.

[3] Department of Statistics Malaysia Official Portal 2020 [updated 2020 July 15; cited 2021 Aug 15]. Available from: https://www.dosm.gov.my/v1/index.php?r=column/cthemeByCat&cat=155&bul_id=OVByWjg5YkQ3MWFZRTN5bDJiaEVhZz09&menu_id=L0pheU43NWJwRWVSZklWdzQ4TlhUUT09

[4] Statistics - Malaysia Energy Information Hub [updated 2021; cited 2021 Aug 15]. Available from: https://meih.st.gov.my/statistics?p_auth=bLmIkSA0&p_p_id=Eng_Statistic_WAR_STOASPublicPortlet&p_p_lifecycle=1&p_p_state=maximized&p_p_mode=view&p_p_col_id=column-.

[5] Performance and Statistical Information on Electricity Supply Industry in Malaysia 2018

Putrajaya, Malaysia, 2018 [updated 2018; cited 2021 Aug 15]. Available from: https://www.st.gov.my

[6] FAOSTAT: Production/Yield quantities of Rice, paddy in World + (Total) *Food and Agriculture Organzation of the United Nations* [updated 2019; cited 2021 Aug 15]. Available from: http://www.fao.org/faostat/en/#data/QC/visualize.

[7] Lim J S , Abdul Manan Z, Wan Alwi S R, Hashim H 2012 A review on utilisation of biomass from rice industry as a source of renewable energy *Renew. Sustain. Energy Rev.* **16** 3084–3094

[8] Kadam K L, Forrest L H, Jacobson W A 2000 Rice straw as a lignocellulosic resource: Collection, processing, transportation, and environmental aspects *Biomass and Bioenergy* **18** 369–389

[9] Junpen A, Pansuk J, Kamnoet O, Cheewaphongphan P, Garivait S 2018 Emission of air pollutants from rice residue open burning in Thailand, 2018 *Atmosphere.* **9** 449

[10] Chakrabarti S, Khan M T, Kishore A, Roy D, Scott S P 2019 Risk of acute respiratory infection from crop burning in India: Estimating disease burden and economic welfare from satellite and national health survey data for 250 000 persons *Int. J. Epidemiol.* **48** 1113–1124

[11] Mofijur M, Mahlia T M I, Logeswaran J, Anwar M, Silitonga A S, Ashrafur Rahman S M, Shamsuddin A H 2019 Potential of rice industry biomass as a renewable energy source *Energies.* **12** 4116

[12] Mohiuddin O, Mohiuddin A, Obaidullah M, Ahmed H, Asumadu-Sarkodie S 2016 Electricity production potential and social benefits from rice husk, a case study in Pakistan *Cogent Eng.* **3** 1–13

[13] Mahlia S M 2015 Paddy residue based power generation in Malaysia: Environmental assessment using LCA approach *ARPN J. Eng. Appl. Sci.* **10** 1–6.

[14] Mutsaers H 2019 The challenge of the oil palm: Using degraded land for its cultivation *Outlook Agric.* **48** 190–197

[15] Production of Oil Palm Products 2019 [updated 2020 June 15; cited 2021 Aug 15]. Available from: http://bepi.mpob.gov.my/index.php/en/ production/production-2019/production-of-oil-palm-products-2019.html.

[16] Chiew Y L, Shimada S 2013 Current state and environmental impact assessment for utilizing oil palm empty fruit bunches for fuel, fiber and fertilizer - A case study of Malaysia *Biomass and Bioenergy.* **51** 109–124

[17] Tenorio C, Moya R, Valaert J, Forestal A, Carlos S, Alajuela C R 2016 Characterisation of pellets made from oil palm residues in Costa Rica *J. Oil Palm Res.***28** 198–210

[18] Obuka Nnaemeka Sp, Onyechi Pius C, Okoli Ndubuisi C 2018 Palm Oil Biomass Waste a Renewable Energy Resource for Power Generation *Saudi J. Eng. Technol.* **3** 680–691

[19] Shrivastava P, Khongphakdi P, Palamanit A, Kumar A, Tekasakul P 2020 Investigation of physicochemical properties of oil palm biomass for evaluating potential of biofuels production via pyrolysis processes *Biomass Convers. Biorefinery.* 1–15

[20] Bajo P O, Acda M N 2017 Fuel pellets from a mixture of rice husk and wood particles *BioResources.* **12** 6618–6628

[21] Rajput R K 2010 *Thermal Engineering (8th Edition)* (S.Chand & Company Limited, 7361, Ram Nagar, New Delhi) chapter 19 pp 802–890

Acknowledgments
The authors would like to thank Universiti Teknologi Petronas (UTP) for the academic advice and space given for this study.

IOP Conf. Series: Earth and Environmental Science **897** (2021) 012012 doi:10.1088/1755-1315/897/1/012012

Renewable Hydrocarbon Production via Rapeseed Oil Hydrotreatment Over Palladium Catalysts

I Malina[1,*], K Malins[1], M Strods-Vavilovs[1] and V Uleiskis[1]

[1]Riga Technical University, Institute of Applied Chemistry, Paula Valdena Str. 3, Riga, Latvia

*Ilze.Malina_1@rtu.lv

Abstract. The effect of SiO_2-Al_2O_3 (Pd5%/SA), activated carbon (Pd5%/C) and Al_2O_3 (Pd5%/A) supported palladium (5%) catalysts on renewable hydrocarbon synthesis *via* rapeseed oil hydrotreatment was investigated. The hydrotreatment experiments were carried out in solvent free medium under initial H_2 pressure 100 bar and at 340 °C temperature for 120 min using catalyst amount 5%. Gas chromatography-mass spectrometry (GC/MS) analysis were used for estimation of hydrocarbon content in the obtained samples. Pd5%/SA catalyst provided complete conversion of rapeseed oil into marketable liquid renewable hydrocarbons without presence of oxygen containing substances under studied hydrotreatment conditions. Moreover, all tested Pd catalysts gave narrow range of linear saturated hydrocarbons (*n*-C15-C19). Pd5%/C and Pd5%/A catalysts gave partial feedstock conversion into hydrocarbons even in long residence time. Overall liquid hydrocarbon yields were from 55.3% to 82.3%.

1. Introduction

Growing population and industrialization has increased the world's demand for energy and fuels. As fossil energy resources are limited, development of alternative green energy products are with great importance. In parallel, the development of agriculture in many countries has shown rapid growth and produced amounts of vegetable oils have increased. For example, EU have produced ~3.6 billion kg of rapeseed oil (RO) in the last year [1]. These renewable raw materials can be used as feedstock for first (1G) and second (2G) generation biofuel production. 1G can be manufactured by transesterification and esterification of glycerides and fatty acids, which lead to product – biodiesel. Although, biodiesel is known as alternative to petroleum diesel, it has some drawbacks as poor cold-flow properties and oxidation stability. It is less energy dense than liquid fossil fuels due to oxygen in fatty acid ester molecule. Another green biofuel obtainable from the same vegetable oils with similar properties to fossil fuels is "green diesel" or "hydrotreated vegetable oil" (HVO). HVO technologies include various catalytic reactions, such as, hydrogenation, hydrocracking, hydrodecarboxylation, hydrodecarbonylation, and hydrodeoxygenation of vegetable oils or animal fats under hydrotreating conditions at high temperature and pressure. The hydrotreating pathways and selectivity to particular products – renewable hydrocarbons - mainly depend on operating conditions and catalyst type (active metal, support, and promoter) [2].

Green diesel attracted wide attention as renewable fuel for diesel vehicles as well as sustainable aviation fuel [3]. HVO capacity within EU reached 3.5 Million tons in 2020 and it is forecasted to grow to 11.3 Million tons by 2025 [4]. Such market grow is initiated by EU new climate and energy policies according to which 14% of the energy consumed in road and rail transport must be from renewable sources by 2030 [5]. Green diesel presents excellent fuel properties, including higher cetane number, higher heating value, and negligible oxygen content than biodiesel, therefore research devoted to parameter optimization of HVO technology, different biomass feedstock utilization, selective and active hydrotreating catalyst synthesis is one of top priorities for EU to boost renewable energy shares in the transport sector.

Ni, Ni-Mo, and noble metals supported on activated carbon, SiO_2, Al_2O_3, SiO_2-Al_2O_3 etc. are typical hydrotreating catalysts for HVO synthesis from fatty acids and their glycerides containing feedstocks. Ni catalysts are low cost and relatively easy to synthesize and recycle, however, they also easily deactivate, showing low reusability [6]. On the other hand, some of noble metal catalysts like Pd and Ru possess high catalytic activity under relatively mild operating conditions even at low metal loadings and catalyst amounts towards feedstock. Unlike Ru catalysts, Pd supported on SiO_2-Al_2O_3 has comparatively high impact on hydrocarbon production by hydrodeoxygenation. This lead to increased hydrocarbon yield which is an important aspect for industrial application [7].

Present work demonstrates the effect of commercial Pd (active metal loading 5%) catalysts with three different supports (activated carbon, Al_2O_3, SiO_2-Al_2O_3) on hydrocarbon production from RO under similar hydrotreating conditions to evaluate their potential for practical application in HVO production.

Table 1. The list of abbreviations used in the text.

1G	first generation
2G	second generation
DG	diglyceride
n-C15	*n*-pentadecane
n-C16	*n*-hexadecane
n-C17	*n*-heptadecane
n-C18	*n*-octadecane
n-C19	*n*-nonadecane
n-C22	docosane
FT-IR	Fourier transform infrared spectroscopy
GC/MS	Gas chromatography-mass spectrometry
GC/FID	Gas chromatography- flame ionization detector
HVO	hydrotreated vegetable oil
MG	monoglyceride
RME	rapeseed oil methyl ester
RO	rapeseed oil
TG	triglyceride

Table 2. The main characteristics of rapeseed oil (RO).

	RO
Monoglycerides, wt.%	0.2
Diglycerides,wt.%	0.6
Triglycerides, wt.%	98.7
Saponification value, mg KOH/g	190.4
Acid value, mg KOH/g	0.1
Calorific value, MJ/kg (d.b)	39.37
Water content, wt%	0.03
N, S	≤0.3[a]
Fatty acid composition, wt.%	
Palmitic acid (C16:0)	5.0
Stearic acid (C18:0)	1.5
Oleic acid (C18:1)	63.6
Linoleic acid (C18:2)	20.8
α–Linolenic acid (C18:3)	7.5
Other fatty acids	1.6

[a]– Method detection limit

2. Materials and Methods

2.1. Materials

Supported palladium catalysts - Pd 5% on activated carbon powder, standard, reduced, nominally 50% water wet (Pd5%/C), Pd 5% on alumina, powder (Pd5%/A), and Pd 5% on silica-alumina, powder (Pd5%/SA) - were supplied from *Alfa Aesar*, *Sigma-Aldrich*, and *Riogen*, respectively. *n*-Pentadecane *n*-C15 (99%), *n*-hexadecane *n*-C16 (99%), *n*-heptadecane *n*-C17 (99%), *n*-octadecane *n*-C18 (99%), *n*-nonadecane *n*-C19 (99%) and docosane *n*-C22 (99%), toluene (99.8%), derivatization reagent *N*-methyl-*N*-(trimethylsilyl)trifluoroacetamide (MSTFA), analytical standards tricaprin, 1,2,4-butanetriol, methyl heptadecanoate, methyl esters of palmitic, stearic, oleic, linoleic and α–linolenic acid for gas chromatography-mass spectrometry (GC/MS) and gas chromatography-flame ionization detector (GC/FID) analysis were purchased from *Sigma-Aldrich*.

The refined RO delivered from local food grade vegetable oil producer Iecavnieks&Co was used in experiments. The main characteristics of RO is given in Table 2.

2.2. *Characterization of rapeseed oil (RO) and hydrocarbon samples*

The RO was analysed by gas chromatography (GC) system 7890A (Agilent Technologies) equipped with two capillary columns, two flame ionization detectors (FID) and 7683B automatic liquid sampler. HP-INNOWax (30 m × 0.25 mm × 0.25 μm) column was utilized for determination of fatty acid composition according to a modified EN 14103 standard method: carrier gas H_2 flow rate 5 ml/min; detector temperature 390 °C; temperature program – 200 °C (hold 25 min). GC-FID data of rapeseed oil methyl ester (RME) analysis (99.4 %, prepared from RO and methanol in the presence of alkaline catalyst) was used for estimation of fatty acid composition. It was assumed that RO have the same fatty acid composition as RME. Methyl heptadecanoate was used as internal standard. Fatty acid methyl ester peaks were identified by comparing retention times of particular standards. DB5-ht (15 m × 0.32 mm × 0.10 μm) column was used for determination of monoglyceride (MG), diglyceride (DG), and triglyceride (TG) content in RO according to a modified EN14105 standard method: carrier gas H_2 flow rate 2 ml/min; detector temperature 390 °C; temperature program – 50 °C (hold 5 min) → 180 °C (15 °C/min) → 230 °C (7 °C/min) → 370 °C (10 °C/min, hold 5 min). Derivatization reagent MSTFA and two internal standards tricaprin, 1,2,4-butanetriol were used in GC analysis.

Composition of hydrocarbon samples were determined by GC/MS analysis, using gas chromatograph GCMS-QP2010 Ultra (Shimadzu) equipped with RTX-XLB capillary column (60 m × 0.25 mm × 0.25 mm), a quadrupole mass selective analyzer and secondary electron multiplier detector. The flow rate of carrier gas (He) was 2.4 ml/min. The injector and ion source temperatures were 300 and 220 °C, respectively. Temperature program: 40 °C (hold 2 min) → 320 °C (10 °C/min, hold 10 min). Total program time was 40 min and split ratio 300:1. Injection volumes of the liquid samples for GC analysis were 1.0 μl. The hydrocarbons and other compounds were identified by comparing their mass spectra with the NIST08 database and retention times with reference compounds. *n*-Alkanes *n*-C15-C19 were quantified using docosane (*n*-C22) as internal standard in the toluene solutions. Calibration curves for each compound C_n were generated by plotting the ratio of peak areas A_{Cn}/A_{C22} against the concentration ratio (c_{Cn}/c_{C22}) of the calibration samples, where n=15-19. The unknown hydrocarbon *n*-C15-C19 concentration in the sample were determined from obtained calibration curves. Each experiment and analytical procedure was repeated twice and the maximum relative standard deviation of the method is 3.5%.

Fourier transform infrared (FT-IR) stretching vibration band (1750 - 1700 cm^{-1}) was used for detection of the unconverted carboxyl and carbonyl compounds in the hydrocarbon samples. FT-IR spectrometer PerkinElmer Spectrum 100 equipped with accessory 100T was used. Method detection limit for the presence of carboxyl and carbonyl compounds in hydrocarbon samples is 0.5% and 1%.

Water content in RO was determined using METTLER TOLEDO DL39 Karl Fischer coulometer according to the standard method ISO 12937. Acid value was determined according to the EN 14104 standard method. Saponification value was determined according to the ISO 3657 standard method. Calorific value was determined by C 200 (IKA) oxygen-bomb calorimeter according to the standard method DIN 51900-3:2005.

Overall liquid hydrocarbon yields (Y_{CH}) were calculated by equation (1).

$$Y_{CH}=(m_p \cdot C_{CH})/m_f \tag{1}$$

where m_p is the mass of product and m_f is the mass of feedstock, C_{CH} hydrocarbon content determined by GC.

2.3. *Catalytic tests*

Hydrotreating process was investigated utilizing initial H_2 pressure of 100 bar, operating temperature of 340 °C, residence time 120 min, mixing speed of 300 rpm and 5% of catalyst amount.

Each experiment was conducted using 40 g of RO mixed with catalyst in a batch type stainless steel autoclave-reactor (*Parker Autoclave Engineers*). The reactor was equipped with magnetically coupled mechanical mixer (6 blade agitator impeller) and Sentinel Series Controller. Filled and sealed autoclave-reactor was purged with H_2 (flow rate 10 ml/s) for 15 min to fully eliminate air atmosphere from the pressure vessel. Then H_2 pressure was increased to necessary initial value. Hydrotreating process was terminated after 120 min by switching off the mixer and rapid cooling of pressure vessel with air fan was performed. The 95-97% of liquid hydrocarbons and other substances produced in the process was easily separated from reaction mixture by centrifugation at 3000 rpm for 3 min. The rest of products were extracted with acetone. The acetone was recovered by distillation using rotary evaporator. Then thin layer of the sample in petri dish was dried at 120 °C for 20 min to remove water produced in hydrotreating process. The hydrocarbon samples isolated from reaction mixture by both methods were collected and used for determination of liquid hydrocarbon yields.

3. Results and Discussion

3.1. The effect of catalysts on hydrocarbon yield and composition

Feedstock used in hydrotreatment experiments was rapeseed oil with fatty acid composition given in Table 2. RO contains mainly unsaturated free fatty acids TG of carbon chain length C18 (93.4%). The feedstock also contains moderate amount of unsaturated C16 (5%) and low amount of fatty acids with hydrocarbon chain length higher than C18 (up to 1.6%). Low amount of other glycerides, water and fatty acids in RO indicates to excellent quality of the rapeseed oil. Theoretically only dominant linear saturated hydrocarbons with carbon chain length analogous to fatty acid composition of feedstock or by one carbon atom shorter can be produced by main oxygen removal reaction pathways: hydrodecarboxylation, hydrodecarbonylation, and hydrodeoxygenation.

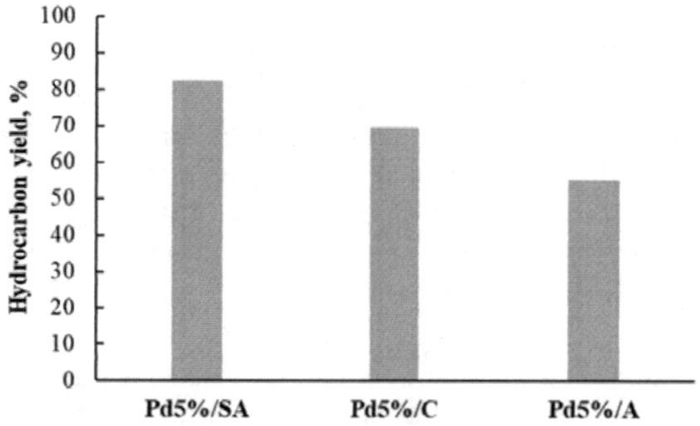

Figure 1. The effect of catalysts on hydrocarbon yield in RO hydrotreatment. Hydrotreating conditions: initial H2 pressure 100 bar, mixing speed 300 rpm, overall residence time 120 min, operating temperature 340 °C, catalyst amount 5%.

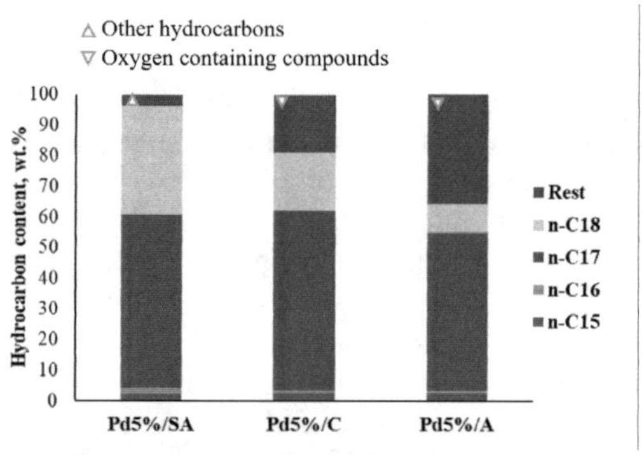

Figure 2. The effect of catalyst type on dominant hydrocarbon (n-pentadecane, n-hexadecane, n-heptadecane, n-octadecane) composition. Hydrotreating conditions: initial H2 pressure 100 bar, mixing speed 300 rpm, overall residence time 120 min, operating temperature 340 °C, catalyst amount 5%.

All Pd catalysts used in this study are completely different by support and specific surface area. The specific surface area of particles is a function of porosity, pore size distribution, shape, and size [8], and it is directly related to catalyst performance. Pd5%/C supported on activated carbon has the largest specific surface area (980 m^2/g) among other two tested catalysts. Activated carbons as catalyst supports typically are highly porous materials with large specific surface area 950-2000 m^2/g [9]. Despite great textural properties Pd5%/C showed average performance in comparison to other two catalysts (Figures 1-3). The hydrocarbon yield over Pd5%/C reached 69.7% in 120 min of residence time. The catalyst mostly produces *n*-C17 (58.4%) and *n*-C18 (18.9%) from C18 fatty acid glycerides of RO. Obtained hydrocarbon sample also consist partially converted oxygen containing intermediates, such as, alcohols and carbonyl compounds detected by GC/MS and FT-IR. Partial

conversion of feedstock at relatively long residence time can be related with catalyst structure and metal dispersion. Catalysts prepared from activated carbon commonly have lower bulk powder densities than SiO_2, Al_2O_3, and SiO_2-Al_2O_3. Large active metal particles and/or increased distance between each other, especially at low Pd loadings like 5% might be the reasons related with reduced performance. Excessive Pd dispersion in low density of highly porous Pd5%/C catalyst support can restrict the accessibility of feedstock molecules and other compounds produced during hydrotreating process to active metal phase on catalyst surface hindering their fast conversion into renewable hydrocarbons. Reduced performance of Pd5%/C also can be contributed to significant presence of water (50% of catalyst mass recommended by supplier for storage) in catalyst. Twice amount of wet Pd5%/C catalyst was used for hydrotreating test because half of catalyst mass is water. Elevated amount of water in reaction mixture from catalyst can negatively affect reaction equilibrium, reaction rate, and overall conversion of feedstock into desired product – pure hydrocarbons without presence of oxygen containing compounds.
The lowest yield (55.3%) of renewable hydrocarbons was obtained in the presence of Pd5%/A. The catalyst has almost ten times lower specific surface (101 m^2/g) area than Pd5%/C. This could be one of the main reasons for reduced performance to achieve high conversion of feedstock into pure renewable hydrocarbons in short residence time. The catalyst produces 51.2% of *n*-C17 and 9.2% of *n*-C18. The hydrocarbon sample similarly to that obtained over Pd5%/C also contains considerable amount of partially converted oxygen compounds (Figure 3).

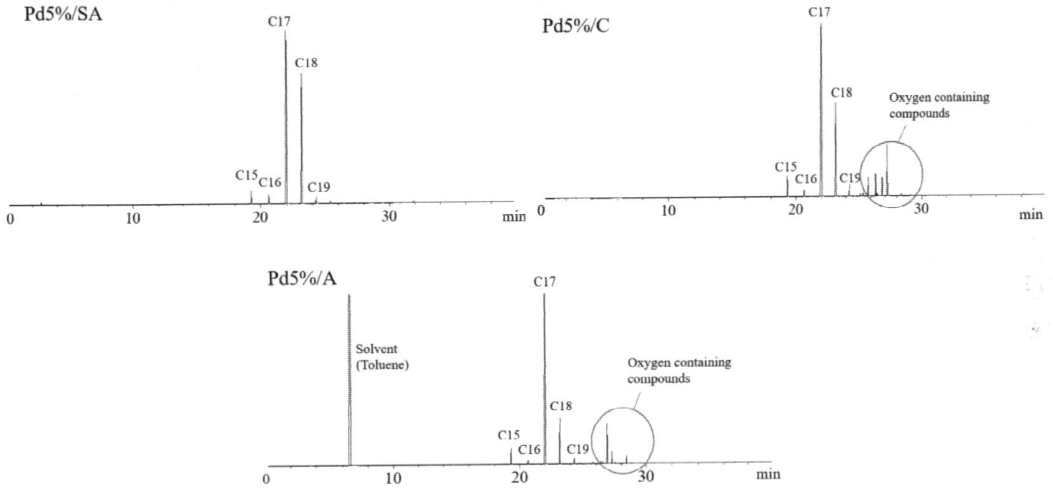

Figure 3. GC/MS chromatograms of hydrocarbon samples produced from RO by hydrotreatment over commercial Pd5%/SA, Pd5%/C and Pd5%/A catalysts. Hydrotreating conditions: initial H2 pressure 100 bar, mixing speed 300 rpm, overall residence time 120 min, operating temperature 340 °C, catalyst amount 5%.

The highest hydrocarbon yield (82.3%) was achieved over Pd5%/AS catalyst. The catalyst has large specific surface area 460 m^2/g. In the presence of Pd5%/SA catalyst under studied experimental conditions complete RO conversion into pure marketable hydrocarbons was accomplished. The hydrocarbon composition was *n*-C15 (2.5%) *n*-C16 (1.9%) *n*-C17 (56.3%) *n*-C18 (35.5%) and *n*-C19 (3.8%). All three catalysts produce low amount of *n*-C19 from C20 fatty acid glycerol esters, which is present in rapeseed oil in small quantities.
Many factors can affect catalyst performance and selectivity to particular hydrocarbons in high temperature and pressure processes like HVO production. Active metal type, loading, dispersion, support type, textural properties, particle size, surface morphology, and composition are the key

factors related with catalyst structure. Experimental conditions are the second major factor which affects the product yield and composition. Tested catalysts exhibit different performance and selectivity in hydrocarbon synthesis from RO. The catalysts by performance can be arranged in following order - Pd5%/SA > Pd5%/C > Pd5%/A. Hydrodeoxygenation is most advantageous reaction pathway for hydrocarbon production. The maximum liquid hydrocarbon yield can be obtained only by hydrodeoxygenation. *n*-C17/*n*-C18 ratio show the effect of catalysts on hydrodeoxygenation. The *n*-C17/*n*-C18 ratios produced over Pd5%/SA, Pd5%/C, and Pd5%/A are 1.6, 3.1, and 5.6, respectively. The order of catalysts by selectivity to hydrocarbons produced by hydrodeoxygenation is analogous to their performance. According to literature data and our previous study [6] the selectivity is directly related with active metal type, support and its acidic sites. Evidently, the SiO_2- Al_2O_3 and activated carbon of catalysts Pd5%/SA and Pd5%/C are supports with abundant acidic sites, which have pronounced impact on hydrodeoxygenation reaction. All tested Pd catalysts possess certain activity in HVO production from RO. However, the best results were achieved over Pd5%/SA due to high yield and full conversion of feedstock into desired product at relatively mild solvent-free one-pot operating conditions. Produced pure linear saturated renewable hydrocarbons with high calorific value 47.15 MJ/kg can be used in different areas similarly to those obtained from fossil sources.

4. Conclusions

Pd as active metal supported on three different carriers Al_2O_3 (Pd5%/A), SiO_2-Al_2O_3 (Pd5%/SA), and activated carbon (Pd5%/C) shoved sufficient activity in hydrocarbon synthesis from rapeseed oil. The catalysts by performance and selectivity to hydrocarbons produced by hydrodeoxygenation can be arranged in following order: Pd5%/SA > Pd5%/C > Pd5%/A. Only Pd5%/SA converted RO into pure marketable hydrocarbons with high calorific value 47.15 MJ/kg, yield (82.3%), and composition: *n*-C15 (2.5%), *n*-C16 (1.9%), *n*-C17 (56.3%), *n*-C18 (35.5%), and *n*-C19 (3.8%) without presence of oxygen containing compounds. Pd5%/C and Pd5%/A have significantly lower activity and impact on hydrodeoxygenation reaction. Depending on support type and catalyst characteristics Pd as active metal has potential for practical application in solvent-free one-pot renewable hydrocarbon synthesis from vegetable oils using relatively mild hydrotreating reaction conditions.

5. References

[1] Rapeseed oil production volumes in EU. https://www.amisyfoodmachine.com/application/European-rapeseed-oil-production-industry.html (accessed 30.04.2021)
[2] Vásquez MC, Silva EE and Castillo EF 2017 *Biomass Bioenergy* 105 197
[3] Panoutsou C, Germer S, Karka P, Papadokostantakis S, Kroyan Y, Wojcieszyk M, Maniatis K, Marchand P and Landalv I 2021 *Energy Strategy Rev.* 34 100633
[4] Greenea Horizon 2030. Which investments will see the light in the biofuel industry. https://www.greenea.com/publication/greenea-horizon-2030-which-investments-will-see-the-light-in-the-biofuel-industry/ (accessed 30.04.2021)
[5] European Commission. Directive on the promotion of the use of energy from renewable sources 2018/2001. PE/48/2018/REV/1
[6] Malins K 2021*Fuel* 285 119129
[7] Chen S, Zhou G and Miao C 2019 *Renewable Sustainable Energy Rev.* 101 568
[8] Amador C and de Juan LM 2016 *Comput. Aided Chem. Eng.* 39 509
[9] Lu J-Da and Xue J *Critical Care Nephrology* (Third Edition), Elsevier 2019

Acknowledgements

This research is funded by the Latvian Council of Science, project “Renewable hydrocarbon production by hydroprocessing from alternative sources over selective noble metal catalysts”, project No. lzp-2020/2-0027.

IOP Conf. Series: Earth and Environmental Science **897** (2021) 012013 doi:10.1088/1755-1315/897/1/012013

Reuse of organic waste from *Eucalyptus globulus* extract with high reducing potential in the green synthesis of silver nanoparticles

Asmat-Campos, D.[1, 2, *], Nazario-Naveda, R.[1], Delfín-Narciso D.[1], Juárez-Cortijo, L.[1]

[1] Grupo de Investigación en Ciencias Aplicadas y Nuevas Tecnologías, Universidad Privada del Norte (UPN), Trujillo, Perú.
[2] Universidad Privada del Norte, Dirección de Investigación e Innovación, Trujillo, Perú

E-mail: davidasmat88@hotmail.com , david.asmat@upn.edu.pe

Abstract. The research provides a new and sustainable methodology for the synthesis of silver nanoparticles, using *Eucalyptus globulus* extract, this due to the fact that it presents metabolites capable of acting as a reducing potential of our silver nitrate precursor, and thus obtaining nanostructured material. This is also associated with the reuse of this type of organic material, which currently abounds as waste in the Peruvian highlands. In the specific case of this research, the effect on the stability over time of the biosynthesized silver nanoparticles was evaluated by varying the pH, with values of 4.82, 8.05 and 10.15. It was observed that as the pH increases the production of nanoparticles is higher, having a saturation threshold close to pH 8. It was also found that for alkaline pH close to 10 a more complete reaction of the reducing agent occurs, but with a high dispersion.

1. Introduction

Nanobiotechnology is a growing field whose objective is to create and improve nanoscale structures for use in various applications, these nanoparticles have a high area/volume ratio which gives them unique physical and chemical properties compared to larger structures [1].

In recent years the biosynthesis of metallic nanoparticles has generated a lot of interest due to the growing need to develop more environmentally friendly synthesis methods. Metallic nanoparticles are of great importance due to the wide spectrum of applications that goes from bactericidal products to pharmaceutical products [2]. Some of the most studied metallic nanoparticles include gold, platinum, copper, and silver.

Silver nanoparticles have been used in sensor technology, disease diagnostic systems, and many other biomedical applications. Silver nanoparticles can be synthesized by various chemical and physical methods, the same ones that use toxic products such as sodium citrate or sodium borohydride (reducing agents) in the synthesis procedure [3], the use of these products brings with it a negative impact on the environment, as well as limitations in its use in the biomedical field. At present, it is necessary to develop biocompatible and ecological processes incorporating green chemistry methods, where various types of natural extracts are used, whose biological agents have reducing, stabilizing or both properties [4–6].

To carry out the biosynthesis of silver nanoparticles, several plants have been used by means of extracts obtained with various solvents, these contain chemical compounds such as essential oils, phenols, flavonoids, carbohydrates, proteins, alkaloids, glycosides, oligosaccharides, quinine, saponins, steroids, tannins, among others [7]. This work reports the biosynthesis of silver nanoparticles using eucalyptus to take advantage of the reducing properties of its components. Esmaili et al. they biosynthesized silver nanoparticles using ethanolic extract from dried *Eucalyptus comadulensis* leaves [8]. Thakur et al. reported the antibacterial activity of silver nanoparticles biosynthesized with *Eucalyptus globules* leaf extracts, using chloroform, acetone, and methanol as solvents, among others [9]. On the other hand, other researchers prefer to use aqueous extracts of eucalyptus leaves since, due to their applications, they turn out to be much safer for health and the environment than chemical solvents [10–16].
In the synthesis process, different parameters must be controlled for the formation of nanoparticles. The pH is one of the important parameters that allow controlling characteristics such as shape, size and stability in the biosynthesis of silver nanoparticles [2]. That is why the objective of this research was to study the stability in time according to the pH of the nanoparticles in suspension biosynthesized with aqueous extract of eucalyptus leaves.

2. Materials and methods

2.1. Process of obtaining the extract
The aqueous extract of eucalyptus leaves was prepared following the following procedure: Eucalyptus globulus leaves were harvested, washed with distilled water and then naturally dried under the sun for three days. The dried leaves were cleaned until they were free of impurities and then cut into small pieces and then shredded. 60 grams was weighed and placed in a beaker with 1 liter of distilled water. the mixture was heated for 30 minutes at 90 °C while stirring and then allowed to cool to room temperature. Finally, the mixture was filtered using whatman filter paper number 42. the extract was stored in a refrigerator at 4 °C for subsequent syntheses.

2.2. Silver nanoparticle synthesis process
The synthesis was carried out using the green chemistry method (biosynthesis), having as a precursor 50 mL of silver nitrate ($AgNO_3$), from Merck, Germany (CAS: 7761-88-8) at a concentration of 1 mM. The solution was brought to a hotplate with stirring keeping it at 60 °C for 10 min. at 300 rpm, subsequently 2.5 mL of aqueous extract of eucalyptus leaves was added drop by drop, finally in the solution the pH was controlled by adding sodium hydroxide (CAS number 1310-73-2) drop by drop under magnetic stirring at 300 rpm. Aliquots with a pH of 4.82, 8.05 and 10.15 were collected to later be analyzed by UV-vis spectrophotometry.

2.3. Experimental tests
The colloidal samples were characterized by UV-vis spectrophotometry (Shimadzu, UV 1900, Tokyo, Japan) to evaluate the presence of the plasmon peak typical of silver nanostructures and in turn evaluate the stability of the colloid over time. The data from the spectrophotometry results was processed with Origin software, the same one that precisely defined the values of the optical absorbance peaks and their respective wavelengths.
For the analysis in TEM, they were carried out in a JEOL equipment (JEM 2011 model) operated with a voltage acceleration of 120 Kv.
NP Ag samples for pH 10 were analyzed using a Nicolet iS50 FT-IR infrared spectrometer (Thermo Fisher Scientific). A scan was performed in the range of 500 - 4000 cm-1. Spectra were analyzed with OMNIC 8.1 software (OMNIC Series 8.1.10, Thermo Fisher Scientific).

3. Results and discussion

The results presented below show graphs of UV-vis spectrophotometry to find the location of the plasmon resonance peak and thus confirm the formation of silver nanoparticles manufactured under a biosynthesis process using aqueous extract of eucalyptus leaves. With the same characterization technique, the stability behavior of silver nanoparticles has been evaluated over time, observing their behavior on days 1 and 17 and thus confirming the influence of pH in the synthesis processes, understand day 1 like the day of manufacture of the nanoparticles.

Figure 1 shows the results of UV-vis spectrophotometry for silver nanoparticle colloids at different pH (4.82, 8.05 and 10.15) on day 1. Nanoparticle formation is confirmed in all cases by the presence of the resonance peak. characteristic plasmon. The colloid with a pH 4.82 has a peak around 441 nm and an optical absorbance of 0.94 u.a., which still implies a low production of nanoparticles. By increasing the pH to 8.05, a greater production of nanoparticles is observed with a plasmon resonance peak located at 421 nm and 1.36 u.a. absorbance, also the shift towards the blueshift (shorter wavelength) would indicate the formation of smaller nanoparticles. Finally, a decrease in the absorbance peak is observed when pH 10.15 is reached, the absorbance fell to 0.72 u.a. with the peak at 413 nm. These results confirm the influence of pH on the production of nanoparticles, and also reveal the presence of a reducing agent saturation threshold that would be reached around pH 8. These results are confirmed by Veerasamy et al., In their study with extract of mangosteen leaf, reported that synthesis at acid pH favors the aggregation of silver nanoparticles on nucleation. However, a higher pH facilitates the nucleation and subsequent formation of a large number of smaller diameter nanoparticles [17].

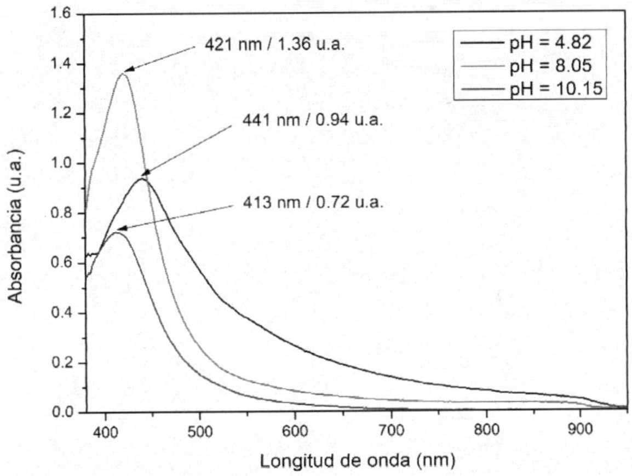

Figure 1. UV-vis spectrophotometry of silver nanoparticles biosynthesized with aqueous extract of eucalyptus at different pH for day 1.

Figure 2 shows the UV-vis spectra of the silver nanoparticle colloids after 17 days of synthesis. The changes are significant for the sample with pH 4.82, a significant decrease in the absorbance peak is observed from 0.94 to 0.65 u.a. and a shift towards 438 nm, which implies that at acidic pH the reaction is incomplete and unstable. As the pH increases, the peaks remain at the same wavelength, highlighting a small increase in absorbance from 1.36 to 1.42 u.a. for the samples with pH 8.05, which means that the reaction was incomplete despite the high number of nanoparticles produced on day 1, however, stability with respect to the size of the nanoparticles is achieved. Dubey et al. reported that silver nanoparticles demonstrate a lower potential zeta value at strongly acidic pH compared to alkaline pH solutions, which indicates for the latter, greater stability and smaller size of the nanoparticles [18]. For colloids with pH 10.15 no changes in absorbance or peak shifts were observed, but a broadening which could be attributed to high dispersity. This same result is confirmed by Andreescu et al, who concluded that an increase in pH leads to the formation of highly dispersed nanoparticles. This phenomenon could be related to electrostatic repulsion at high pH or attributed to the high magnitude of the potential zeta [19].

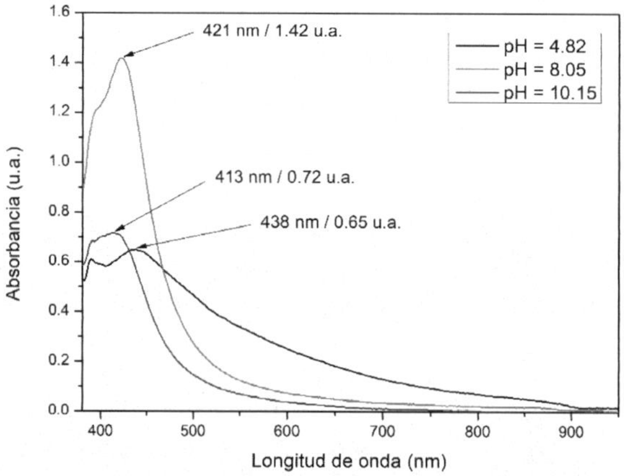

Figure 2. UV-vis spectrophotometry of silver nanoparticles biosynthesized with aqueous extract of eucalyptus at different pH for day 17.

According to the results obtained on stability in size of the silver nanoparticles, the colloids with pH 10 present an excellent resistance against agglomeration, however, in the case of pH 8.05, as the days pass there are still processes of reduction of the nitrate precursor of silver due to increased absorbance. Furthermore, the plasmon peaks are well defined and due to their bandwidth, they show high monodispersity and smaller diameter nanoparticles [20], the plasmon peak is relatively unchanged, a unique characteristic in relation to all other samples. With the increase in pH to 4.82, a tendency for the plasmon peak to widen is observed, which possibly implies processes of aggregation and cluster formation, generating even larger nanoparticles.

The variation of the pH in the synthesis process implies important changes. In this sense, it is significant to highlight that sample that is stable over time, the sample synthesized with pH 10.15, is the one characterized by being stable over time, the other two samples have a tendency to form agglomerations and have a variation in its absorbance.

Figure 3 shows the transmission electron microscopy (TEM) image of Ag nanoparticles at pH 10.15 17 days after the synthesis has been carried out, where its spherical morphology can be highlighted with an average size at 20 nm., without agglomerations, which to a certain extent justifies what was observed in the results by UV Vis spectrophotometry. The sample in question has a high tendency to be monodisperse, this linked to the fact that the majority of nanoparticulate material has values of average size to the aforementioned, this can be partially contrasted based on the absorbance spectra indicated in figure 2, in where the graph of the sample pH 10.15 has a leptokurtic-like bandwidth configuration.

Figure 4 shows results by FT-IR of the NP Ag sample at a pH of 10.15 after 17 days of having carried out the synthesis, the spectrum confirms that the carbonyl group of the amino acid residues and protein peptides has the greatest capacity to bind to the metal. In this sense, it is most likely that the proteins could form a layer that covers the metallic silver nanoparticles, to avoid agglomeration and therefore stabilize them in the middle. This evidence suggests that biological molecules could possibly perform the function of formation (reduction) and stabilization of silver nanoparticles in aqueous medium through carbonyl groups through free amine or cysteine groups in proteins. The proteins present on the surface of the silver nanoparticle act as protective agents for amino acid residues and the peptides have a great capacity to bind to the silver ion [21].

IOP Conf. Series: Earth and Environmental Science **897** (2021) 012013 doi:10.1088/1755-1315/897/1/012013

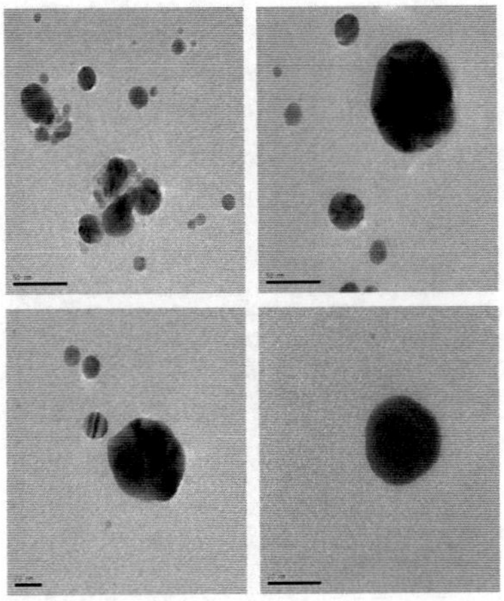

Figure 3. Transmission electron microscopy of silver nanoparticles (NP Ag) synthesized by the green route method at pH 10.15, after 17 days.

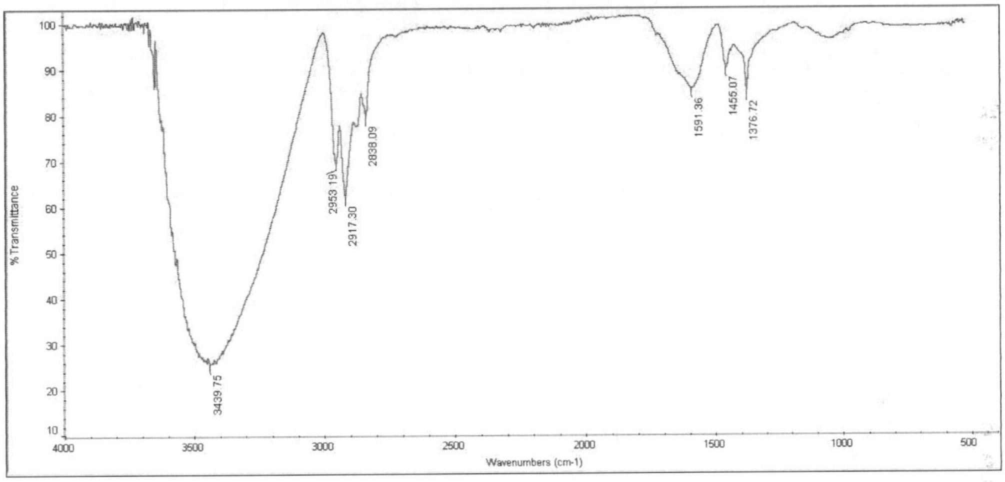

Figure 4. FT-IR of colloidal silver nanoparticles (NP Ag), synthesized by the green route method at pH 10.15, after 17 days.

4. Conclusions

The reducing capacity of the aqueous extract of eucalyptus leaves for the biosynthesis of silver nanoparticles has been demonstrated and its stability in time for different pH's was studied. It was observed that as the pH increases the production of nanoparticles is higher, having a saturation threshold close to pH 8. At this point smaller and monodisperse nanoparticles were manufactured, the increase in the absorbance peak also indicates that the reaction was incomplete. It was found that for alkaline pH close to 10 a more complete reaction of the reducing agent occurs, but with a high dispersity. It is suggested that future work may be focused on having a better approach to the precursor reduction mechanisms, through techniques such as nuclear magnetic resonance.

5. References

[1] HF A, H K and AD W 2019 Synthesis of Silver Nanoparticles Using Aqueous Extract of Medicinal Plants' (Impatiens balsamina and Lantana camara) Fresh Leaves and Analysis of Antimicrobial Activity *Int. J. Microbiol.* **2019**

[2] Akhtar M S, Panwar J and Yun Y-S 2013 Biogenic Synthesis of Metallic Nanoparticles by Plant Extracts *ACS Sustain. Chem. Eng.* **1** 591–602
[3] ZQ T and B R 2004 Adsorption and reaction at electrochemical interfaces as probed by surface-enhanced Raman spectroscopy *Annu. Rev. Phys. Chem.* **55** 197–229
[4] Yulizar Y, Ariyanta H A and Abduracman L 2017 Green Synthesis of Gold Nanoparticles using Aqueous Garlic (Allium sativum L.) Extract, and Its Interaction Study with Melamine *Bull. Chem. React. Eng. Catal.* **12** 212–8
[5] SP C, M C, R P, A A and M S 2006 Synthesis of gold nanotriangles and silver nanoparticles using Aloe vera plant extract *Biotechnol. Prog.* **22** 577–83
[6] IM C, I P, K S-H, M T and G R 2016 Plant-Mediated Synthesis of Silver Nanoparticles: Their Characteristic Properties and Therapeutic Applications *Nanoscale Res. Lett.* **11** 1–14
[7] M R-N, SM P, SA S, F A and H B 2014 Green synthesis of silver nanoparticles using Eucalyptus leucoxylon leaves extract and evaluating the antioxidant activities of extract *Nat. Prod. Res.* **28** 1964–9
[8] Esmaili N, Mohammadi P, Abbaszadeh M and Sheibani H 2020 Green synthesis of silver nanoparticles using Eucalyptus comadulensis leaves extract and its immobilization on magnetic nanocomposite (GO-Fe3O4/PAA/Ag) as a recoverable catalyst for degradation of organic dyes in water *Appl. Organomet. Chem.* **34** e5547
[9] Thakur K, Bala I, Rajeshwer D M and Bhatt A K 2017 Evaluation of Effectiveness of Biologically Synthesized Silver Nanoparticles of Eucalyptus globules Leaf Extract against Pathogenic and Acne-inducing Bacteria *J Nanomed Nanotechnol* **8** 443
[10] SM P, M T, V M and M R-N 2015 Procedure optimization for green synthesis of silver nanoparticles by aqueous extract of Eucalyptus oleosa *Spectrochim. Acta. A. Mol. Biomol. Spectrosc.* **136 Pt C** 1249–54
[11] Mohammed A E 2015 Green synthesis, antimicrobial and cytotoxic effects of silver nanoparticles mediated by Eucalyptus camaldulensis leaf extract *Asian Pac. J. Trop. Biomed.* **5** 382–6
[12] Sila M J, Nyambura M I, Abong'o D A, Mwaura F B and Iwuoha E 2019 Biosynthesis of Silver Nanoparticles from Eucalyptus Corymbia Leaf Extract at Optional Conditions *Nanohybrids Compos.* **25**
[13] Salgado P, Mártire D O and Vidal G 2019 Eucalyptus extracts-mediated synthesis of metallic and metal oxide nanoparticles: current status and perspectives *Mater. Res. Express* **6** 082006
[14] Alghoraibi I, Soukkarieh C, Zein R, Alahmad A, Walter J-G and Daghestani M 2020 Aqueous extract of Eucalyptus camaldulensis leaves as reducing and capping agent in biosynthesis of silver nanoparticles *https://doi.org/10.1080/24701556.2020.1728315* **50** 895–902
[15] Balamurugan M, Saravanan S, Balamurugan M and Saravanan S 2017 Green Synthesis of Silver Nanoparticles by using Eucalyptus Globulus Leaf Extract *JIEIA* **98** 461–7
[16] Mo Y, Tang Y, Wang S, Lin J, Zhang H and Luo D 2015 Green synthesis of silver nanoparticles using eucalyptus leaf extract *Mater. Lett.* **C** 165–7
[17] Veerasamy R, Zi Xin T, Gunasagaran S, Foo Wei Xiang T, Fang Chou Yang E, Jeyakumar N and Arumugam Dhanaraj S 2011 Biosynthesis of silver nanoparticles using mangosteen leaf extract and evaluation of their antimicrobial activities
[18] Dubey S P, Lahtinen M and Sillanpää M 2010 Tansy fruit mediated greener synthesis of silver and gold nanoparticles *Process Biochem.* **45** 1065–71
[19] Andreescu D, Eastman C, Balantrapu K and Goia D V. 2007 A simple route for manufacturing highly dispersed silver nanoparticles *J. Mater. Res.* **22** 2488–96
[20] M S, K S, SW W, CW C, S K and YS Y 2009 Cinnamon zeylanicum bark extract and powder mediated green synthesis of nano-crystalline silver particles and its bactericidal activity *Colloids Surf. B. Biointerfaces* **73** 332–8
[21] DS B, S B, R D, DB M, BK P and A V 2009 Extracellular biosynthesis of functionalized silver nanoparticles by strains of Cladosporium cladosporioides fungus *Colloids Surf. B.*

Biointerfaces **68** 88–92

Acknowledgments
The researchers thank the Research and Innovation Directorate of the Universidad Privada del Norte for their full support in the development of this research, as well as a special thanks to the reviewers for their contributions and suggestions.

IOP Conf. Series: Earth and Environmental Science **897** (2021) 012014 doi:10.1088/1755-1315/897/1/012014

Ecological and sustainable synthesis of silver nanoparticles from alcoholic extract of *Eucalyptus globulus*: Evaluation of alcoholic solvent influence (70° and 96°)

D. Asmat-Campos [1,2]*, D. Delfin-Narciso [2], L. Juárez-Cortijo [2] and R. Nazario-Naveda [2]

[1]Universidad Privada del Norte, Dirección de Investigación e Innovación, Trujillo, Perú
[2]Universidad Privada del Norte, Grupo de Investigación en Ciencias Aplicadas y Nuevas Tecnologías, Trujillo, Perú

*Corresponding author's e-mail: david.asmat@upn.edu.pe

Abstract. The present study provides an ecological and sustainable methodology for obtaining nanostructured material from *Eucalyptus globulus* leaf extract, as a potential value-added alternative and a contribution to circular economy. Silver nanoparticles (NP Ag) were synthesized, through the reducing action of the alcoholic extracts of eucalyptus on the precursor silver nitrate ($AgNO_3$) evaluating the influence of alcoholic solvent (70 ° and 96 °) and pH in the synthesis. The silver colloids obtained were evaluated by UV-vis spectrophotometry, which shows the formation of nanoparticles through the plasmon resonance peak; showing that for pH values 9.9 and 10 with alcohol extract of 70 ° and 96 ° respectively, silver nanoparticles with plasmon resonance peaks at 410 nm and 412.5 nm are obtained. While for pH values 3.86, 11.8 (96°) and 4.7, 8.2 (70°) nanoparticles with higher polydispersity and in a lower proportion are obtained. The results suggest that the alcoholic extracts of eucalyptus can act as reducing agents and that the optimum pH value for the synthesis of silver nanoparticles corresponds to 10.

1. Introduction

Eucalyptus, in different varieties commonly used as wood resource, leaves and bark are used to a lesser extent, and are usually left on the ground. Current research leads to a new use of these resources by developing extracts that are used in the synthesis of various nanoparticles, thereby giving them both economic and environmental added value [1,2].

Nowadays metallic nanoparticles have become an important object of study due to their new and unique properties that arise at this scale. Studies on their magnetic, electrical, and medical properties, among others, provide knowledge for future applications based on these nanometric materials [3,4].

There are various methods for the synthesis of nanoparticles; however, current methods are aimed at green synthesis, a method in which chemical reducing agents, harmful to the environment, are replaced by natural elements, such as plant extract for example. The biological components of these extracts such as polyphenols, alkaloids, flavonoids, reducing sugars, tannic acids, among other components, act as bioreducing agents [5–10]. This green synthesis has the advantage of not generating polluting chemical by-products, and they are lower cost by presenting less energy expenditure and requiring less infrastructure.

Silver nanoparticles are of great interest due to their various applications, among which their bactericidal properties stand out. Various authors have reported the green synthesis of these nanoparticles in which different plant extracts are used, such as blueberries, tomatoes, onions, grapes, cocoa, among many others. These works coincide in the reducing capacity of the extracts, showing characteristic plasmon resonance peaks between 400 nm and 450 nm, and showing nanoparticle size between 9 nm and 50 nm [5,6,11].

Using the aqueous extract of eucalyptus leaves, iron nanoparticles between 20 and 80 nm were synthesized [2]; likewise, it was possible to synthesize zinc oxide nanoparticles with sizes between 20 nm and 40 nm approximately [1]. Works related to the synthesis of silver nanoparticles show us the reducing capacities of this extract; three different varieties of eucalyptus were used (*Eucalyptus urophylla, Eucalyptus citriodora* and *Eucaliptus robusta*), achieving evidence of the formation of these nanoparticles through the characteristic surface plasmon resonance peak at 447 nm, 445 nm and 482 nm for each extract respectively, and with mean size 23.25 nm [7]. The ethanolic extract of Eucalyptus citriodora was also investigated as a possible reducing agent in the formation of silver nanoparticles with a spherical shape between 8 and 15 nm, whose UV-Vis absorption peak is at 416 nm [12].

In this study, the results of the green synthesis process of silver nanoparticles are shown, specifically evaluating the influence of the type of solvent (70 ° and 96 ° alcohol) to obtain the extract, and pH in the synthesis process of nanoparticles, to evaluate the reducing potential of the extract and the process of formation of nanostructured material. Results are presented by UV Vis spectrophotometry evaluating the presence of the surface plasmon peak corresponding to the NP Ag.

2. Materials and methods

2.1. Preparation of eucalyptus extract

60 g of dried leaves of eucalyptus (Eucalyptus globulus) was used; the same ones that were washed and dried in an oven at 40 °C. Later, these leaves were ground with the help of a mortar until obtaining a fine powder, which was mixed with 1 L of alcohol (70 ° and 96 °) at room temperature with the help of a magnetic stirrer. Then we proceeded to filter with which the eucalyptus extracts were obtained in both 70 ° and 96 ° alcohol and stored at 4 ° C for later use.

2.2. Green synthesis of silver nanoparticles

The synthesis was carried out using the green chemistry method (biosynthesis), having as a precursor silver nitrate ($AgNO_3$), from Merck, Germany (CAS: 7761-88-8) at a concentration of 1M. The solution was brought to a hotplate with stirring keeping it at 60 ° C for 10 min. at 400 rpm; subsequently, the extracts were added dropwise for evaluation, and finally the pH of the solution was varied by adding sodium hydroxide (CAS number 1310-73-2) dropwise and stirring at 400 rpm.

2.3. Silver nanoparticle characterization

The colloidal samples were characterized by UV-vis spectrophotometry (Shimadzu, UV 1900, Tokyo, Japan) to evaluate the presence of the plasmon peak typical of silver nanostructures, the effect of pH and the stability of the colloid over time. Spectrophotometry measurement was performed for each of the samples, on the day of synthesis (day 1), 17 and 65 days later.

3. Results and Discussion

The silver nanoparticles were synthesized by green synthesis, evaluating the effects of the type of solvent used in the preparation of the extract (70 ° and 96 ° alcohol), as well as the pH values, 4.7, 8.2 and 9.9 for the extracts in alcohol of 70 °, and pH of 3.86, 10 and 11.8 for alcohol of 96 °.

Figure 1 shows the collides obtained for each of the parameters indicated above. As mentioned by various authors, the change in colour towards brown by the collides would indicate the formation of silver nanoparticles [13–15].

IOP Conf. Series: Earth and Environmental Science **897** (2021) 012014 doi:10.1088/1755-1315/897/1/012014

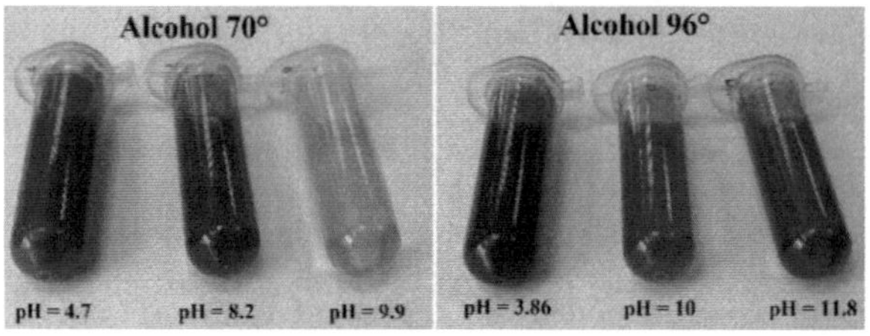

Figure 1. Colloids of silver nanoparticles under the influence of different solvents and pH values.

These same parameters are analysed through their UV-Vis absorbance spectrum (Figure 2), where it is shown that their resonance peaks are between 410 nm and 430 nm, indicative of the formation of silver nanoparticles [16,17]. This graph allows us to note that there are indeed phytochemical compounds capable of the reduction process towards Ag^0, which is evidenced in the previous literature, where polyphenols and flavonoids are indicated as responsible agents for this reduction process [13,15,18].

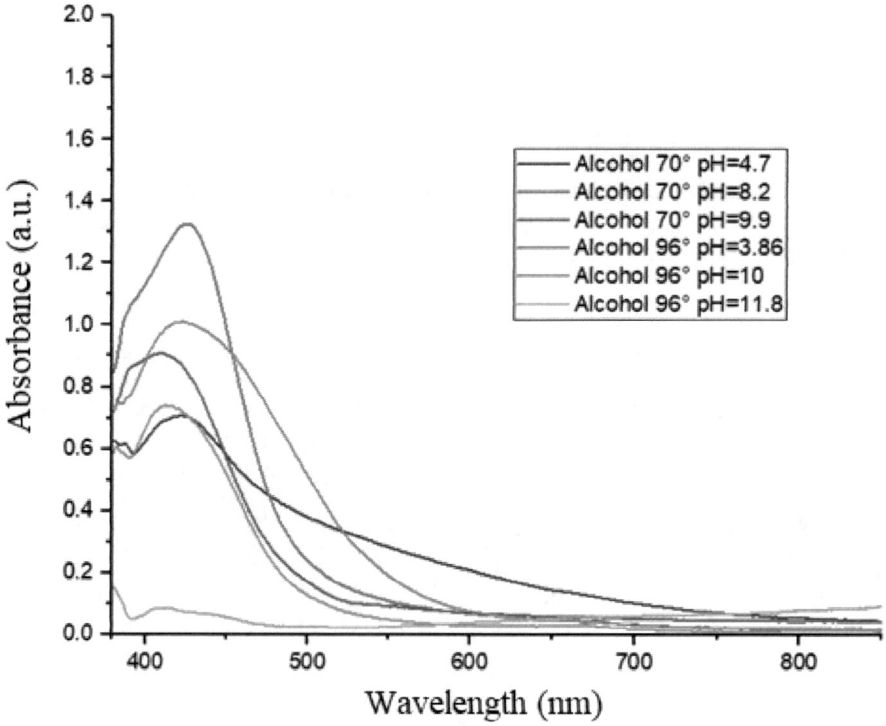

Figure 2. UV-vis spectrophotometry of the influence of type of solvent and pH on the production of Ag nanoparticles.

The pH values intervene in the reaction kinetics with effects on the size and shape of the nanoparticles [19,20]. It is graphically evident (Fig. 2) that the values of pH 9.9 and 10 for the synthesis with alcohol extracts of 70 ° and 96 ° respectively, provide us with a better formation of nanoparticles as their plasmon resonance peaks are better defined, narrow, and with values of 410.9 nm and 412.5 nm.

Regarding the stability analysis, this was carried out from 1 to 65 days. In Figure 3, the UV-Vis absorbance spectra are presented for each of the pH values and type of extract solvent. It can be observed that, over time, the peaks have an increase in absorbance and a shift towards the "red shift".

IOP Conf. Series: Earth and Environmental Science **897** (2021) 012014 doi:10.1088/1755-1315/897/1/012014

The increase in the absorbance peak is due to the aggregation processes, originating larger silver nanoparticles. The foregoing is a consequence of the still presence of Ag^+ charges, whose surface allows the adherence of other nanoparticles, generating clusters, which in turn provide us with broader peaks, indicative of greater polydispersity in terms of nanoparticle size [21,22].

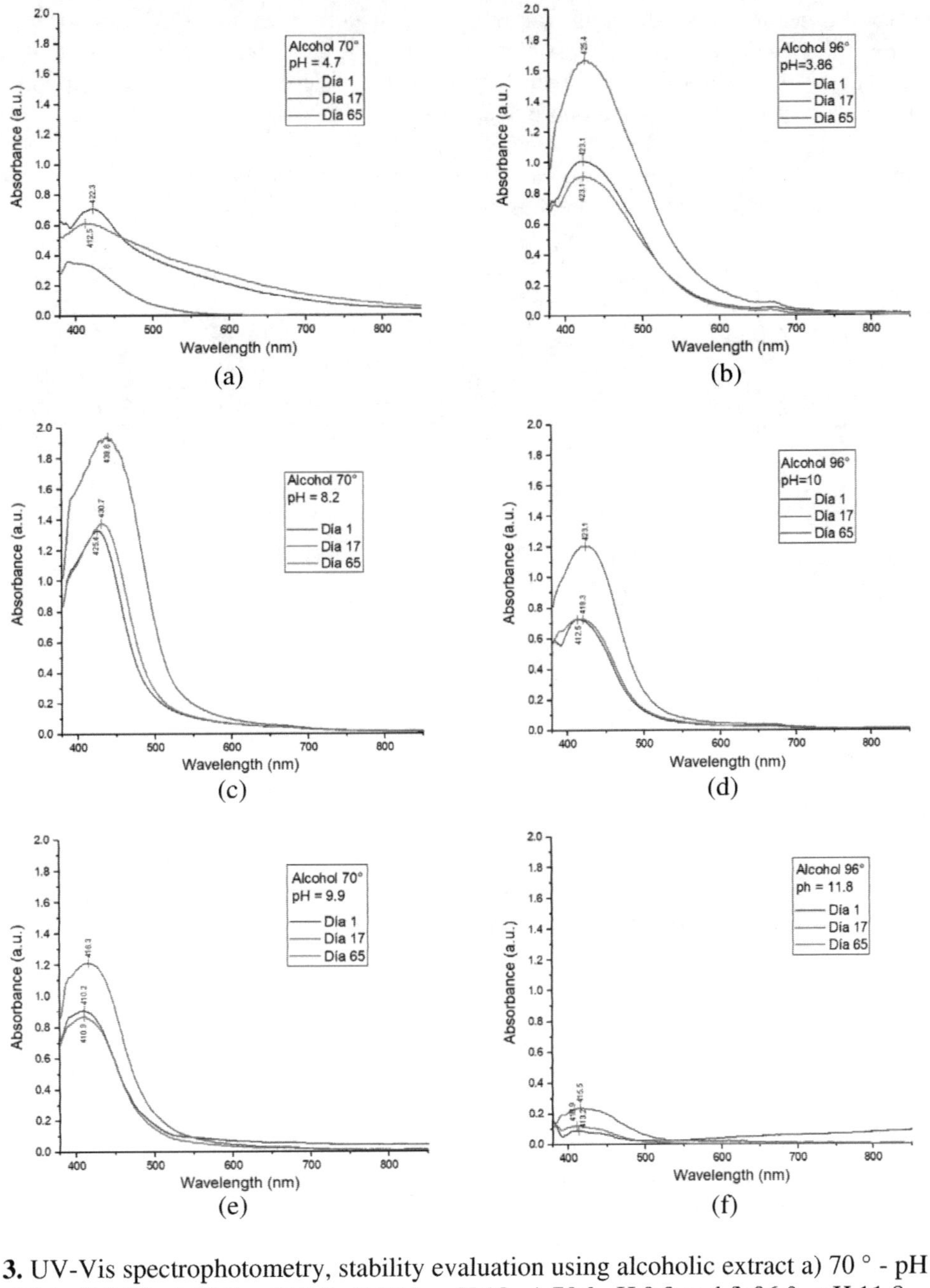

Figure 3. UV-Vis spectrophotometry, stability evaluation using alcoholic extract a) 70 ° - pH 4.7, b) 96 ° pH 3.86, c) 70 ° - pH 8.2, d) 96 ° - pH 10, e) 70 ° pH 9.9 and f) 96 ° - pH 11.8.

Note that although in Figures 3b and 3c the absorbance peaks are higher compared to the rest of the graphs, they also have a higher bandwidth, which would indicate that the nanoparticle sizes are less uniform. The 3d and 3e graphs on the other hand, provide us with more defined resonance peaks with absorbance values of 0.90991 u.a. - 1.21361 u.a. (day 1- day 65) and 0.73981 u.a. - 1.21888 u.a. (day 1 - day 65) respectively, indicative of the formation of nanostructures with less dispersion in size due to having a narrower bandwidth than those mentioned above.

Figure 3 (a) for a pH of 4.7 with eucalyptus extract in 70° alcohol, shows us the little formation of silver nanoparticles, based on the presence of absorbance peaks that are not very well defined, especially over time; in which the aggregation process occurs, leading to the disappearance of the resonance peak due to the low presence of nanostructures. At the other extreme, Figure 3(f), gives us evidence that at these pH values the synthesis by means of the alcoholic extract of eucalyptus provides us with a very low production of nanoparticles, so none of the above parameters is recommended. mentioned.

Comparing all the absorbance spectra, we can see that the best syntheses through the parameters under study occur with approximate pH values equal to 10.

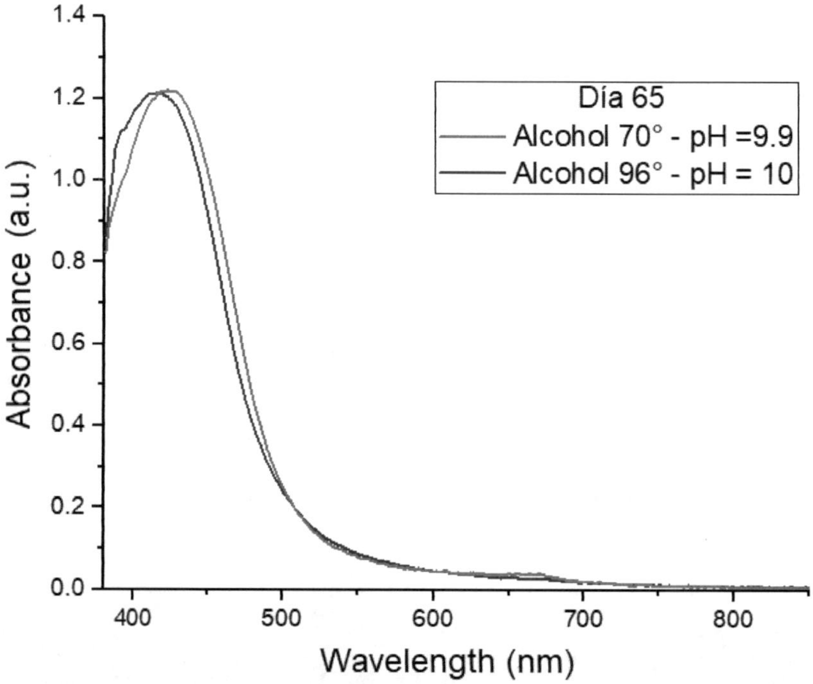

Figure 4. UV-Vis spectrophotometry, stability comparison for synthesis with alcoholic extract of 70 ° and 96 ° with pH values of 9.9 and 10 respectively.

Figure 4 shows the absorbance results 65 days after the synthesis for the pH values 9.9 and 10, in both it can be noted that the absorbance values are very similar 1.21361 u.a. and 1.21888 u.a. for eucalyptus extracts in alcohol of 70 ° and 96 ° respectively. However, the wavelength of the resonance peak for 70 ° alcohol synthesis is 416.3 nm, while for 96 ° it corresponds to 423.1. The above indicates the formation of smaller nanoparticles through the first extract due to the shift of this peak towards the "blue shift" although not with a marked difference.

4. Conclusions

The reuse of eucalyptus leaves is a potential alternative as a reducing agent within the silver nanoparticle synthesis process. However, it is important to establish an adequate methodology for the extraction of reducing compounds; therefore, the solvent plays a fundamental role in the process of making these extracts. In this sense, silver nanoparticles were able to synthesize using the alcoholic extract of eucalyptus, both at 70 ° and 96 °, this process being economic and ecological showing the

IOP Conf. Series: Earth and Environmental Science **897** (2021) 012014 doi:10.1088/1755-1315/897/1/012014

characteristic plasmon resonance peak for silver nanoparticles around 416.3 nm. Likewise, the management of the pH variable in the precursor reduction process had a significant influence, this after the analysis of the absorbance spectra by means of the surface plasmon resonance peaks, it becomes evident that for pH values of 4.7 or 3.86 for example, although there is nanoparticle production, this is very low or they are large, in the same way with a pH of 11.8. However, it was also possible to determine the optimum pH value, which corresponds to values approximately equal to 10, under which the synthesized nanoparticles show a well-defined characteristic plasmon peak and which remains with good stability over time.

5. References

[1] Chauhan A K, Kataria N and Garg V K 2020 Green fabrication of ZnO nanoparticles using Eucalyptus spp. leaves extract and their application in wastewater remediation *Chemosphere* **247** 125803

[2] Wang T, Lin J, Chen Z, Megharaj M and Naidu R 2014 Green synthesized iron nanoparticles by green tea and eucalyptus leaves extracts used for removal of nitrate in aqueous solution *J. Clean. Prod.* **83** 413–9

[3] Zhang Y, Ni S, Wang X, Zhang W, Lagerquist L, Qin M, Willför S, Xu C and Fatehi P 2019 Ultrafast adsorption of heavy metal ions onto functionalized lignin-based hybrid magnetic nanoparticles *Chem. Eng. J.* **372** 82–91

[4] Dakshayani S S, Marulasiddeshwara M B, Sharath S K, Golla R, P. R K, Devaraja S and Hosamani R 2019 Antimicrobial, anticoagulant and antiplatelet activities of green synthesized silver nanoparticles using Selaginella (Sanjeevini) plant extract *Int. J. Biol. Macromol.* **131** 787–97

[5] Asmat-Campos D, Abreu A C, Romero-Cano M S, Urquiaga-Zavaleta J, Contreras-Cáceres R, Delfín-Narciso D, Juárez-Cortijo L, Nazario-Naveda R, Rengifo-Penadillos R and Fernández I 2020 Unraveling the Active Biomolecules Responsible for the Sustainable Synthesis of Nanoscale Silver Particles through Nuclear Magnetic Resonance Metabolomics *ACS Sustain. Chem. Eng.* **8** 17816–27

[6] Ranoszek-Soliwoda K, Tomaszewska E, Małek K, Celichowski G, Orlowski P, Krzyzowska M and Grobelny J 2019 The synthesis of monodisperse silver nanoparticles with plant extracts *Colloids Surfaces B Biointerfaces* **177** 19–24

[7] Mo Y Y, Tang Y K, Wang S Y, Lin J M, Zhang H B and Luo D Y 2015 Green synthesis of silver nanoparticles using eucalyptus leaf extract *Mater. Lett.* **144** 165–7

[8] Ahn E Y, Jin H and Park Y 2019 Assessing the antioxidant, cytotoxic, apoptotic and wound healing properties of silver nanoparticles green-synthesized by plant extracts *Mater. Sci. Eng. C* **101** 204–16

[9] Uddin I, Ahmad K, Khan A A and Kazmi M A 2017 Synthesis of silver nanoparticles using Matricaria recutita (Babunah) plant extract and its study as mercury ions sensor *Sens. Bio-Sensing Res.* **16** 62–7

[10] Ahmed S, Ahmad M, Swami B L and Ikram S 2016 A review on plants extract mediated synthesis of silver nanoparticles for antimicrobial applications: A green expertise *J. Adv. Res.* **7** 17–28

[11] Chand K, Cao D, Eldin Fouad D, Hussain Shah A, Qadeer Dayo A, Zhu K, Nazim Lakhan M, Mehdi G and Dong S 2020 Green synthesis, characterization and photocatalytic application of silver nanoparticles synthesized by various plant extracts *Arab. J. Chem.* **13** 8248–61

[12] Wintachai P, Paosen S, Yupanqui C T and Voravuthikunchai S P 2019 Silver nanoparticles synthesized with Eucalyptus critriodora ethanol leaf extract stimulate antibacterial activity against clinically multidrug-resistant Acinetobacter baumannii isolated from pneumonia patients *Microb. Pathog.* **126** 245–57

[13] Mosaviniya M, Kikhavani T, Tanzifi M, Tavakkoli Yaraki M, Tajbakhsh P and Lajevardi A 2019 Facile green synthesis of silver nanoparticles using Crocus Haussknechtii Bois bulb

extract: Catalytic activity and antibacterial properties *Colloids Interface Sci. Commun.* **33** 100211

[14] Das P, Ghosal K, Jana N K, Mukherjee A and Basak P 2019 Green synthesis and characterization of silver nanoparticles using belladonna mother tincture and its efficacy as a potential antibacterial and anti-inflammatory agent *Mater. Chem. Phys.* **228** 310–7

[15] Raota C S, Cerbaro A F, Salvador M, Delamare A P L, Echeverrigaray S, Da Silva Crespo J, Da Silva T B and Giovanela M 2019 Green synthesis of silver nanoparticles using an extract of Ives cultivar (Vitis labrusca) pomace: Characterization and application in wastewater disinfection *J. Environ. Chem. Eng.* **7** 103383

[16] Katta V K M and Dubey R S 2020 Green synthesis of silver nanoparticles using Tagetes erecta plant and investigation of their structural, optical, chemical and morphological properties *Mater. Today Proc.*

[17] Kambale E K, Nkanga C I, Mutonkole B P I, Bapolisi A M, Tassa D O, Liesse J M I, Krause R W M and Memvanga P B 2020 Green synthesis of antimicrobial silver nanoparticles using aqueous leaf extracts from three Congolese plant species (Brillantaisia patula, Crossopteryx febrifuga and Senna siamea) *Heliyon* **6** e04493

[18] Alghoraibi I, Soukkarieh C, Zein R, Alahmad A, Walter J G and Daghestani M 2020 Aqueous extract of Eucalyptus camaldulensis leaves as reducing and capping agent in biosynthesis of silver nanoparticles *Inorg. Nano-Metal Chem.* **50** 895–902

[19] Mora-Godínez S, Abril-Martínez F and Pacheco A 2020 Green synthesis of silver nanoparticles using microalgae acclimated to high CO2 *Mater. Today Proc.*

[20] Parthiban E, Manivannan N, Ramanibai R and Mathivanan N 2019 Green synthesis of silver-nanoparticles from Annona reticulata leaves aqueous extract and its mosquito larvicidal and anti-microbial activity on human pathogens *Biotechnol. Reports* **21** e00297

[21] Tamilarasi P and Meena P 2020 Green synthesis of silver nanoparticles (Ag NPs) using Gomphrena globosa (Globe amaranth) leaf extract and their characterization *Mater. Today Proc.* **33** 2209–16

[22] Khan M, Ahmad F, Koivisto J T and Kellomäki M 2020 Green synthesis of controlled size gold and silver nanoparticles using antioxidant as capping and reducing agent *Colloids Interface Sci. Commun.* **39** 100322

IOP Conf. Series: Earth and Environmental Science **897** (2021) 012015 doi:10.1088/1755-1315/897/1/012015

Optimal load dispatch strategies for a hybrid system using renewable energy sources

Ronald Ayala Ramírez[1], Javier Tenesaca Chacaguasay[1]*, and Juan Lata García[1]
[1]Universidad Politécnica Salesiana, Electrical Engineering Department, Guayaquil, Ecuador

*Corresponding author's e-mail: jtenesacac6@est.ups.edu.ec

Abstract. Recently, the idea of hybrid power systems (HES) has attracted interest for the electrification of isolated or energy efficient areas. This document examines the modelling and optimal dimensions of a hybrid microgrid using different dispatch strategies. The sizing of the HES components such as Photovoltaic panels, Batteries, Inverter, a Diesel generator has been optimized by three strategies: (i) load tracking, (ii) cycle load, and (iii) combined dispatch. The location of the case study is in a rural community in Ecuador whose load profile is 17 kW. By utilizing HOMER software, optimization for the HES was achieved with the Combined Dispatch strategy (CD) which presented the minimum levels in the net annual cost (NPC), initial capital, levelized cost of energy (LCOE) of $ 90,073.10, $ 21,208 and $ 0.2016 / kWh, respectively. The conclusions offer a guide to consider the resources and generation combination essential for the optimal operation of an island microgrid with different dispatch scenarios.

1. Introduction

There is an abundance of isolated places around the world in which power supply through the central electrical Grid is neither economical nor technically feasible due to their geographic conditions. These systems are normally referred as an autonomous Microgrid. [1] Most of these microgrids usually have poorly optimized systems in which they normally operate with a diesel generator which is not only a terrible option money-wise due to the volatility of oil prices but also negatively affecting the environment due to gas emissions produced when burning through the fuel. In these cases, Off-grid systems involving renewable energy can be a feasible solution. [2] Generation units in Microgrids can be Diesel or Thermal Generators, and Renewable Energy Sources (RES), can be Photovoltaic (PV) Panels, Wind powered Turbines (WT), Battery Energy Storage Systems (BESS) or Fuel Cells. These technologies offer feasible energy solutions in remote locations but due to their stochastic nature, their operation may or not be conditioned. [3] This depends on whether the system contains a reliable generation unit and batteries as backup units. Hence a hybrid system is preferable, requiring optimal sizing and dispatch strategy to offer a higher reliability, and quality energy than a system based on a fossil fuel energy source. [4]

Recent research has successfully implemented different hybrid energy systems. PV panels, Wind turbines, Diesel Generators are the most used resources in these investigations. Also, Battery banks, Fuel cells and Hydro pumps are the common denominators for energy storage.

IOP Conf. Series: Earth and Environmental Science **897** (2021) 012015 doi:10.1088/1755-1315/897/1/012015

A. Toopshekan [5] In his research he presents the dispatch of renewable and sustainable energy on a system composed of solar panels / WT / diesel / battery, the researchers come up with a novel control strategy through MATLAB and HOMER, the results show a 9% improvement in cost over two other dispatch strategies by load tracking and by cycle load. The simulations were carried out for different sensitivities and electrical loads in Iran.
N. Sifakis [6] Presents the optimization analysis of a RES with connection to the electrical grid in a seaport, in the research they compare the impact of three storage technologies, and two dispatch strategies in 17 scenarios with respect to possible combinations taking into account the energy cost, future employability in addition to carbon footprint, best results were obtained with vanadium redox flow batteries offering up to 10h autonomy, Supplied energy cost is at 12.9 c € / kWh In all the important presentations presented, the cost of energy is lower than the rate of the conventional network with one point reducing the environmental impact.
K. Liu [7] used particle swarms' optimization to optimize the micro-grid operation of a system containing Gas turbines, Fuel Cells, wind power and PV power generation, In their control strategy, the first step is to choose whether PV or Wind power is the one with the least costs and less pollution next the fuel cell power when the supply is insufficient, and lastly the Gas turbine Power. Results show that the particle swarms' optimization technique used is a fast solution which can greatly improve power quality, reduce costs operation and maintenance costs while also lowering environmental pollution. K. Kuzakana [8] used a simulation dispatch method based on MATLAB in a rural residential home and a transmitter-receiver station case study, he compares the obtained results to a setting in which the Diesel Generator is the sole energy supplier. The results show noteworthy energy savings utilizing the modelled optimal energy control strategies.
J. Lu [9] sought to substitute an old Diesel generator in a remote sea Island with a Photovoltaic-Eolic-Diesel-Batteries HES using HOMER as a method to optimize and simulate the system, using different combinations of hybrid energy sources and storage to establish the optimal model of the system, through simulations they concluded that the HES can get rid of the Diesel-based Generation dependency to operate in a more cost-efficient manner. X Liu [10] Optimized an AC/DC hybrid microgrid of independent operation by means of a modified artificial BCA, building an optimal model with an economic approach. The objective function was built around reducing environmental emissions, Finally the optimal sizing model demonstrates that it has a lower cost when compared to the traditional model. S. Oh [11] Proposed a predictive Dispatch strategy and did a real-time simulation to prove near optimal performance on the management algorithm. The case study was an islanded microgrid, their optimization considered energy costs, Generator and battery life cycles utilizing normal parameter values. The simulations show that the algorithm provides an optimal control for the system with a minimum impact from erroneous load demand prediction to the algorithm's performance.
This paper analyses a case study on an isolated community located in Colta-Ecuador, which is currently using a sole Diesel generator as their means of energy supply and presents an optimized off-grid model using renewable energy sources in conjunction with a diesel generator.
People living in this area have very low income, so the main focus will be utilizing the dispatch strategy that allows the lowest energy cost throughout the project's life. Three different dispatch strategies are considered for a comparative performance analysis. Section II Consists of the design and modelling of the micro-grid system in HOMER. Section III contains the simulation and optimization results. And finally, section IV Draws the conclusions.

2. Methodology

This review analyses the viability towards a new hybrid microgrid structure for a rural isolated community, considering solar energy resource as its main source of power, while also including several components such as a battery bank and a DC/AC converter.

The correct analysis of a certain renewable energy project requires the use of proper criteria in the selection of the system's location to ensure the accurate operating behaviour before the different scenarios. For this purpose, the subsequent analysis frameworks are adopted.
Please follow these instructions as carefully as possible so all articles within a conference have the same style to the title page. This paragraph follows a section title so it should not be indented.

2.1. Case Study's characteristics and load profile
The case study is placed in the "Castug Alto" community located in Colta – Ecuador. This community is composed of 40 households with an approximate population of 120 people. Electrical Energy generation is delegated to a single conventional Diesel Generator with a 10-kW throughput which is also near the end of its life cycle and with an Operation and Maintenance cost of around $ 1137.33 per year.
The cost of diesel fuel in Ecuador has seen a price increase from 1.48 USD per gallon to 1.52 USD per gallon. This increase also impacts the energy production costs [12]. Given the increased Diesel price, the cost of fuel consumption per year equates roughly to $ 4240.
The Diesel Generator has been replaced every time it ceases to function in its entirety, so the time between replacements varies greatly.
Residential households in this community have a very low load demand on their electronics and illumination. Total daily load profile is illustrated in figure 1. Demand is low during the (00:00 – 5:00 hrs) period, given that the local population is resting during this time. The profile shows an increase in demand between 05:00 a 7:00 am due to the population is getting ready to go to school or their jobs.
The rest of the day presents a slight decrease in energy consumption and maintains this energy flow until approximately 16:00 hrs. After 16:00 there is a considerable increase in load demand, with its peak being around 19:00 to 21:00 due to most villagers being awake and present at their households. Finally, the load starts diminishing as the resting hours start getting closer. The data shows that the community presents a total daily load plaint of 109 kWh, with a 10 kW power peak and a mean of 4,54 kW.

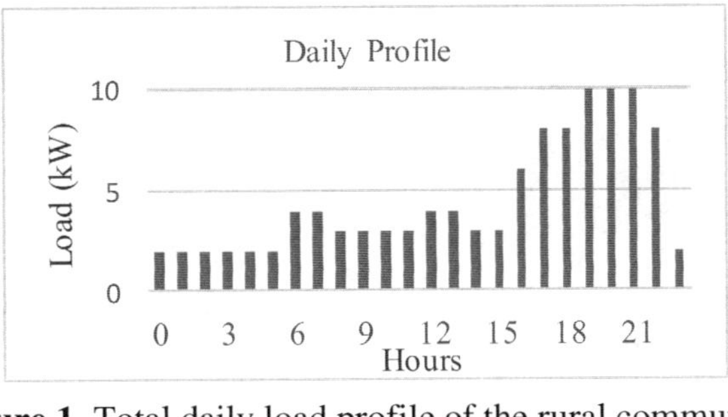

Figure 1. Total daily load profile of the rural community.

2.2. Solar Resource and Temperature
To determine the PV power output, both the temperature in the zone and the solar radiation are required like critical reference. The HOMER software requires an average horizontal global monthly radiation and on-site temperature as input parameters.
Solar radiation and on-site temperature on the selected location (Latitude 1 ° 46.9 'S, Longitude 78 ° 42.3'W) which were obtained on the NASA website [13] are detailed bellow.
The figures of the radiation index and solar clearness for the rural community are given at figure 2. Median solar radiation per year is 4.48 kWh/m2 per day, also the highest value radiation is 4,7 kWh/m2 per day is registered in November and the lowest valuer radiation 4,23 kWh/m2 per day registered within June. With these parameters, it is established that the solar radiation quantity in the area is quite high, which makes Solar energy an attractive alternative.

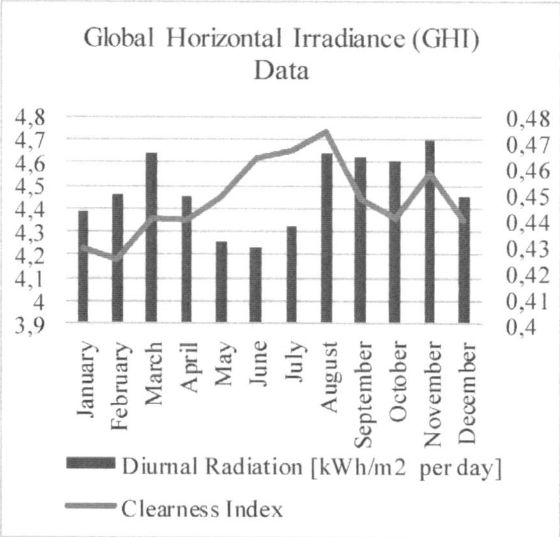

Figure 2. Clearance indication and mean solar global radiation at month of rural Community.

The surrounding temperature presents itself as a critical parameter for determining the efficiency of PV cells. Hence, surrounding temperature must have a precise measurement. Figure 3 gives details in mean ambient temperature of this area by month. Winter season shows that its higher temperature at: 11.43 ◦C in April, and summer season shows that its higher temperature is at: 9.7 ◦C in July.

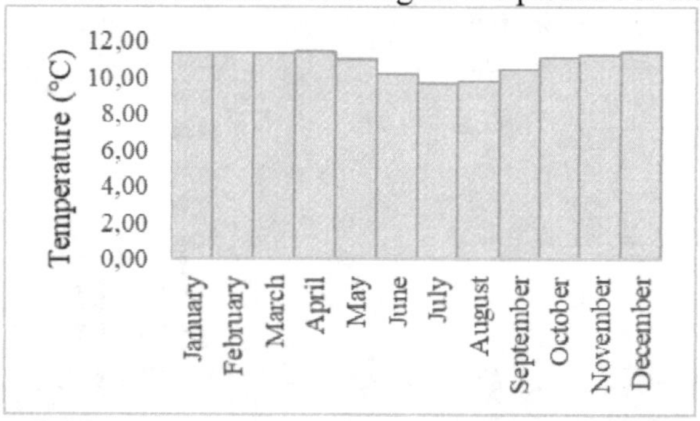

Figure 3. Mean temperature at month of the rural community.

2.3. Proposed Hybrid structure description

In the hybrid system shown in figure 4, the components such as: PV panels and Battery bank are by a direct route attached to the DC bar, while Diesel Generator is brought together to the AC bar. The DC/AC converter interconnects the AC and DC bars.

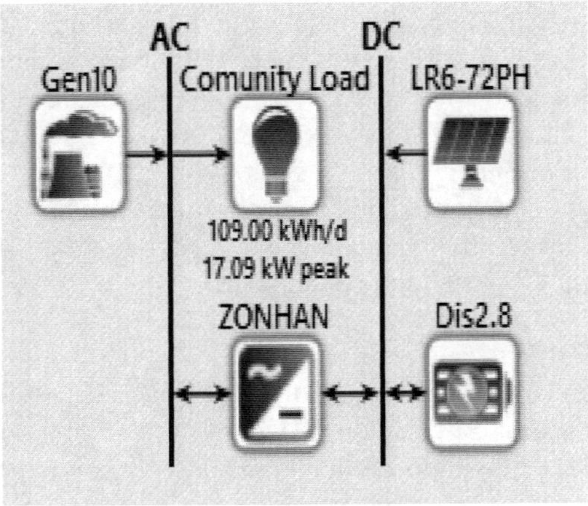

Figure 4. Proposed hybrid system

2.3.1. Photovoltaic Panel. Given the location conditions, Solar resource is abundant, which is why PV panels from the HOMER commercial catalog are used. The LONGI SOLAR LR36 is selected with a nominal capacity of 0.370 KW, with an initial cost of 180$, replacement cost of 180$, an efficiency of 19.1%, Operation and Maintenance at 10$ per year.

2.3.2. Diesel Generator. Hybrid systems usually work with a Diesel generator as a backup unit in front of unexpected energy production reduction or a sudden increase in demand. [14]. In this system, the optimal choice is a diesel generator, it has a rated Power of 10 kW, Operation life of 15000 Hours, minimum load ratio of 25%, beginning cost of $ 5000, Operation and maintenance costs set at $ 0.3 per hour, Diesel price is set at $ 0.40 per liter and replacement cost of $ 5000.

2.3.3. Battery bank. The battery size is considered as one of indispensable components in hybrid systems, this is due to their main purpose is to lower the Renewable energy generation's intermittence [14 -16]. In this simulation, Battery models commonly available are used, such as the Discover AES (24Vdc, 100 Ah, 2.64 kWh). Each battery possesses price initial of $840, O & M cost of $ 10 annual and replacing cost of $ 840.

2.3.4. DC/AC Converter. An electronic converter unit is necessary on a system in which DC components are being used to supply an AC load or vice versa. Initial cost is set at $ 672 and so is replacement cost. O&M cost is $ 10 annual. And it has a 15-year life cycle expectancy with a 90% efficiency.

2.4. Dispatch strategies

Performing an optimal load dispatch on a hybrid system supposes a technical challenge in which the controller set to control the hybrid system must comply with load requirements. The set of constraints that dictate the behaviour of generators and batteries is known as load dispatch [17].

For this system, three dispatch strategies are considered: Load Following, Cycle Charging and Combined Method. Establishing the optimal dispatch strategy follows a set of constraints, which include Generator sizing, Fuel price, Battery quantity, operation and maintenance cost, to name a few. HOMER software simulates the three strategies, and the result is taken to make the optimal system evaluation.

2.4.1. Load Following (LF). In this load dispatch strategy, every time a Generator is required, it then provides enough but no more than the necessary energy to satisfy the load. This control is mostly used when there are a lot of available renewable resources that are normally above the load [18].

IOP Conf. Series: Earth and Environmental Science **897** (2021) 012015 doi:10.1088/1755-1315/897/1/012015

2.4.2. Cycle Charging (CC)). In this load dispatch method, the energy supplier functions at its entire capacity in order to meet the demand, and surplus energy produced is then used to fill up the batteries. This strategy is better used when renewable energy resource is scarce. This ensures the battery uses less charge cycles to improve battery life, delaying time needed for battery replacement [19].
Combined Dispatch (CD). In Combined Dispatch strategy the controller decides whether to use LF or CC depending on the system's conditions. This controller uses the CC strategy when the present time demand is small, oppositely when the demand is on a high degree, the controller uses the Load Following method [20]. section IV Draws the conclusions.

3. Results and discussion

For this hybrid system the simulation is performed using the HOMER software. Optimal dimensioning values are calculated for each dispatch strategy, which allows the comparison of critical data such as: NPC, COE and O&M. The strategies used in this article show innovative results. The simulation was optimized using the following parameters an yearly discount rate of 7.84 % and a 25 year project duration of life. Furthermore, it was considered a 2% as maximum capacity shortage for a get better confiability.
Table 1 gives the differentiation in NPC, Capital, COE and O&M, to different dispatch control in case study. The NPC and COE costs are more economics for CD strategy as sample $ 90,073.10 and 0.2106 $/kWh respectively and the high-cost for LF strategy as 109,319.9 $ and 0.2549 $/kWh respectively. The CD strategy presented a capital cost of $ 21,208, the lowest of all, followed by the CC strategy $ 25,474. The LF strategy obtained a capital cost of $ 54,174.68, because the size of the renewables is the highest. The LF strategy present the O&M cost and CO2 emission lowest 5,096.89 $/yr. and 7,292 Kg/yr. respectively, making more environmentally friendly, but with a high NPC cost.

Table 1. Comparison of NPC, Capital, COE and O&M for different controller.

Homer Controller	NPC ($)	*Initial Capital ($)*	COE ($/kWh)	*O&M ($/year)*
Cycle Charging (CC)	91,953	25,474	0.214	6,144
Load Following (LF)	109,319.90	54,174.68	0.2549	5,096.89
Combined Dispatch (CD)	90,073.10	21,208	0.2106	6,365.01

The size of each component of the hybrid structure is compared and details in table 2. Under the same loading profile, a clear difference in the size of all the components is appreciated. Depending on the different control strategies, the optimal size of PV, converter and battery are different from each other, except for the diesel generator, which will be a fixed size of 10kW.
In LF strategy, the Battery size is extremely high, its capacity is about 86.8 MW. Due the power produced from PV fully satisfy the load. Thus, to lay up the overage energy requires a bigger storage for later use.

Table 2. Optimum Dimension established in Homer evaluation.

Dispatch Strategies	*PV (KW)*	*Diesel Generator (KW)*	*Battery (kW-h)*	*Converter Size (KW)*	*Load Profile*
Cycle Charging	18.1	10	33.6	7.2	109 kWh/d

IOP Conf. Series: Earth and Environmental Science **897** (2021) 012015 doi:10.1088/1755-1315/897/1/012015

(CC)					10 kW peak
Load Following (LF)	41.4	10	86.8	13.3	109 kWh/d 10 kW peak
Combined Dispatch (CD)	15.7	10	22.4	8.19	109 kWh/d 10 kW peak

In the LF strategy, an outstanding use of the renewable source (solar radiation) can contribute with an approximate 89% of annual energy production, the highest standard contrasted to the CC and CD strategies. The renewable component (PV) creates a greater dependence on the hybrid system compared to the other strategies CC (48.4%) and CD (42.4%).
The contribution of energy by the Diesel generator in the CC, LF and CD strategies were 51.6%, 11% and 57.6%, respectively. Figure 5 shows the annual electrical energy production of the three dispatch strategies.

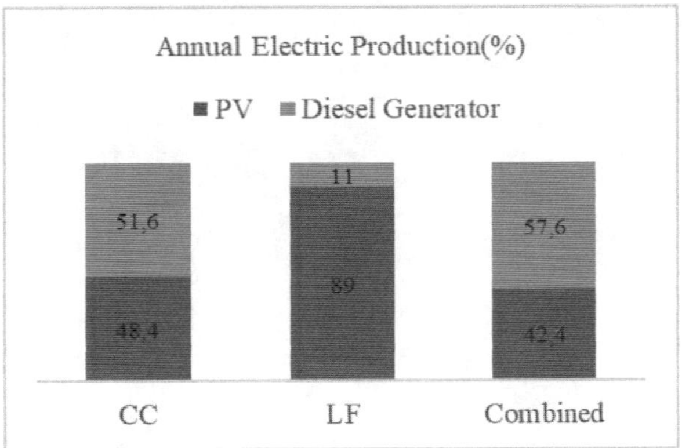

Figure 5. Annual electric production comparison by source under different dispatch strategies

In terms of pollution, Diesel consumption has a negative effect on harmful to human health and the environment, since various way of polluting agent are being throwing throughout the operation. These discharges contain residue as carbon monoxide (CO) and carbon dioxide (CO2). In addition to other polluting gasses are details shown in table 3.
A hybrid energy source has positive effects on the environment [21]. The combination of photovoltaic energy with diesel and batteries is capable of significantly minimizing emissions. Annual emission releases for HES have been defined and demonstrated under different control tactics to contrast cases from an environmental perspective. In the table 3, it details the emissions produced by the hybrid configuration (PV / diesel / battery) employing LF, CC and CD strategies.

Table 3. Dispatch strategies and the pollution produced.

Quantity (Kg/yr.)	CC	*LF*	CD
Carbon Dioxide	22,154	7,292	24,438
Carbon Monoxide	168	55.2	185
Unburned Hydrocarbons	6.11	2.01	6.74

Particulate Matter	10.2	3.34	11.2
Sulfur Dioxide	54.4	17.9	60
Nitro Oxides	190	62.7	210

The system with LF has been the biggest environmentally amiable alternative for bring about the lowest proportion of CO2 release (7,292 kg/yr.), in contradiction to the configuration that employs DC strategy, whose had CO2 expulsion is 24,438 kg /yr. Although this system is not the most economical since NPC and Capital cost high as shown in table 1.
In figure 6, we can observe the state of charge of the batteries. This indicates that the CC strategy can store a greater amount of energy with respect to the LF and DC strategies, because the diesel generator works at its greatest range to satisfy the user requirements, relegating the batteries only for emergencies. The state of charge in the DC strategy is the lowest because the diesel generator, the solar panels and the batteries are continuously feeding the load.

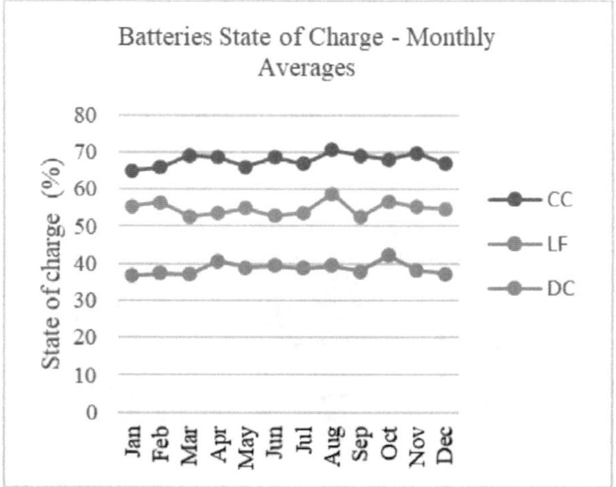

Figure 6. Batteries State of Charge - Monthly Averages

4. Conclusions

Optimization of load dispatch strategies is essential in evaluating the feasibility of hybrid energy systems. For this study, three different control strategies were used such as: Cycle charging (CC), Load following (LF) and combined dispatch (CD). HOMER platform was employed to estimate the general study, which includes the sizing optimization and sensitivity analysis. The microgrid proposed has a PV system, battery bank, a DC / AC converter, and a Diesel generator for a rural community in Ecuador.
According to the cost evaluation, the outcome details that the mix of 15.7 kW Photovoltaic Panels, 8 batteries, a diesel generator with 10 kW, and 8.19 kW power converters controlled by the CD method is the most favourable solution. since it has the lowest cost NPC ($ 90,073.10), capital cost ($ 21,208) and a COE ($ 0.2106 / kWh) which were 2% and 17.6% cheaper than the systems with CC and LF strategies, respectively. However, the type of load and the future increase in demand could vary the performance of the hybrid system. This study plays an important function in decision making that leads to superior energy administration methods. By decreasing the Shortage of Capacity by 2%, the size of the system will be greater resulting in a higher cost of energy.

5. References

[1] Rishi K S, Shivraman M, Sukumar M 2018 Power management and economic load dispatch-based control of hybrid pv-battery-diesel standalone ac system *IEEE Access*

IOP Conf. Series: Earth and Environmental Science **897** (2021) 012015 doi:10.1088/1755-1315/897/1/012015

[2] Patarau T, Petreus D, Etz R, Lazar E 2018 Techno-economic feasibility study on an off-grid renewable energy microgrid for an isolated greenhouse in Romania *IEEE Access*

[3] Santillán-Lemus F D, Hertwin M P, Aguilar-Mejía O 2019 Optimal economic dispatch in microgrids with renewable energy sources *Energies*

[4] Sajjad A, Mazhar A, Elena G 2018 Optimal energy management for off-grid hybrid system using hybrid optimization technique *IEEE Access*

[5] Toopshekan A, Yousefi H, Astaraei F R 2020 Technical, economic, and performance analysis of a hybrid energy *J. Energy*

[6] Sifakis N, Konidakis S, Tsoutsos T 2021 Hybrid renewable energy system optimum design and smart dispatch for nearly zero energy ports *J. Cleaner Prod*

[7] Liu K, Zhang Q, Tian B, Li W, Xuan W, Zhang F, Li J 2016 The Research on an Optimal Dispatch Model of Micro-Grid

[8] Kusakana K 2016 Energy dispatching of an isolated diesel-battery hybrid power system *IEEE Access*

[9] Jiaxin L, Weijun W, Yingchao Z, Sheng Y 2017 Techno-economic feasibility of pv-wind-diesel-battery hybrid energy system in a remote island in the south China sea *AMSE J.*

[10] Liu X, Du Z, Yan X 2016 The optimal sizing for ac/dc hybrid stand-alone microgrid based on energy dispatch strategy *IEEE Power Energy Technol. Syst. J.*

[11] Seaseung O, Suyong C, Jason N, Jongbok B, Marvin C 2017 Efficient model predictive control strategies for resource management in an islanded microgrid *Energies*

[12] Diario El Comercio, [Online]. Available: https://www.elcomercio.com/actualidad/negocios/precios-gasolina-combustibles-diesel-ecuador.html. (2021).

[13] Power Data Access Viewer, [Online]. Available: https://power.larc.nasa.gov/data-access-viewer/. (2021).

[14] Lu J, Wang W, Zhang Y, Ye S 2017 Techno-economic feasibility of pv-wind-diesel-battery hybrid *IIETA*

[15] Oloo F, Molefyane B, Rampokanyo M 2020 Modelling and optimisation studies for generator dispatch strategies for deployment of an off-grid micro-grid in South Africa *IEEE-ENERGYCON*

[16] Kassa Y, Zhang J, Zheng D 2019 Optimal energy management strategy in microgrids with mixed energy resources and energy storage system *IET Journals*

[17] Ishraque F, Shezan A, Ali M, Rashid M 2021 Optimization of load dispach strategies for an island microgrid connected with renewable energy sources *Appl. Energy*

[18] Arevalo P, Benavides D, Leonardo J, Hernandez L, Jurado F 2020 Optimal energy management strategies to reduce diesel consumption for a hybrid off-grid system *Redin - Universidad de Antioquia*

[19] Saleh A, Naim M, Rafi M, Ramli M, Mekhilef S 2019 Energy management and optimization of a pv/diesel/battery hybrid energy system using a combined dispatch strategy *Sustainability*

[20] Shezan A, Hasan K, Datta M 2019 Optimal sizing of an islanded hybrid microgrid considering alternative dispatch strategies *Australian University Power Engineering Conference*

[21] Fazelpour F, Soltani N, Rosen M 2016 Economic analysis of standalone hybrid energy systems for application in Tehran, Iran *Int. J. Hydrogen Energy*

IOP Conf. Series: Earth and Environmental Science **897** (2021) 012016 doi:10.1088/1755-1315/897/1/012016

A review of voltage and reactive power control algorithms in medium voltage distribution networks

S Šantaras[1]

[1] Lithuanian Energy Institute, Smart Grids and Renewable Energy Laboratory, Breslaujos g. 3, 44403, Lithuania

simonas.santaras@lei.lt

Abstract. Current power network was built in mind that there will be mostly one-way power transmission, but as there is increasing penetration of distributed generation (DG), power network evolves and changes. Thereof old control strategies and algorithms used in power distribution do not fully utilize ever evolving power network in ensuring minimum active power loses, best distribution power generation for optimal voltage and frequency control.
The main objective of this paper is review and comparison of heuristic algorithms in medium-voltage (MV) distribution networks with high distributed generation (DG) penetration. The specific goal is integrating reviewed algorithms in distributer power network control to minimise power loss, control active and reactive power flow between transmission and distribution power networks and utilization of current power network structure with growing number distributed generation penetration and thus minimizing required cost for its upgrade. Due to complexity of different power network structures and control methods used in managing it, algorithms are reviewed used in ensuring fast and reliable calculations by guaranteeing as minimum active power loss as possible and by improving power system performance over existing centralised as well as decentralised control methods. For both control methods distributed generators requires to be endowed with communication capabilities, as it is effective in driving the voltages within the admissible intervals and, additionally, it exploits the cooperation among the distributed generators and controllers to reach power network objectives of minimal power loss, voltage stability as mentioned before.

1. Introduction

As the world is dealing with climate change, carbon dioxide-generating energy sources have been classified as detrimental to the atmosphere, and green energy sources as a potential energy source has gained more attention. A agreement has been achieved, laying the foundations for tackling environmental challenges effectively and paving the way for more ambitious problems. As a result, implementing policies to improve energy conservation and alternative energy (such as solar energy) would significantly shorten the time required to meet net greenhouse gas emissions goals. However, the scarcity of renewable energy has resulted in additional technological and economic issues, limiting its appeal. Renewable energy plants, on the other hand, require unique areas where resources are typically available, as opposed to traditional power plants. This condition necessitates the development of a mathematical optimization method to aid in the planning and decision-making process, especially in terms of the scale and location of renewable energy sources (RES) platforms. Utility companies have created distributed generator (DG) to stimulate competition and generate flexibility in the clean energy industry in order to reduce the cost of large-scale construction. Indeed, in a liberalized

electricity market, the number of generators related to power distribution will grow, particularly in terms of size and location. The CEOs of green energy firms are streamlined and they have a significant effect on reducing technological challenges associated with renewable energy incorporation (such as energy waste). According to records, inadequate estimation of the position and scale of DG would result in increased network losses [1], [2]. The optimum location and scale of DG will not only minimize errors in the electrical grid, but will also improve network voltage and efficiency because it will dictate the maximum level of RES DG that can be connected to the grid. Also, with the ongoing development in materials and energy system control technologies, the cost of RES DG is expected to fall and so ease DG integration in in power network. In comparison to massive traditional power plants, the reduced scale reduces construction time. Furthermore, DG can be used as a backup solution in the event of an emergency and provides the device's islanding feature. Their benefit is that there is less construction, the current systems can be modified, allowing the electricity grid to be transformed from centralized to decentralized.
There is increasing global interest in the use of device optimization strategies to deploy DG RES: more than half of all research undertaken in developing countries (such as Europe) is currently under investigation. However, developing countries like China, Iran, and India conduct roughly 30% of all studies. The emphasis on developed and emerging countries derives from international pressure on policies and incentives to minimize carbon dioxide emissions and encourage the use of renewable energy. The decision has been made to view green energy as a feasible alternative to conventional energy sources in order to satisfy rising domestic demand while still ensuring national security and protection. Among the numerous advantages of installing DG RES, studies have shown that as system losses rise, utilities can face new difficulties, owing to a lack of utility development. The utility companies can solve the problem using applied optimization technology.

2. Drivers and challenges of distributed generation growth

This segment provides an overview of the various factors that have contributed to the increasing interest in DG implementation, with an emphasis on DGs from RES. This article would focus on the real context of the change to more aggressive power system maintenance and smart grid implementation. A summary of the obstacles to be resolved will also be included.

2.1. Definitions

A distributed generator (DG) is an energy generating source that is connected to a power grid. DG are typically connected to distribution networks, but using International Council on Large Electric Systems (CIGRE) definitions, its power is specified as up to 50-100 MW [3], of which some countries can approve DG development only if it is connected to a transmission network. Using the Electric Power Research Institute (EPRI) definition, DG power ranges from a few kilowatts to 50 MW [4].

2.2. Drivers

It is no secret that the power network was built without considering the impact of human action on global temperature, and that humans would need to evolve and transition power production to one created by renewable energy. Due to atmosphere change at an accelerated pace, the race has already begun with a single goal in mind: to generate electricity with the smallest possible carbon footprint. Renewable energy sources (RES) come into play since the electricity generated does not contain carbon, and the carbon is only emitted in the manufacturing and assembly of renewable energy resources, but its footprint is just a fraction of the footprint of traditional power resources as compared to generating fuel.

2.3. Challenges

While there are several benefits and reasons supporting the extension of DG, fast integration of DG may also trigger economic and technological issues. The following are some of the most pressing issues confronting DG. The first is that within a certain voltage spectrum, grid current has shifted from unidirectional to bidirectional. As a result, the successful deployment of the DG unit would have an effect on network reliability and electrical device efficiency. As a result, the selection of DG installation capability is influenced not just by the expense and benefits of each technology, but also by the optimum position and scale, which can greatly minimize overall system loss [5]. This is one of the issues with DG installation. The competition in the conventional non-liberalized power structure is typically distinguished by a vertically integrated monopoly. Investment in DG is compensated by the increased competition brought on by new prospects.

3. Power network models objectives and constrains

This segment covers the steps and causes that must be precisely coordinated when planning real estate. Power flow models, general objective features, limits, and involuntary data sets are also worth investigating.

3.1. Power flow models

The fundamental element of network control is energy flow. Methods of energy transfer vary from optimum power flow (OPF) to continuous power flow (CPF) to probabilistic power flow (PPF) (PPF, etc.). The power failure conditions, on the other hand, are determined by the form of generator source. For instance, consider the energy moment of a diesel generator. Provide energy through PV or WT. Similarly, if it is single-phase or three-phase, the structure of the energy distribution can shift. Figure 1 [6] depicts the design diagram of the single-phase power supply.

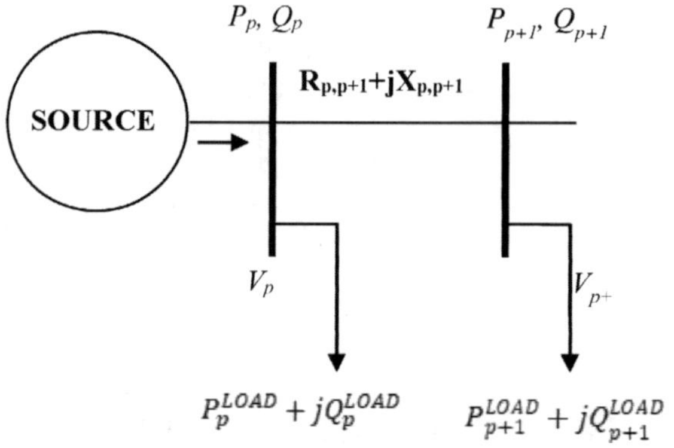

Figure 1. Design diagram of the single-phase power supply [6].

3.2. Common objectives functions

In fact, the objective function is a digital restriction created to replicate the system's properties. The most well-known aim of ideal switching is to reduce real energy loss, which is accompanied by improving voltage stability, improving voltage distribution and minimizing costs and increasing profits.

3.2.1. Power loss minimization. Power grid outages are almost unavoidable, but efforts to mitigate them cannot be overestimated. This objective is regarded as an optimal combined target, and it is mostly used as the primary target for solving calculations (such as voltage curves). Kirchhoff's rule, which states that introducing the number of the results of each transmission will fix the loss issue in the whole power grid, can affect power loss. Excess energy is inefficient in terms of money.

3.2.2. Voltage stability improvement. Voltage reliability is an important element in distribution network load control. Power outages may be caused by sudden voltage surges or variations. Uncompensated obtained power, unfortunately, may also trigger voltage instability [6], especially in microgrids with limited range. The Voltage Stability Index (VSI) is a network stability and bandwidth measurement tool that can only be used to address issues when designing traffic networks. Rejection methods, on the other hand, are seen in just a few techniques [7]. Line stability index (Lmn), line stability index (Lp), current line stability index (NLSI), line voltage stability index (LVSI), voltage stability index (FVSI), and overvoltage proximity indicator are all part of the proposed technologies (long time). To coordinate repeated networks, the word VSI has been paired with meta-heuristic calculations.

3.2.3. Load variance minimization. A mixture of electric cars will help to achieve this aim. Given the stacking system of electric cars, it is critical to ensure a fair charging schedule for electric vehicles. If this target is achieved, power outages can be limited. Based on its position, the average goal load shift is determined. In theory, this aim can also take into account dumping and smoothing costs in order to increase network performance. Changes in load can increase network efficiency (not related to VSI)[8]. This objective is linked to the advent of electric cars. Given how electric vehicles interact with one another, a realistic strategy would minimize network volatility. The perfect retreat will help save resources.
To increase network reliability, this aim could theoretically include excellent shaving and load balancing. Controlled load variations have the potential to increase network reliability in all ways (synonymous with VSI).

3.2.4. Peak load shaving. Peak load shaving is a valuable objective of V2G efficiency, since it allows you to fulfill performance expectations by using usable resources rather than stretching frame boundaries.

3.2.5. Cost minimization. The financial aspect of the optimal mix should take compatibility and approval into account. If such an idea does not exist, a significant portion of the programs will not be finished. As a result, the costs of carrying out an optimal pooling of DG unit programs must be kept to a minimum. Most studies calculate the absolute costs as the amount of the DG unit's production costs and operation and service costs [9],[10].
Charging prices for electric cars must be reduced at some stage. This demonstration is very beneficial to all lattice and EV drivers. A heap movement strategy is used, in which lower cost functions are used to promote fewer time consuming operations.

3.2.6. Operating profit maximization. Investing in upgrading the electricity grid could result in a profit for investors. For example, when developing an electric vehicle, the aggregator should have an appealing benefit expense compared to the profit. With this in mind, partners in joint venture growth must expect a predictable return [9].

3.3. Constraints
Investing in upgrading the energy structure can provide financial backers with returns. For example, if you use EV booking, the aggregator's benefit should have low charging costs [9].

3.3.1. Bus voltage limits. To keep the network stable, the bus voltage cap is used. Each bus is required to provide an allowable voltage for the power supply of a DG or EV device, with a maximum deviation of 5%. The highest permissible variation in certain situations is 10%.

3.3.2. Bus capacity limits. The bus capacity limit is used to change the bus's overall allowed load. This limitation mostly extends to EV charging, where multiple EVs may only be charged from a single bus. This inequality cap must be equal to or less than the total amount of electric vehicles attached to the same bus plus the aggregate of traditional electric vehicles (residential or commercial vehicles).

3.3.3. DG penetration limit. The overall nominal performance of the deployed DG units is proportional to the DG penetration limit (such as PV and WT). Electric vehicle intrusion is often

IOP Conf. Series: Earth and Environmental Science **897** (2021) 012016 doi:10.1088/1755-1315/897/1/012016

limited, whether gradual measures are permitted [11] or the full amount of intrusions is used, as in [12].

3.3.4. Power flow balance. The power flow balance ensures that the combined reactive and responsive power of the network equals the amount of the active and reactive power flowing from the DG unit, the substation, and the power network's active and reactive losses.

3.4. Non-arbitrary data-set

The active and reactive power of the buses, as well as the resistance and reactance of the branches linking the buses, was used to optimize the dimensioning and positioning of power supply units and power electronics. These parameters influence critical goal functions including power failure, voltage profile, and voltage. The IEEE dataset is a benchmark for solving size and position issues, and it was chosen as a high priority in the analysis for this study. Passages from the literature have also been considered. Climate, human behaviour, and other factors may also have an impact on other necessary data such as demand, PV output, and electricity prices. These variables may cause inconsistencies in the results. As a result, various models for dealing with such cases have been created. While recent studies in the literature have taken complexities into consideration, this research focuses solely on the optimization mechanism that employs metaheuristic algorithms in a single objective or multi-objective context.

4. Algorithms used in power network

4.1. Algorithm classification

The key issues with renewable energy are uncertainty and variability, especially in terms of the availability of wind, solar, and hydro power resources. Energy-related approaches, such as standard and intelligent search methods, are also included in the literature. In most cases, the DG bandwidth is modeled as a nonlinear mathematical optimization challenge. Determine some objective limits and functions first. Optimization technology aids decision-making by producing the best option or collection of best solutions, or producing output variables from a set of reduced initial input variables

Many optimization approaches have been proposed in the literature to date in attempt to solve these models. These approaches are classified as heuristic methods, empirical methods, gradient and second-order methods, and iterative methods, as seen in Fig. 2. This paper is being focused only on heuristic methods.

There are two kinds of optimization approaches based on the objective function: single-objective methods and multi-criteria methods. The most popular aim of the analysis is to minimize electrical device failure. Other strategies, too, are aimed at lowering overall costs.

In reality, the various objectives of the optimization dilemma would inevitably lead to some form of conflict, since no one approach will satisfy all of the needs. The maximum growth in DG capability would clash not only with the increase in production line depletion, but also with the potential increase in investment costs and public interest in lowering carbon dioxide emissions [13]. Multiple target functions are normally minimized or expanded concurrently in optimization problems with multiple conditions [14].

The weighted sum approach, in which a multi-criteria problem is translated into a single objective problem using a fixed weight, is one of the most commonly applied techniques for solving multi-criteria optimization problems. The weighted sum system has several drawbacks due to its nature. On the one side, the weighted sum procedure cannot be used to solve non-convex problems and additional goals cannot be added. The suggested solution, on the other side, is limited to a range of weights (priority) chosen for the objective purpose.

4.2. Heuristic methods

IOP Conf. Series: Earth and Environmental Science **897** (2021) 012016 doi:10.1088/1755-1315/897/1/012016

Many meta-heuristic algorithms are built on the idea of a characterized agent or group of agents that work without human intervention to accomplish an objective or set of objectives.
For large complex problem solving heuristic methods often help by reducing calculation time and producing good enough results. Since there is a lot of new heuristic algoritms beeing invented, classification of these methods becomes significant in comparing one to another. Methology on how to heuristic methods work is inspired by the world around us. Due to that, meta-heuristic optimization algorithms can be divided into five categories from which 3 are physics, evolution and swarm inspired techniques, other 2 categories come from combining algorithms. For classification of different meta-heuristic algorithms see figure 3.

4.2.1. Evolutionary-based algorithms. For optimal solutions, these evolutionary based heuristic algorithms employ Darwinian evolution and natural selection theory. GA, Differential Evolution (DE), and Evolution Strategy are examples of this type (ES). A generation, like the evolutionary mechanism, may consist of parents and children who inherit the characteristics of their parents. Selection, crossover, and mutation are the three main operators used by evolutionary algorithms. The crossover operator manages the mating mechanism to produce offspring, while the selection operator requires two parents. Finally, the mutation operator is often used to create stronger pupils. Many different forms of study have used evolutionary algorithms to optimize the integration of electrical units, such as in [14], [15], [16], [17], [18].
The genetic algorithm [14] provides the best knowledge for studying the trend. It updates the power flow response to optimally position the capacitor bank in the distribution line's critical region. Saudi Electricity Company (SEC) placed it through its paces. Consider [20] the ideal BESS solution for the ideal transportation company in the Virtual Power Plant in this direction (VPP). Genetic algorithms have been used to reduce the energy loss and power instability induced by photovoltaic systems, taking into account the susceptibility of impact and the characteristics of load growth.
The developer [15] increased the objective of reducing power loss in the ideal capacitor location in the uneven transportation frame by using an enhanced GA. GA has been evaluated on IEEE 4, 123, and 85 transmission feeders and is primarily used to approximate ideal transmission response efficiency. In [16], preferably weigh and place the BESS device in the grid photovoltaic system's circulating frame using a BA-based two-stage piping system. The aim is to minimize voltage variations generated by PV module control by supplying active power and BESS units in the event of a power loss. Compare the suggested equation to other meta-heuristic estimates for prediction, assembly period, and calculation time.

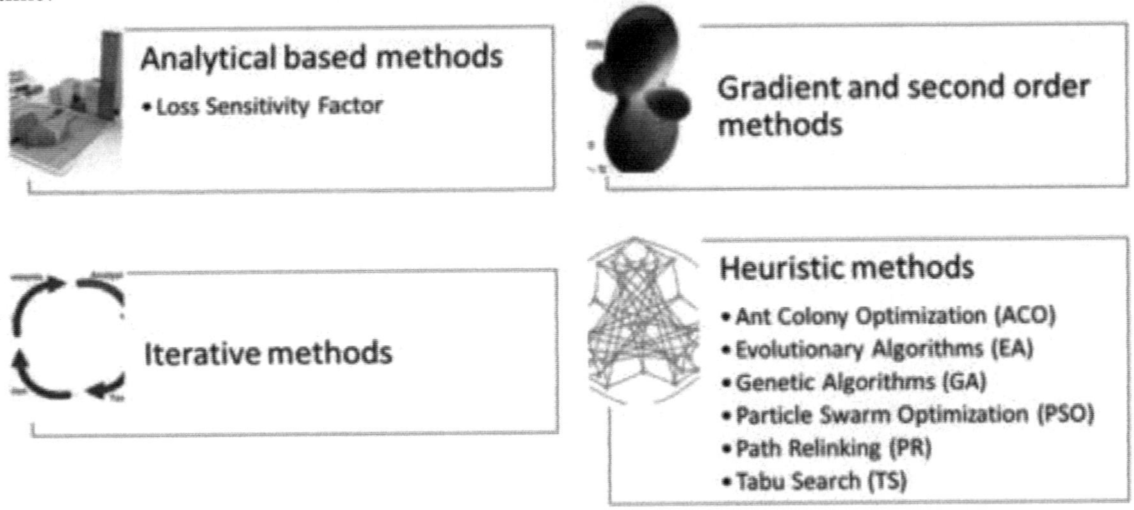

Figure 2. Classification of different methodologies to solve DGs placement and sizing problem [19].

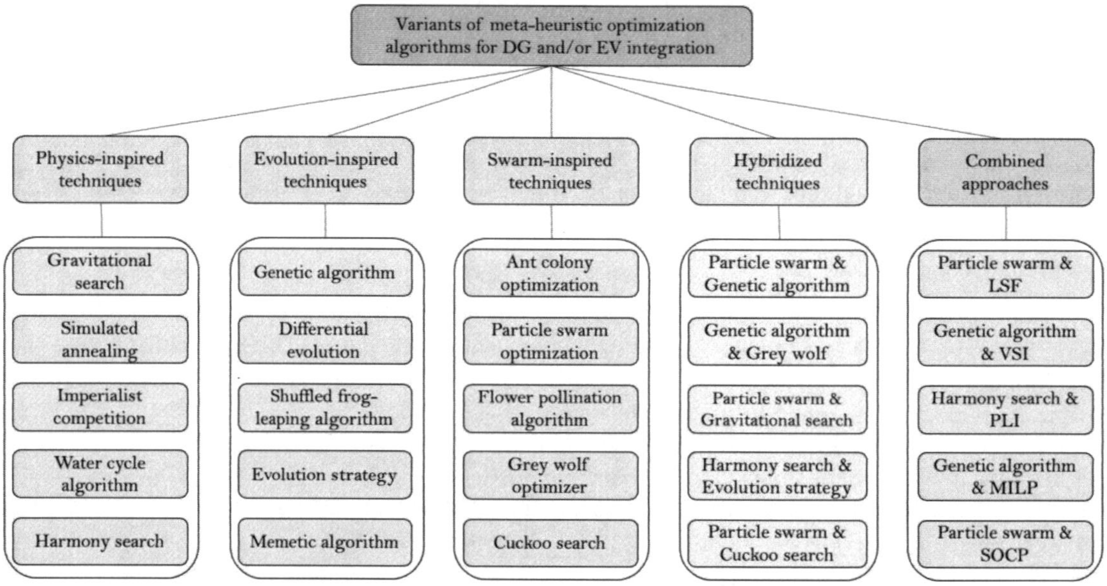

Figure 3. Classification of different meta heuristic algorithms [14].

[21] DE operations generate objective and contrast vectors, which are then used to generate pre-vectors by standard EA managers (such as Hybrid and Transformable). [22] Using isolated and continuous variables In [22], DE was used to determine the optimal area, scale, and output parameters of the DG system in the required influencing organization in order to minimize energy loss. The differential search algorithm (DSA) is used in [23] to increase the absolute power failure, voltage divergence, and overall workload in the transmission network by comparative calculations. In comparison to AG, the final outcome may be obtained by improving DSA.
For the optimal allocation and calculation of DG and D-STATCOM units in the allocation network, the modified Shufled Frog Leaping Algorithm (MSFLA) algorithm is proposed [23]. The aim is to reduce failure in the distribution string in order to enhance voltage reliability and effect efficiency. This measurement was checked and compared to the GA calculation in the IEEE 33 transmission. MSFLA is observed to produce fewer GD and STATCOM bandwidth than GA. MSFLA also increases stress distribution more effectively than GA. MSFLA has also been extended to the energy framework of dashboards in high-performance delivery companies by the authors of [24]. DG and BESS been paired for the best switches and priorities. [25] developed a GA-based simplification approach to simplify the mixture of DG, STATCOM, PHEV, and FACTS blocks and reduce total real energy usage in the delivery organization. In GA calculates each chromosome's health score based on the status of each transport intensity. The measurement is performed in the IEEE 37 Transport Dispersion Organization, assuming that the block alignment of DG, STATCOM, and PHEV will conduct real and responsive maintenance on the improved strength and power factor of the extended frame. Other prototype cars can be used to validate the current standard. As a result, different simulations may be run to boost the model.
Coercion is a significant component of the ideal union dilemma. As a result, the management is a significant issue. The majority of study has focused on computational approaches that effectively solve the issue. GA, on the other hand, will successfully solve this dilemma. The Penalty-Free Genetic Algorithm (PFGA), which deals with restrictions and exclusions, was developed in [17]. In the capacitor circuit, choose the direction of the penalty limit and measurement challenge. Since the opportunity or desire to be disciplined for breaching prohibitions is restricted, it is necessary to procure two chromosomes to deal with the conditions: if two chromosomes do not violate the requirements, choose one chromosome depending on its suitability; if both chromosomes do not

violate the requirements, choose a chromosome. If one of the chromosomes follows the order, the other chromosome is chosen to go on with the measurement. And, since both chromosomes break the constraints, the chromosome with the lowest harm score is chosen. To pick the nearest quality of another chromosome, PGFA employs single-point mixing and unequal likelihood rounds. The greatest praise is always reserved for the first class of the next cycle, regardless of dialect. This loop eliminates the ambiguity that can occur by using total restrictions in the transformation planner. PFGA is used to minimize control and energy loss and has been studied on transport frames 18, 68, and 141.
In [26], neither the hybrid manager nor the exchange manager are able to correctly allocate the BESS block. They employed a dual boss who selected rewards depending on the qualities of the two subordinates. This technique finds a good compromise between exploring and abusing high-quality environments. Your transition planner employs a heterogeneous method focused on loops and polynomial probability transformations. Since, unlike [17], the amount of extraction is determined by the mass number, one output extraction is assumed for each chromosome, thus extending the conversion period increases measurement time. [27] Formalized paraphrase To accelerate power loss and voltage drift, normal distribution crossover (NDC) is used to produce new chromosomes. It has been discovered that NDC-based GA calculations are quicker than PSO and PSO-GA calculations.
In transportation organisations, GA efficiency has been tailored for perfect BESS connections [18]. OPF technology controls BESS reserves and reduces network errors, while GA has the potential to boost net present value (NPV). This saves time and reduces the amount of age ranges needed to tackle the perfect coordination dilemma. It has previously been evaluated on the IEEE 33 bus platform and in previous GA races. The findings indicate that by shortening the measurement period, the amount of GA ages can be increased.

4.2.2. Swarm-based algorithms. This type of algorithm is focused on simulating the interaction of groups of animals (referred to as quest agents or particles) in order to accomplish a shared target. The parameter added to the simulation has the greatest effect on the algorithm's efficiency. To boost voltage reliability and mitigate power loss, the optimum integration of electrical units has been refined and modified, and parameters have been reduced, resulting in improved overall network efficiency. Some experiments have also altered the algorithm engine to improve computational performance. [28] considered the charge delivery scheme, and numerous PSO topologies were suggested to improve capacitor positioning. It generates random numbers using Gaussian and Cauchy probability distributions rather than a uniform probability distribution. The diffusion function is extended alternately to the social and cognitive elements, yielding fifteen different types of PSO mechanisms. The GA and Taboo search algorithms are used to evaluate all PSO groups. Changing the social variable of PSO will obviously increase the algorithm's efficiency as compared to other permutations. They have a tendency to collide so much. To stop this setback, In [29] suggested a sustainable PSO to attain the lowest power and generation costs while incorporating renewable energy into the distribution network. The location of the particles is modified by a rivalry mechanism applied to randomly chosen pairs of particles, rather than by their personal and global best output [29]. The PSO algorithm is used in [30] to evaluate the optimum location and size of the DG block while also taking into account the estimation of the harmonic power flow of various harmonics dependent on the nonlinear load in the 31-bus distribution network. The key constraint in terms of units, pollution, and fuel is maintaining a reasonable degree of harmonic distortion. In [31], he suggested an enhanced PSO for optimum parallel capacitor usage and scale in radial distribution systems. The algorithm is applied by lowering the high current induced by the voltage decrease, thus decreasing real power loss and lowering the capacitor cost. According to estimates, as comparison to the PSO implementation, the proposed PSO implementation has a lower capacitor expense [32]. Similarly, in [33], PSO is used to solve the objective function to find the best location and scale of capacitor banks in a delivery network of 10, 33, and 69 buses. The aim is to minimize the lack of active and reactive power while saving the most money for the distribution business. Another consideration is to incorporate a feature to inflict line losses to ensure minimum voltage interference.

[34] suggested an artificial bee colony (ABC) system for optimizing the delivery of dispersed BESS blocks in the distribution network, which enhanced by removing power loss, voltage drift, and linear load. According to studies, this algorithm outperforms PSO in terms of reducing real output loss.
The BESS device is optimally inserted into the microgrid in [35] to input and consume real resources as electricity rates adjust. Improve the voltage curve and the voltage reliability of the microgrid to reduce power loss. The scale of the study can be broadened by running the algorithm through its paces on a test bus grid and contrasting it to other algorithms. PSO was suggested in [36] to maximize the scale and position of capacitors in the delivery network in order to minimize real electricity loss and economic costs. The algorithm was tested on 34 and 85 bus networks, and its performance was compared to that of WOA.
Any scholars concentrate mostly on the benefits of these two algorithms. In [37], for example, there are two group-based meta-heuristics: Bat Algorithm (BA) and Cuckoo Search (CS). The analogy was rendered in order to solve the issue of determining the ideal size and position of the capacitor. In the 34 and 85 bus networks, the target is to minimize real output loss while maximizing network savings. It is discovered that although CS outperforms the consistency of the BA solution, its convergence speed is slower than that of the Swar algorithm.
The arbitrary specification of parameters is one of the key drawbacks of community algorithms. Incorrect parameter settings can trigger premature convergence [38], which can result in local optimized outcomes. Finding an easy way to limit or remove parameter changes as far as possible is a basic trick. This idea was investigated in [39], which established a symbiotic chaotic organism (CSOS) search algorithm with less parameters to locate and decide the size of the DG unit in the distribution network. A disorderly vector is added in place of parameters to render it less vulnerable to the effects of local optima. In terms of convergence, the algorithm outperforms the initial SOS on 33, 69, and 118 RDS buses. The Hybrid Grey Wolf optimizer was created by the creator of [40]. Other than the population scale, no other parameters need to be adjusted for this optimizer. This algorithm is used to solve the issue of determining the optimal RG to allocate in the delivery network in order to reduce active and reactive power losses. The DG unit's height, power factor on each bus, and bus voltage are all within the acceptable range.
Meta-heuristic algorithms, such as WOA, ALO, and JAYA, have been designed to address the realization of a wide number of parameters. [41] used an ALO-based optimization strategy to position fixed shunt capacitors in the power distribution system optimally thus minimizing overall power. ALO is often used in [42] to optimize the placement of simultaneous capacitors in the delivery network when optimizing goal functions like energy loss and stress conditions. ALO has been tested in IEEE bus networks 33 and 69. To achieve the optimal flow in the existing renewable energy delivery network, use the JAYA algorithm [43]. The most difficult integration challenge [44].
In [45], WOA was used to optimize the size and position of capacitors in RDN. Developed to mitigate energy loss, increase the voltage curve, and reduce costs. WOA was tested on IEEE 34 and 85 bus systems and contrasted to other meta-heuristic algorithms such as PSO and BFOA. [46] investigated the strongest incorporation result of BESS and PV-DG blocks in various scenarios. In comparison to PSO and FA, WOA is added to a 25-bus network to reduce overall energy loss. The output of WOA and PSO is considerably higher than that of FA, with WOA doing marginally better than PSO.
Other algorithms, such as [47], which uses the discrete ray search algorithm (DLSA) to optimize the location of the capacitor, have also been applied to the optimal integration issue. Inside the windmill. The suggested target is to reduce administrative expenses while mitigating electricity waste. Contrast DLSA with the GA and the Discrete Harmony Search Algorithm (DHSA). The moth algorithm (MFA) has been used to maximize the scale and placement of the capacitor bank delivery network since [48]. To minimize the deficit, an iterative algorithm is used to solve the energy flux. Simultaneously, the responsive and reactive load curves are simulated at 15-minute intervals.
[49] Formalized paraphrase Implemented a Flower Pollination Algorithm (FPA) focused on a thick-tailed probability distribution to find the best global approach that can minimize the overall cost of installed capacitors thus determining the best capacitor size power network. The Butterfly Search

Algorithm (MSA) employs the Levy Flight approach in a related manner and has suggested [50] to change the bus voltage while positioning and calculating DG units in the distribution network.

4.2.3. Physics-inspired algorithms. These equations mimic the functions of actual miracles, which may be technological breakthroughs, human collaboration, or some other distinguishing occurrence. This classification can be assigned to population-based calculations to develop, especially in terms of composition. Instead of a vast number of quest experts, a search specialist may be included in material science equations to engage in the search space to locate answers, and vice versa. An Improved Harmony Algorithm (IHA) is suggested in [51] for ideal capacitor calculation in transportation organizations, where the power loss index (PLI) distinguishes the feasible transportation methods for the ideal capacitor installation. This measure was checked with 69 transmissions and contrasted to others including PSO, ABC, DE, and HSA. The expanded organisation found the stochastic principle of electric cars and the dubious energy performance of PV-DG systems in [52]. Use the Gravity Search Algorithm (GSA) layering problem to maximize the advantages of the PV portfolio while limiting annual energy loss. Display the electric car's battery to change the charging level at each level and charge/discharge level. The hypothetical region and scale of the PV-DG device are built on the first layer, taking into account the charge/release potential of electric vehicles, and an ideal area is formed on the next step. Load and post the results profile. HSA is another successful meta-heuristic approach built on an elegant sound mix. This approach can be found in [53] and [54] for a detailed review of its usefulness in improving energy structure. In [55], an improved HSA was developed for the optimal combination of units in the circulation organisation to reduce overall labor expense.
These algorithms mimic the behavior of physical processes such as chemical reactions, human experiences, or findings of some natural occurrence. Distribute the system and evaluate it to other algorithms (such as PSO, ABC, DE, and HSA). [52] investigates the randomness of electric cars and the effect of instability on the success of PV-DG systems in the delivery network. The Gravity Search Algorithm (GSA) is a two-step problem designed to optimize the advantages of photovoltaic incorporation while minimizing annual energy loss. The battery of an electric vehicle has been modeled, and the state of charge at each stage and charge/discharge state can be changed. To reduce overall cost of ownership, Advanced HSA will better incorporate the drive into the power delivery network. In [56], HSA hybridizes with FA, as will be addressed in the following segment.

4.2.4. Hybrid algorithms. A hybrid algorithm is an algorithm that combines two or more meta-heuristic algorithms. Algorithms may be merged to solve the whole optimization problem in a single algorithm, or they cannot be combined to solve separate sub-problems. Any instances of implementation can be found in literary works. A hybrid algorithm was proposed in [57] to reconfigure the distributor feeder via ESS, DG, and solar PV to maximize capacitor distribution. To achieve minimized power loss and decreased voltage drift, improved PSO and MSFLA were used in conjunction with VSI technology. Since the whole thing is split into two sections, each algorithm will solve the same problem. Then, IPSOMSFLA selects the best health attribute again and again until the full amount of iterations is achieved. This algorithm, according to sources, will have greater voltage stability and lower running costs than previous algorithms.
Taking complexity into account, a hybrid PSO-GA algorithm was suggested in [58] to optimally position the energy storage device in the wind farm network. Designed to reduce maintenance costs and greenhouse pollution while improving voltage deflection. The algorithm was put through its paces on a grid of 30 IEEE buses. According to studies, PSO-GA outperforms GSA in all goals excluding voltage suppression.
[59] optimizes and allocates DG units throughout the delivery network using GA and Intelligent Water Drops (IWD). By defining applicant buses, VSI is used to minimize real power loss. IWD computes fitness values, thus GA operators build a new generation of chromosomes. The hybrid algorithm was evaluated on a device with 33 buses and 69 buses, and it demonstrated excellent computation time that increased linearly with the amount of DG blocks.

In [60], the aim room created a hybrid algorithm (including community optimization and TBLO) to increase network reliability while allocating charging stations to the IEEE33 bus network. Voltage stability metrics, efficiency measures, and power failure indicators are also part of network stability. CSO and TLBO combine explicitly to accelerate convergence and reduce the risk of premature convergence. The INV parameters in CSO-TLBO, like those in CSO, are user-definable and must be modified appropriately to prevent local optimum solutions. According to sources, CSO-TLBO has a quicker convergence speed than the evolutionary general decomposition algorithm (MOEA/D) and NSGA-II, but the computation period is longer.

The author [56] combined HSA and FA (CHSFA) by lowering running costs and increasing delivery network sales to increase distribution network business revenues. To cope with the volatility of load demand, a rigorous paradigm has been applied that employs deterministic methods rather than probabilistic methods based on confidence intervals or forecasts based on historical evidence. CHSFA employs the HSA engine to locate the harmony of the best goal value in memory, and the FA engine to display the random value. To get the right answer, repeat this step again. CHSFA has been measured on 38 bus delivery networks and is stated to have a higher convergence pace than HSA. Although there is no study comparing calculation times, it is estimated that CHSFA would take longer to calculate than HSA due to its complicated mechanism.

[61] created a hybrid DE and Harmony Search (DEHS) algorithm to find the best place for multiple STATCOMs in an IEEE 30 mesh bus network. The DE mutation operator is widely used in this analysis to locate the best answer in the harmonic memory without modifying the HSA search engine. The opposing theory is to shorten the quest method and hence the convergence period. DEHS, as an HSA, raises the network stability index. It also has a higher convergence rate than HSA.

[62] combines a number of evolutionary algorithms, including GA, DE, and the Pareto Force Evolutionary Algorithm (SPEA-II), to incorporate RES units (solar and wind energy), capacitor banks, and electric cars into the delivery network all at once. The algorithm employs a vector cross feature to choose between three mutation strategies at random. The aim is to increase voltage efficiency while lowering gas pollution and installation costs.

4.2.5. Combined techniques. This group, in relation to cross-computer grouping, is a hybrid of two distinct enhancement methods. Scientific/metaheuristic and numerical/metaheuristic methods are included in the simulations. Both techniques should be used to reduce the sophistication of the optimal electrical device combination and solve the undeniable degree of simplex estimation. They are often the first stage in the process of upgrading the newly redesigned model. Calculations are typically paired with different methods of changing output systems to improve productivity. [63] investigated the optimal location and height of the ESS and PV-DG blocks in the distribution network. The region of the GD block was determined using an in-depth effect study of 33 buses and a fire spread model. Reduce power outages and boost power efficiency.

The power loss reduction factor and enhanced optimization of the multi-purpose golden ratio was used in [64] to optimally distribute and scale capacitors and STATCOMs in a different power distribution network. IEEE 69 and 118 bus systems were allocated to restrict output loss, react to impact assumptions, and increase voltage curves. The authors of [65] suggested an ABC-PSO hybrid approach that can smoothly quantify the load flow based on the load flow to minimize energy loss and increase voltage distribution during the monitoring phase. The Loss Sensitivity Factor (LSF) approach is used to classify vehicles that are vulnerable to loss of control and then pick the ideal capacitor set using a soft-tap frame. The hybrid ABC-PSO algorithm is used for capacitor ranking and estimation. 34 RDN hubs were checked. Similarly, in [66], a half-calculated HSAABC was built to increase the size and position of capacitors in RDN when taking into account various load models. PLI, VSI, and LSF were introduced to measure the overall company's power loss, discern lower voltage quality on the hub, and identify high dynamic impulse loss on the hub under a specified capacitor condition. In a 69-bus device with varying loads, an MFA-based algorithm was designed to optimally quantify capacitors in order to minimize energy loss. Since the output of LSF ALO is similar to defined and

IOP Conf. Series: Earth and Environmental Science **897** (2021) 012016 doi:10.1088/1755-1315/897/1/012016

calculated preferably in [67], an inexhaustible DG should be used on a separate microgrid. The suggested equations were tested on the RDS 69 bus and contrasted to other calculations, demonstrating that the total energy deficit was minimized and the net reserve was increased. [68] adds VSI to the LSF method to optimally position the capacitor in the delivery organization. The fluffy symmetric technique improves the estimation of enhanced bacterial absorption (IBFOA). The most rigorous approaches for limiting power loss and improving voltage reliability have been applied and validated in an enterprise of 33, 69, and 141 hubs.

When arranging the cycle, the maker of [69] sacrificed the perfect condenser region and condenser height. The LSF procedure was used in this case to choose opposing conveyors for capacitor placement in order to achieve GSA for ideal capacitor calculation. In the chosen truck. The FPA equation is used to optimally calculate the size of the capacitor, with the primary goal of limiting real power usage. The effects of the calculations are analyzed in the 10, 33, and 69 bus systems.

[70] suggested a BA enhancement strategy to limit the ratio of energy average interest to energy expense in a multi-level improvement scheme. The architecture employs phased curve programming to merge power demand reservations, ESS blocks, and PV-based DG blocks for households. The turbulence relationship is linked to the bee settlement miscalculation in [71]. This perfect condition is ideal for BESS transportation in the PV and WT IEEE 33 bus system organisation. The aim is to reduce power loss and voltage drift while also expanding the linear stack.

5. Conclusion

Since the beginning of power network, it has always evolved and now it is facing yet again new changes, this time due to increasing penetration of distributed generation. Due to power network complexity and growing number of uncertainties current control strategies and algorithms need to be updated to ensure minimum active power loses, best distribution power generation for optimal voltage and frequency control.

The usage of meta-heuristic algorithms to solve the optimum integration problem and its complex realization to solve the objective function problem was discussed in this paper. Evolutionary algorithms, swarm-based algorithms, physical algorithms, hybrid algorithms, and combined algorithms have all been commonly debated and known as metaheuristic algorithms. In the literature, each of these definitions has been explored in depth using descriptions of program modes, judgment variables, and objective functions. Since the process of optimum integration is dependent on achieving several objectives, researchers may make additional choices, such as selecting the best yet correct answer. Furthermore, depending on the network circumstance, the simultaneous integration and management of uncertainties in the delivery network may significantly complicate the model. Ensuring that algorithms are properly integrated into the system requires the system to be adapted for rapid management and monitoring. The selection of the most appropriate algorithm for medium voltage power network management depends on the network management structure. Thus, the use of different control algorithms like in hybrid and combined techniques in medium voltage power network may help making them more universal, easier to manage and more resilient to upcoming network changes.

In the future, proposed work on the subject of optimum incorporation into the distribution network could be needed. For example, the development of new meta-heuristic algorithms necessitates minimal user interaction; the development of the optimal number of parameters; the development of a hybrid algorithm with fast convergence and short calculation time; the development of a better support method for the selection of candidate buses for the best position of the electrical unit in the distribution network; and the development of a better support method for the selection of candidate buses for the best position of the electrical unit in the distribution network.

6. References

[1] Anaya K L and Pollitt M G 2015 Integrating distributed generation: Regulation and trends in three leading countries *Energy Policy* **85**

[2] Griffin T, Tomsovic K, Secret D and Law A 2000 Placement of dispersed generations systems for reduced losses *Proceedings of the Hawaii International Conference on System Sciences*
[3] CIGRE W G and Working Group 37.23 1998 Impact of increasing contribution of dispersed generation on the power system-Final Report *Electra, Sept.*
[4] Ackermann T, Andersson G and Söder L 2001 Distributed generation: A definition *Electr. Power Syst. Res.* **57** 195–204
[5] Lopes J A P, Hatziargyriou N, Mutale J, Djapic P and Jenkins N 2007 Integrating distributed generation into electric power systems: A review of drivers, challenges and opportunities *Electr. Power Syst. Res.* **77**
[6] Mohamed Imran A, Kowsalya M and Kothari D P 2014 A novel integration technique for optimal network reconfiguration and distributed generation placement in power distribution networks *Int. J. Electr. Power Energy Syst.* **63**
[7] Modarresi J, Gholipour E and Khodabakhshian A 2016 A comprehensive review of the voltage stability indices *Renew. Sustain. Energy Rev.* **63**
[8] Adetunji K E, Hofsajer I W, Abu-Mahfouz A M and Cheng L 2021 A Review of Metaheuristic Techniques for Optimal Integration of Electrical Units in Distribution Networks *IEEE Access* **9**
[9] Grover-Silva E, Girard R and Kariniotakis G 2018 Optimal sizing and placement of distribution grid connected battery systems through an SOCP optimal power flow algorithm *Appl. Energy* **219**
[10] Ju C, Wang P, Goel L and Xu Y 2018 A two-layer energy management system for microgrids with hybrid energy storage considering degradation costs *IEEE Trans. Smart Grid* **9**
[11] Ginigeme K and Wang Z 2020 Distributed Optimal Vehicle-To-Grid Approaches with Consideration of Battery Degradation Cost under Real-Time Pricing *IEEE Access* **8**
[12] Adetunji K, Hofsajer I and Cheng L 2020 A coordinated charging model for electric vehicles in a smart grid using whale optimization algorithm *Proceedings of 2020 23rd International Conference on Information Fusion, FUSION 2020*
[13] Harrison G P, Piccolo A, Siano P and Wallace A R 2007 Exploring the tradeoffs between incentives for distributed generation developers and DNOs *IEEE Trans. Power Syst.* **22**
[14] Reddy M J and Kumar D N 2006 Multi-Objective Optimization using Evolutionary Algorithms *Water Resour. Manag.* **20**
[15] de Araujo L R, Penido D R R, Carneiro S and Pereira J L R 2018 Optimal unbalanced capacitor placement in distribution systems for voltage control and energy losses minimization *Electr. Power Syst. Res.* **154**
[16] Babacan O, Torre W and Kleissl J 2017 Siting and sizing of distributed energy storage to mitigate voltage impact by solar PV in distribution systems *Sol. Energy* **146**
[17] Vuletić J and Todorovski M 2016 Optimal capacitor placement in distorted distribution networks with different load models using Penalty Free Genetic Algorithm *Int. J. Electr. Power Energy Syst.* **78**
[18] Xiao J, Zhang Z, Bai L and Liang H 2016 Determination of the optimal installation site and capacity of battery energy storage system in distribution network integrated with distributed generation *IET Gener. Transm. Distrib.* **10**
[19] Abdmouleh Z, Gastli A, Ben-Brahim L, Haouari M and Al-Emadi N A 2017 Review of optimization techniques applied for the integration of distributed generation from renewable energy sources *Renew. Energy* **113**
[20] Liu Y, Yang J, Tang Y, Xu J, Sun Y, Chen Y, Peng X and Liao S 2019 Bi-level fuzzy stochastic expectation modelling and optimization for energy storage systems planning in virtual power plants *J. Renew. Sustain. Energy* **11**
[21] Storn R and Price K 1997 Differential Evolution - A Simple and Efficient Heuristic for Global Optimization over Continuous Spaces *J. Glob. Optim.* **11**
[22] Das S and Suganthan P N 2011 Differential evolution: A survey of the state-of-the-art *IEEE*

IOP Conf. Series: Earth and Environmental Science **897** (2021) 012016 doi:10.1088/1755-1315/897/1/012016

Trans. Evol. Comput. **15**

[23] Injeti S K 2018 A Pareto optimal approach for allocation of distributed generators in radial distribution systems using improved differential search algorithm *J. Electr. Syst. Inf. Technol.* **5**

[24] Azizivahed A, Arefi A, Ghavidel S, Shafie-Khah M, Li L, Zhang J and Catalao J P S 2020 Energy management strategy in dynamic distribution network reconfiguration considering renewable energy resources and storage *IEEE Trans. Sustain. Energy* **11**

[25] Singh B and Singh S 2019 GA-based optimization for integration of DGs, STATCOM and PHEVs in distribution systems *Energy Reports* **5**

[26] Celli G, Pilo F, Pisano G and Soma G G 2018 Distribution energy storage investment prioritization with a real coded multi-objective Genetic Algorithm *Electr. Power Syst. Res.* **163**

[27] Huiling T, Jiekang W, Fan W, Lingmin C, Zhijun L and Haoran Y 2020 An optimization framework for collaborative control of power loss and voltage in distribution systems with DGs and EVs using stochastic fuzzy chance constrained programming *IEEE Access* **8**

[28] Lee C S, Ayala H V H and Coelho L D S 2015 Capacitor placement of distribution systems using particle swarm optimization approaches *Int. J. Electr. Power Energy Syst.* **64**

[29] Deveci K and Güler Ö 2020 A CMOPSO based multi-objective optimization of renewable energy planning: Case of Turkey *Renew. Energy* **155**

[30] HassanzadehFard H and Jalilian A 2018 Optimal sizing and location of renewable energy based DG units in distribution systems considering load growth *Int. J. Electr. Power Energy Syst.* **101**

[31] Mosbah M, Mohammedi R D, Arif S and Hellal A 2017 Optimal of shunt capacitor placement and size in Algerian distribution network using particle swarm optimization *Proceedings of 2016 8th International Conference on Modelling, Identification and Control, ICMIC 2016*

[32] Elsheikh A, Helmy Y, Abouelseoud Y and Elsherif A 2014 Optimal capacitor placement and sizing in radial electric powe r systems *Alexandria Eng. J.* **53**

[33] Karimi H and Dashti R 2016 Comprehensive framework for capacitor placement in distribution networks from the perspective of distribution system management in a restructured environment *Int. J. Electr. Power Energy Syst.* **82**

[34] Das C K, Bass O, Kothapalli G, Mahmoud T S and Habibi D 2018 Optimal placement of distributed energy storage systems in distribution networks using artificial bee colony algorithm *Appl. Energy* **232**

[35] Mostafa M H, Ali S G, Aleem S H E A and Abdelaziz A Y 2019 Optimal Allocation of Energy Storage System for Improving Performance of Microgrid Using Symbiotic Organisms Search *2018 20th International Middle East Power Systems Conference, MEPCON 2018 - Proceedings*

[36] Mendoza G E, Vacas V M and Ferreira N R 2019 Optimal Capacitor Allocation and Sizing in Distribution Networks Using Particle Swarm Optimization Algorithm *WCNPS 2018 - Workshop on Communication Networks and Power Systems*

[37] Injeti S K, Thunuguntla V K and Shareef M 2015 Optimal allocation of capacitor banks in radial distribution systems for minimization of real power loss and maximization of network savings using bio-inspired optimization algorithms *Int. J. Electr. Power Energy Syst.* **69**

[38] Rezaee Jordehi A and Jasni J 2013 Parameter selection in particle swarm optimisation: A survey *J. Exp. Theor. Artif. Intell.* **25**

[39] Saha S and Mukherjee V 2016 Optimal placement and sizing of DGs in RDS using chaos embedded SOS algorithm *IET Gener. Transm. Distrib.* **10**

[40] Sanjay R, Jayabarathi T, Raghunathan T, Ramesh V and Mithulananthan N 2017 Optimal allocation of distributed generation using hybrid grey Wolf optimizer *IEEE Access* **5**

[41] George T, Youssef A R, Ebeed M and Kamel S 2018 Ant lion optimization technique for optimal capacitor placement based on total cost and power loss minimization *Proceedings of 2018 International Conference on Innovative Trends in Computer Engineering, ITCE 2018* vol 2018-March

IOP Conf. Series: Earth and Environmental Science **897** (2021) 012016 doi:10.1088/1755-1315/897/1/012016

[42] Boktor C G, Youssef A R, Ali A H and Kamel S 2018 Optimal distribution power flow including shunt capacitor allocation based on voltage deviation and power loss minimization *2017 19th International Middle-East Power Systems Conference, MEPCON 2017 - Proceedings* vol 2018-February
[43] Elattar E E and ElSayed S K 2019 Modified JAYA algorithm for optimal power flow incorporating renewable energy sources considering the cost, emission, power loss and voltage profile improvement *Energy* **178**
[44] Kumawat M, Gupta N, Jain N and Bansal R C 2017 Optimally Allocation of Distributed Generators in Three-Phase Unbalanced Distribution Network *Energy Procedia* vol 142
[45] Prakash D B and Lakshminarayana C 2017 Optimal siting of capacitors in radial distribution network using Whale Optimization Algorithm *Alexandria Eng. J.* **56**
[46] Wong L A, Ramachandaramurthy V K, Walker S L, Taylor P and Sanjari M J 2019 Optimal placement and sizing of battery energy storage system for losses reduction using whale optimization algorithm *J. Energy Storage* **26**
[47] Sirjani R 2017 Optimal capacitor placement in wind farms by considering harmonics using discrete lightning search algorithm *Sustain.* **9**
[48] Ceylan O and Paudyal S 2017 Optimal capacitor placement and sizing considering load profile variations using moth-flame optimization algorithm *Proceedings - 2017 International Conference on Modern Power Systems, MPS 2017*
[49] Abdelaziz A Y, Ali E S and Abd Elazim S M 2016 Optimal sizing and locations of capacitors in radial distribution systems via flower pollination optimization algorithm and power loss index *Eng. Sci. Technol. an Int. J.* **19**
[50] Singh P, Bishnoi S K and Meena N K 2020 Moth Search Optimization for Optimal DERs Integration in Conjunction to OLTC Tap Operations in Distribution Systems *IEEE Syst. J.* **14**
[51] Ali E S, Abd Elazim S M and Abdelaziz A Y 2016 Improved Harmony Algorithm for optimal locations and sizing of capacitors in radial distribution systems *Int. J. Electr. Power Energy Syst.* **79**
[52] Ali A, Raisz D, Mahmoud K and Lehtonen M 2020 Optimal Placement and Sizing of Uncertain PVs Considering Stochastic Nature of PEVs *IEEE Trans. Sustain. Energy* **11**
[53] Geem Z W, Kim J H and Loganathan G V. 2001 A New Heuristic Optimization Algorithm: Harmony Search *Simulation* **76**
[54] Askarzadeh A 2017 Solving electrical power system problems by harmony search: a review *Artif. Intell. Rev.* **47**
[55] Liu C, Abdulkareem S S, Rezvani A, Samad S, Aljojo N, Foong L K and Nishihara K 2020 Stochastic scheduling of a renewable-based microgrid in the presence of electric vehicles using modified harmony search algorithm with control policies *Sustain. Cities Soc.* **59**
[56] Jeddi B, Vahidinasab V, Ramezanpour P, Aghaei J, Shafie-khah M and Catalão J P S 2019 Robust optimization framework for dynamic distributed energy resources planning in distribution networks *Int. J. Electr. Power Energy Syst.* **110**
[57] Lotfi H, Ghazi R and Naghibi-Sistani M bagher 2020 Multi-objective dynamic distribution feeder reconfiguration along with capacitor allocation using a new hybrid evolutionary algorithm *Energy Syst.* **11**
[58] Jani V and Abdi H 2018 Optimal allocation of energy storage systems considering wind power uncertainty *J. Energy Storage* **20**
[59] Moradi M H and Abedini M 2016 A novel method for optimal DG units capacity and location in Microgrids *Int. J. Electr. Power Energy Syst.* **75**
[60] Deb S, Tammi K, Gao X Z, Kalita K and Mahanta P 2020 A Hybrid Multi-Objective Chicken Swarm Optimization and Teaching Learning Based Algorithm for Charging Station Placement Problem *IEEE Access* **8**
[61] Zhang T, Xu X, Li Z, Abu-Siada A and Guo Y 2020 Optimum Location and Parameter Setting of STATCOM Based on Improved Differential Evolution Harmony Search Algorithm *IEEE*

Access **8**

[62] Zeynali S, Rostami N and Feyzi M R 2020 Multi-objective optimal short-term planning of renewable distributed generations and capacitor banks in power system considering different uncertainties including plug-in electric vehicles *Int. J. Electr. Power Energy Syst.* **119**

[63] Tang X, Deng K, Wu Q and Feng Y 2020 Optimal Location and Capacity of the Distributed Energy Storage System in a Distribution Network *IEEE Access* **8**

[64] Noori A, Zhang Y, Nouri N and Hajivand M 2020 Hybrid allocation of capacitor and distributed static compensator in radial distribution networks using multi-objective improved golden ratio optimization based on fuzzy decision making *IEEE Access* **8**

[65] Sharma S and Ghosh S 2020 FIS and hybrid ABC-PSO based optimal capacitor placement and sizing for radial distribution networks *J. Ambient Intell. Humaniz. Comput.* **11**

[66] Muthukumar K and Jayalalitha S 2018 Multiobjective hybrid evolutionary approach for optimal planning of shunt capacitors in radial distribution systems with load models *Ain Shams Eng. J.* **9**

[67] Ali E S, Abd Elazim S M and Abdelaziz A Y 2018 Optimal allocation and sizing of renewable distributed generation using ant lion optimization algorithm *Electr. Eng.* **100**

[68] Kishore C, Ghosh S and Karar V 2018 Symmetric Fuzzy Logic and IBFOA Solutions for Optimal Position and Rating of Capacitors Allocated to Radial Distribution Networks *Energies* **11**

[69] Mohamed Shuaib Y, Surya Kalavathi M and Christober Asir Rajan C 2015 Optimal capacitor placement in radial distribution system using Gravitational Search Algorithm *Int. J. Electr. Power Energy Syst.* **64**

[70] Lim K Z, Lim K H, Wee X Bin, Li Y and Wang X 2020 Optimal allocation of energy storage and solar photovoltaic systems with residential demand scheduling *Appl. Energy* **269**

[71] Das C K, Bass O, Mahmoud T S, Kothapalli G, Mousavi N, Habibi D and Masoum M A S 2019 Optimal allocation of distributed energy storage systems to improve performance and power quality of distribution networks *Appl. Energy* **252**

IOP Conf. Series: Earth and Environmental Science **897** (2021) 012017 doi:10.1088/1755-1315/897/1/012017

Formal modelling of the electricity markets: the example of the load reduction of electricity mechanism "NEBEF"

Florian Selot[1], Bruno Robisson[1], Claire Vaglio-Gaudard[2] and Javier Gil-Quijano[3]

[1] CEA, DRT/LITEN/DTS/SMSP/LSA, 13108, St-Paul-les-Durance, France
[2] CEA, DES/IRESNE/DER/SESI, 13108, St-Paul-les-Durance, France
[3] CEA, DRT/CEATECHR/DSUD, 13108, St-Paul-les-Durance, France

florian.selot@gmail.com

Abstract. The liberalisation of the electricity market initiated at the beginning of the 21st century has opened it to new parties. To ensure the growth of participants' number will support the system's balance, the EU regulation 2019/943 confirms that "all market participants should be financially responsible of the imbalances they cause". In their respective area, the transmission system operators develops the regulation in compliance with this condition. However, as the regulation takes into account the new realities of the market such as renewables, the interactions between the participants become more complex. One of the risks is that the imbalance of an actor may not be due to its own actions, not complying with the EU regulation then. To analyse this kind of implicit condition, we propose a formal approach to model the exchanges of energy. Using the French regulation as a base, we model the participants and their interactions in the form of symbolic equations using the energy-related terms as variables. In this paper, to illustrate the model we will use to analyse the entire electricity market, we apply it to the NEBEF mechanism only. This mechanism is dedicated to the selling of demand response in France and introduces a third party between the final producer and the final consumer: the demand response operator. We model the mechanism and analyse how the mechanism complies with the balancing responsibility. Our results demonstrate that the mechanism complies with the regulation but there are some limits due to the calculation method of the reference consumption.

1. Introduction

Electricity is a singular good that must be consumed at the time it is produced. When there is as much electricity injected as electricity extracted from the grid, the electricity system is balanced. When it is not the case, the frequency deviates from the operating value and if the deviation is too large, it may damage the electricity devices connected to the grid.

The transmission system operators (TSOs) are charged by the government with management of the electricity system in the area they operate. To perform this task, they can propose changes of the national regulation, and/or new rules to take into account the realities of the market. As part of the European internal market for electricity, the French regulation must comply with the European "Network Codes", a framework drafted by the European Network of Transmission System Operators for Electricity (ENTSO-E) with parties relevant to its development: EU institutions, regulators, European Union Agency for the Cooperation of Energy Regulations (ACER). The objective of this framework evolves from the liberalisation of the market at the beginning of the 21st century to the transition to a more sustainable energy system with the last package called the "Clean energy for all Europeans" package. To ensure the support of the system's balance by all actors, the EU Parliament and the EU Council stated that "all markets participants should be financially responsible for the imbalance they cause in the system, representing the difference between the allocated volume and the final position in the market" [1].

At the same time, as more and more actors are willing to take part in this market and the regulation takes into account the new realities of the market such as renewables, demand response or real time control, the interactions between the participants become more complex and the domain of responsibility of each of them less distinct. We think that applying the methodology defined by the TSOs to calculate the position of all actors may create relations that pair the actions of an actor to the imbalance of another.

The entry of new technologies and their impacts on the stability of the grid has been widely study from an economic point of view to the physical management of the electricity transportation. The economic studies have mainly addressed the balancing mechanisms as a whole with studies on the imbalancing price [2] or from a firm point of view with the optimisation of a portfolio by participating in the balancing mechanisms [3]. These approaches are based on the simulation of the behaviour of the actors. Simulation does not suit the analysis of the TSO's methodology because the pairing of actions might be hidden in the data. Moreover, the models that support the simulations are rarely explicit.

Consequently, we propose an approach which consists first to model, from the network codes, the exchanges of energy between the electricity market actors and, second, to analyze, from this model, the possible relations that may pair the actions of an actor to the imbalance of another. To illustrate our approach, we apply it to a recent mechanism dedicated to provide the flexibility of consumers to the energy markets thanks to load reductions of consumption. This mechanism is called "Notification d'Échanges de Blocs d'Effacement" (NEBEF) in French and was questioned for its design.

The paper is divided into 3 sections: the presentation of the electricity system, the description of the model and the analysis of the NEBEF mechanism.

2. Electricity System

2.1. Network & actors

The electricity system refers essentially to two layers: the first one is the component layer and the second one is the business layer.

The component layer represents all the physical entities, from the actors to the devices [4]. It consists of the infrastructures to produce, to consume and to transport electricity. The transmission of electricity over long distance at a high voltage is called the transport system whereas the distribution to residential and small industrial consumers is called the distribution system.

The business layer is the representation of the information exchanges from a business point of view. It consists of the different electricity markets. It is usually subdivided into the energy markets that concerns the wholesale markets in one part and into the balancing mechanisms that ensure the stability of the grid in the other part.

The main several actors involved in the electricity market are:

- The producers generate electricity and sell it to suppliers,
- The electricity suppliers buy electricity from producers or other suppliers and sell it to consumers,
- The consumers use electricity and pay suppliers via their bills,
- The market operators provide a service whereby the offers to sell electricity are matched with bids to buy electricity,
- The distribution system operators (DSO) are paid for delivering electricity to consumer and must ensure the stability of the grid in their area. They are in charge of the data in their area,
- The transmission system operator (TSO) is paid for the long-distance transmission of electricity and is charged by the government to ensure the stability of the grid at a national level, according to [5]
- The balance responsible parties (BRP) are financially in charge of the imbalance of their customers.

IOP Conf. Series: Earth and Environmental Science **897** (2021) 012017 doi:10.1088/1755-1315/897/1/012017

2.2. *System balancing*

The stability translates in a nearly perfect equality between the energy injected and the energy extracted every second. The French TSO, called Réseau de Transport d'Électricité (RTE), manages the process to ensure the stability with the involvement of all the others actors, from the long-term grid investments to the invoicing of imbalances.

Between long-term investments and real-time delivery, actors participate to ensure the stability of the grid. Each actor is responsible for the imbalance of his perimeter, defined as the difference between the sum of its injections and the sum of its extractions over periods of thirty minutes. An actor may delegate this responsibility by joining the perimeter of a balance responsible party [6].

Because the actors cannot forecast their consumption and/or cannot control perfectly the production of their plants, the TSO compensates the global residual imbalance by reserving capacities from power plants and activating them when needed.

The actors pays back both the procurement and the activation of the reserves through the billing of their imbalances. They are encouraged to balance their perimeter as much as possible by a higher cost for the use of reserves than the provision of electricity from the wholesale market. The process to determine the imbalance is divided into 3 phases:

1. The planning period or ex-ante period is the period prior to the real-time period of several years before up to one hour before only. The actors forecast their production and consumption and can trade electricity to balance their perimeter as well as possible. The TSO provisions the reserves that will compensate the forecasted imbalance.
2. The real-time period or delivery is a 30-minute period during which each actor tries to produce or consume as he planned to do. During this period, the TSO is the only actor who can intervene to maintain the global balance of the electricity system: he can call actors to activate the reserves they agree to provide.
3. The invoicing period or ex-post period is the period following the delivery period. The TSO calculates the cost of the actions that he made during the real-time period to balance the grid. He also retro-estimates the imbalance of all parties to charge them for the balancing actions.

2.3. *A market for consumer flexibilities in France: the NEBEF mechanism*

Since 2007 in France, it is possible to sell demand response to the TSO through the adjustment mechanism ("mécanisme d'ajustement"), which is a subpart of the recovery reserve. In 2013, it was enlarged to the markets of energy by the creation of the NEBEF mechanism in order to increase the use of demand response.

The main actor is the demand response operator (DRO) who manages the consumption of his customers and sells the volume that they do not consume thanks to his actions. The DRO operates as a third party between the final producer (the one who produce the electricity for the consumer who offers his flexibility), and the final consumer (the one who buys the volume offered by the operator). The mechanism was designed to respect the process of reserves procurement, i.e. the operator is not necessarily in contract with the same balance responsible party of the consumers that he manages. Consequently, his actions may affect the imbalance of the balance responsible party of the consumers. In this paper, we analyse the relations between the parties involved in the NEBEF mechanism thanks to the model we developed.

3. Modelling

3.1. *Introduction: system studied*

The proposed model is a simplified representation of the management of the electricity system by RTE. Only the business layer of the system is modelled and a special focus on the imbalance of the parties is made.

According to paragraph C.15.1 of [7], the imbalance of an actor equals the difference between the sum of his injections and the sum of his extractions over periods of thirty minutes.

IOP Conf. Series: Earth and Environmental Science **897** (2021) 012017 doi:10.1088/1755-1315/897/1/012017

$$Imb_{Actor\ A} = \sum_{i \in Injections} (I_i^r) - \sum_{e\ Extractions} (E_e^r) \quad (1)$$

Injection and extraction are defined at the same paragraph in [7]. It can be one of the following energy-related term:

- Production of electricity noted P,
- Consumption of electricity noted C,
- Energy activated for FCR noted V_{FCR},
- Energy activated for FRR noted V_{FRR},
- Energy activated for RR noted V_{RR},
- Energy corresponding to load reduction actions (NEBEF) noted V_{LR},
- Energy exchanged through the mechanism designed to declare bilateral exchange of energy. In France, the mechanism is "Programme d'Échanges de Blocs" (PEB). We abbreviate it by V_{PEB},
- Energy exchanged at the future markets noted V_{FM},
- Energy exchanged at the spot markets noted V_{SM}.

The model is designed to study how the interactions may affect the balancing position. Therefore, rather than using numeric variables that may result in singular cases, we use symbolic variables. Consequently, the representation of the system by our model is a set of symbolic equations using the energy-related terms as variables.

3.2. *Temporal modelling*

The RTE's management is a constant monitoring of the system's imbalance within different timeframes (from long-term to real-time). RTE takes actions during the ex-ante period to ensure the stability whereas he takes actions during the ex-post period to allocate the volumes to the actors according to the regulation. At the same time, the actors participate in the markets and exchange volumes of electricity. Although these processes are continuous, our model only considerate the electricity system time base in a discrete way, in accordance with the MA-RE rules [7]. We look at the system at two key times of the RTE's management:

1. The balancing gate closure time, which closes the planning period. At that time, the submission or update of a bid is no longer permitted [8]. The actors have announced their position in the markets and their balance position cannot be modified.
2. The end of the invoicing process for the imbalances of the actors. The actors' data of energy production and energy consumption have been transmitted to RTE. Then, they were processed to determine the balancing position of all parties. At the end of this process, all the electricity produced and consumed during a thirty-minute period has been divided among the actors.

In our model, when the variables refer to their use in the ex-ante period, they are preceded by the term "ea" (ea_Var as example). When they are in the ex-post period, the term "ep" precedes the name of the variables. Some variables are not related to a specific period. In these cases, the variable is not preceded by any term.

3.3. *Actors, Interactions & Mathematical representation for the use case*

Our model represents only two groups of entities defined by the legislation. The first group is the actors and the second one, the markets.

In our use case nine actors participates in the system through two markets. There are two producers, two consumers and a demand response operator. All of them delegate their balancing responsibility to three different balance responsible parties. Some of them participate in a spot market and only the demand response operator participates in the NEBEF mechanism. They also are all supervised by RTE, the transmission system operator. Figure 1 resumes the interactions between the actors.

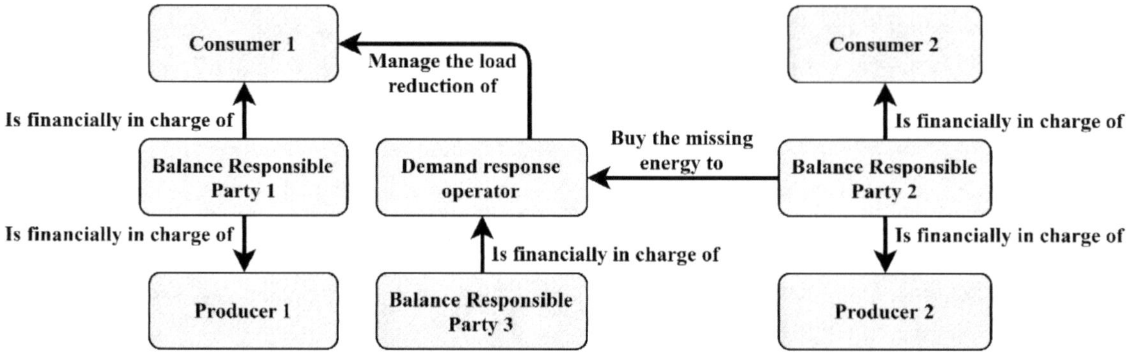

Figure 1. Use case actors and relations

3.3.1. Actors. For each actor, one or several equations represent how the different volumes of energy are accounted in his perimeter and how they are calculated.

When two actors have a different vision of a variable, the abbreviation of the actor is added to the subscript of the variable. The vision of a variable by an actor means the value he gives/knows to/of the variable. For example, if the actor A determines the value of the variable VAR, we refer to it as VAR_A and if the actor B forecasts another value for the variable, we refer to it as VAR_B.

We name the two producers Producer 1 and Producer 2. The balancing position of a producer X is represented by the equation:

$$Imb_{PX} = +PX \tag{2}$$

They produce electricity for the two consumers. The consumer 1 is a consumer who is in charge of several sites. The consumer 1 offers the flexibility that one of his site can provide. The following equation represents the consumer 1:

$$F1 = C1_O + C1_{SAdju} \tag{3}$$

$C1_O$ and $C1_{SAdju}$ represent respectively the physical consumption of the other sites and the adjusted consumption of the site who offers his flexibility. The adjusted consumption represents the sum of the physical consumption and of the calculated adjustments made. An adjustment is a voluntary change in consumption in order to provide balancing resources or to offer flexibility through the NEBEF mechanism. In the use case, we abbreviate the flexibility offered by the consumption site by $C1_F$.

The consumer 2 does not offer the flexibility of his consumption sites to the markets. The variable C2 represents the whole consumption from these sites. The balancing position of both consumers is:

$$Imb_{C1} = -F1 \; for\ the\ consumer\ 1 \tag{4}$$

$$Imb_{C2} = -C2 \; for\ the\ consumer\ 2 \tag{5}$$

The consumer 1 delegates the load-reduction management to a DRO. The DRO, when it is financially attractive, sells the energy not consumed thanks to the load-reduction actions to the energy market. The global volume of load reduction, noted V_{LR}, is considered as an injection and the volume sold, noted V_{SM}, as an extraction in his balancing perimeter. It gives the equation:

$$Imb_{DRO} = V_{LR} - V_{SM} \tag{6}$$

In our use case, the demand response operator is in charge of one site only. Consequently, the volume of energy not consumed thanks to his actions equals the volume of flexibility offered by the site. It means:

$$V_{LR} = C1_F \tag{7}$$

The balancing position of a balance responsible party equal the sum of the imbalances of the parties who delegate it to him.

$$Imb_{BRP} = \sum_{Imb\ X \in actor\ X} Imb_X \tag{8}$$

For the use case, we suppose that three balance responsible parties are involved. The producer 1 and consumer 1 have the same BRP. We name it BRP1. It gives the equation:

$$Imb_{BRP1} = P1 - F1 \tag{9}$$

The producer 2 and consumer 2 have the same balance responsible party, named BRP 2. We suppose that the producer 2 cannot produce as much as the consumer 2 needs, the BRP 2 has to buy the missing volume to the market. We suppose that BRP2 buys the volume V_{SM} to the DRO to this aim. The volume bought counts as an injection. It gives the equation:

$$IMB_{BRP2} = P2 + V_{SM} - C2 \tag{10}$$

We suppose that the BRP 3 is in charge of the demand response operator only. Thus, his balancing position is the same as the demand response operator.

Despite the activation by RTE of the reserves, noted RES, there is still a global residual imbalance, abbreviated NET. From a global point of view, the systems imbalance is summed up by the equation:

$$NET = \sum_{BRP \in BRPs} Imb_{BRP} - RES \tag{11}$$

For the use case, we divide the sum above in two parts. One represents the imbalance of the BRPs participating in the use case. We use the abbreviation Imb_{BRP123} in the model. The other one represents the imbalance of all the other BRPs who are active in one of the electricity markets. We use the abbreviation Imb_{BRPs}. We obtained the following equation:

$$Net = Imb_{BRP123} + Imb_{BRPs} - Res \tag{12}$$

3.3.2. Markets & mechanisms. The second part of our model is the markets. A market defines specific relations between the actors that are modelled by one or several equations.

The first market we look at is the spot markets, throughout which actors can sell load reduction declared by the NEBEF mechanism. On these markets, the producers offer to sell electricity and the consumers bid to buy electricity. Once there is a match between an offer and a bid for a certain quantity at a certain price, the contract cannot be modified In particular, the quantity of energy used for the calculation of the imbalance during the invoicing process is the same as the one contracted. In term of equation, it means:

$$ea\_V_{SM} = ep\_V_{SM} \tag{13}$$

The other market we look at is the NEBEF mechanism. The mechanism links the consumption of the consumers and the volume produced by the operator.

During the ex-ante period, the demand response operator determines the electricity that will be consumed by the site, $C1_S$, according to the reference consumption, $C1_{Rref}$ and the reduction of consumption he plans to do, $C1_F$. It means:

$$ea\_C1_S = ea\_C1_{ref,DRO} - ea\_C1_F \tag{14}$$

For the invoicing process, the calculation is based on the physical data and the reference consumption. It means:

$$ep\_C1_F = C1_{ref} - ep\_C1_S \tag{15}$$

There are three different configurations for the invoicing process of the consumer site. We assume the site is in the "télérelevé au modèle régulé" configuration. "Télérelevé" means that the data are issued from TSO's certified metering installations with a real-time transmission. "Régulé" is the method to compensate the load reduction to the supplier financially.

The configuration also specifies how the load reduction is accounted in the perimeter of the balance responsible party in charge of the consumer. In terms of modelling, it means that the reduction of consumption is added to the physical consumption $C1_s$ and not directly to the imbalance of the BRP. It gives us the equation:

$$ep\_C_{SAdju} = ep\_C_S + ep\_C1_F \tag{16}$$

Moreover, the NEBEF mechanism was designed to ensure the demand response operators can work independently from the supplier. In our use case, one consequence is the information about a load reduction is not transmitted to the supplier. But we suppose that the supplier has access to the historical data of the sites that he supplies.

Figure 2 summarises the relations between the actors from a business point of view.

IOP Conf. Series: Earth and Environmental Science **897** (2021) 012017 doi:10.1088/1755-1315/897/1/012017

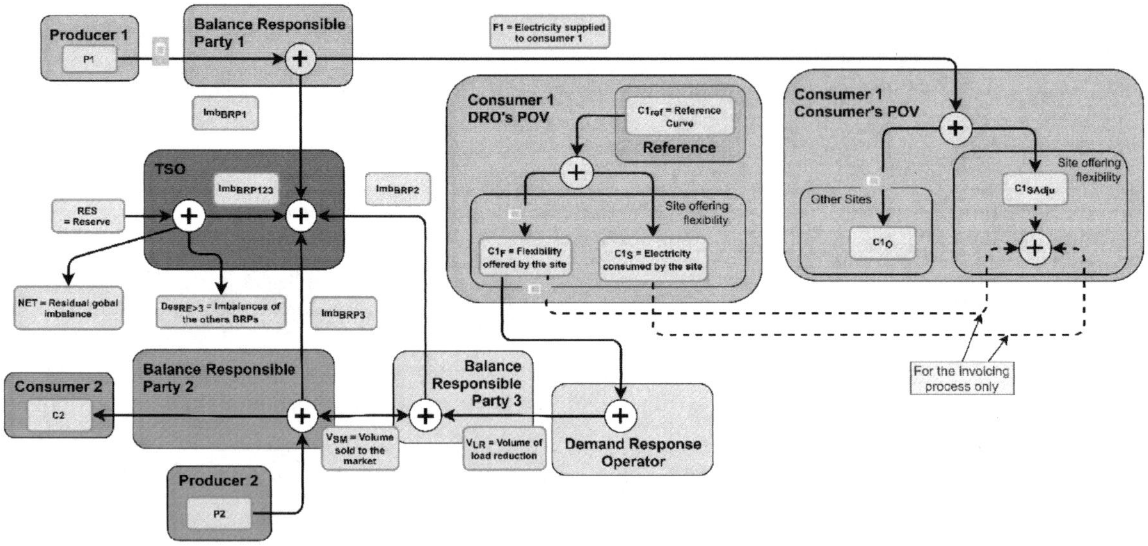

Figure 2. Business representation of the use case

3.4. Method of analysis

The model aims at providing a base for a formal analysis of the exchanges of energy between the parties involved.

We formalise the balancing responsibility of an actor into mathematical conditions that have to be observed.

3.4.1. Formalisation of Balancing Responsibility. The balancing responsibility means that an actor is responsible financially for the actions he takes. We states that it corresponds to two conditions.

The first condition that must be observed is the possibility for the actor to be perfectly balanced at the end of the invoicing process. It means his balancing position must be equal to zero in the case of perfects actions (i.e. he acts "perfectly"). The actions are the forecasts for the ex-ante periods and the production and consumption according to his forecasts/schedules for the real-time period.

The second condition refers to the actions of an actor to another. It means the actions of an actor must not imbalance another one. To verify the condition, we introduce the concept of error. An error is either a forecast different from the consumption physically extracted, either a consumption or production different from the schedule. We state that the condition is verified if the error is contained in the perimeter of the actor and of his balance responsible party. In particular, the error must not spread to other actors.

3.4.2. Steps of the NEBEF analysis. To analyse the compliance of the NEBEF with the balancing responsibility, we broke down the analysis into two phases. In the first one, we assume all the actors act perfectly. If the NEBEF is well designed, the imbalances of the balance responsible parties should equal zero. In that case the NEBEF complies with the first condition.

In the second one, we focus our analysis to the actions of the demand response operator because of his position as a third party. We assume all the actors act perfectly, except the demand response operator. The mechanism complies with the second condition if the error is contained to the imbalance of the BRP 3.

4. Results

To analyse the system, the set of equations that represents the system must be determined. The choice of the variables as parameter imitates the actor's management. For the two periods, the parameters are the injections and extractions of all parties except $C1_S$ for the ex-ante period and V_{LR} for the ex-post

IOP Conf. Series: Earth and Environmental Science **897** (2021) 012017 doi:10.1088/1755-1315/897/1/012017

period. Consequently, the dependent variables are the imbalances of the balance responsible parties and the residual imbalance.
Table 1 summarises the variables that are used for the use case.

Table 1. Abbreviations and their meaning

Variable	Definition	Variable	Definition
P1	Electricity generated by P1's power plants	P2	Electricity generates by P2's power plants
F1	Electricity supplied to consumer 1	C2	Electricity supplied to consumer 2
$C1_O$	Electricity consumed by the sites not offering flexibility	Imb_{BRPs}	Sum of the imbalances of the BRPs who are not involved in the use case
$C1_S$	Electricity consumed by the site providing flexibility	Imb_{BRP123}	Sum of the imbalances of the BRPs 1, 2 and 3
$C1_{SAdju}$	Adjusted consumption of the site providing flexibility	Imb_{BRP1}	Imbalance of the balance responsible party 1
$C1_F$	Volume of flexibility provided by the site offering his flexibility	Imb_{BRP2}	Imbalance of the balance responsible party 2
V_{LR}	Global volume of load reduction made by the DRO	Imb_{BRP3}	Imbalance of the balance responsible party 3
Res	Reserves activated by the TSO to stabilise the grid	Net	Global residual imbalance on the grid

For the ex-ante period, we obtain the relations between the variables presented in the Table 2.

Table 2. System of equations solved for the ex-ante period

Variable	Value at the gate closure time	Variable	Value at the gate closure time
P1	ea_P1	P2	ea_P2
F1	$ea\_C1_o + ea\_C1_{SAdju}$	C2	ea_C2
$C1_O$	ea_C1_o	$C1_{SAdju}$	ea_C1_{SAdju}
$C1_S$	ea_C1_{SAdju}	$C1_F$	ea_V_{LR}
V_{LR}	ea_V_{LR}	V_{SM}	ea_V_{SM}
$C1_{ref}$	$C1_{ref}$	$C1_{ref,DRO}$	$ea\_C1_{ref,DRO}$
Imb_{BRPs}	ea_Imb_{BRPS}	Res	ea_Res
$\mathbf{Imb_{RE1}}$	$\mathbf{- ea\_C1_O - ea\_C1_{SAdju} + ea\_P1}$	$\mathbf{Imb_{RE2}}$	$\mathbf{- ea\_C2 + ea\_P2 + ea\_V_{SM}}$

Variable	Value at the gate closure time
$\mathbf{Imb_{RE3}}$	$\mathbf{ea\_V_{LR} - ea\_V_{SM}}$
Imb_{BRP123}	$- ea\_C1_O - ea\_C1_{SAdju} - ea\_C2 + ea\_P1 + ea\_P2 + ea\_V_{LR}$
Net	$ea\_C1_O + ea\_C1_{SAdju} + ea\_C2 - ea\_Imb_{BRPs} + ea\_Res - ea\_P1 - ea\_P2 - ea\_V_{LR}$

Table 2 shows that the imbalance of the balance responsible parties depends on different variables for each of them at the gate closure time.
For the BRP 1, the electricity supplied to consumer 1 and the electricity production scheduled to producer 1 are the variables that determine the forecast of the imbalance of the balance responsible 1.

IOP Conf. Series: Earth and Environmental Science **897** (2021) 012017 doi:10.1088/1755-1315/897/1/012017

For the BRP2, the electricity supplied to consumer 2, C2, the electricity production scheduled by producer 2, P2, and the electricity bought during the short-market period, V_{SM} are the variables that determine the imbalance of the balance responsible 2.
Finally, the electricity sold to a spot market, V_{SM}, and the reduction of consumption that the DRO plans to do, V_{LR}, are the variables that determine the imbalance of the BRP3.

Table 3. System of equations solved for the ex-post period

Variable	Value at the end of the invoicing process	Variable	Value at the end of the invoicing process
P1	ep_P1	P2	ep_P2
F1	$ep\_C1_o + ep\_C1_{ref}$	C2	ep_C2
$C1_o$	ep_C1_o	$C1_{SAdju,C1}$	ep_C1_{ref}
$C1_S$	ep_C1_S	$C1_F$	$C1_{ref}$ - ep_C1_S
$C1_{ref}$	$C1_{ref}$	$C1_{ref,DRO}$	$ep\_C1_{ref,DRO}$
V_{LR}	ep_C1_{ref} - ep_C1_S	V_{SM}	ep_V_{SM}
Imb_{BRPs}	ep_Imb_{BRPs}	Res	ep_Res
Imb_{BRP1}	**- ep_C1_O - $C1_{ref}$ + ep_P1**	**Imb_{BRP2}**	**- ep_C2 + ep_P2 + ep_V_{SM}**

Variable	Value at the end of the invoicing process
Imb_{BRP3}	**$C1_{ref}$ - ep_C1_S - ep_V_{SM}**
Imb_{RP123}	- ep_C1_O - ep_C1_S - ep_C2 + ep_P1 + ep_P2
Net	ep_C1_O + ep_C1_S + ep_C2 - ep_Imb_{BRPs} - ep_P1 - ep_P2 + ep_Res

Table 3 summarises the relations at the end of the invoicing process. The imbalances of the balances responsible parties do not depend on the same variables.
For the balance responsible party 1, the electricity produced by producer 1, P1, the electricity consumed by the other sites, $C1_O$ and the reference consumption, $C1_{ref}$ are the variables that determines the imbalance of the balance responsible 1.
For the balance responsible party 2, the electricity consumed by consumer 2, C2, the electricity produced by producer 2, P2, and the electricity bought during the short-market period, V_{SM}, are the variables that determines the imbalance of the balance responsible 2.
Finally, the electricity sold to a spot market, V_{SM}, the reference consumption, $C1_{ref}$, and the energy extracted by the site offering flexibility, $C1_{S,C1}$ are the variables that determines the imbalance of the balance responsible 3.

4.1. Feasibility to balance its perimeter
The first phase of the analysis is to verify the first condition. In this phase, we assume the actors act perfectly. It mean that:

- The producer produces the exact quantity contracted with the suppliers.

$$ea\_P1 = ea\_F1 \tag{17}$$
$$ep\_P1 = ea\_P1 \tag{18}$$
$$ea\_P2 = ea\_C2 \tag{19}$$
$$ep\_P2 = ea\_P2 \tag{20}$$

- The consumption of the consumer 2 equals the forecast made in the ex-ante period.

$$ep\_C2 = ea\_C2 \tag{21}$$

- The demand response operator plans to reduce the consumption of its customers as much as he sells to the market.

$$ea\_V_{LR} = ea\_V_{SM} \tag{22}$$

IOP Conf. Series: Earth and Environmental Science **897** (2021) 012017 doi:10.1088/1755-1315/897/1/012017

Moreover, he perfectly reduces the consumption according to his schedule.

$$ep\_C1_s = ea\_C1_s \tag{23}$$

- The BRP 2 buys the quantity of electricity missing between its production and its consumption

$$ea\_C2 = ea\_P2 + ea\_V_{SM} \tag{24}$$

- RTE forecasts the imbalance of the others BRPs perfectly and the procurement and the activation of reserves compensates the sum of their imbalance

$$ea\_RES = ea\_Imb_{BRPs} \tag{25}$$

Table 4 presents the value of the BRPs' imbalance after the simplifications that have been applied to the system.

Table 4. Value of the imbalances in case of perfect actions

Variable	Value at the end of the invoicing process
Imb_{BRP1}	$ea\_C1_{SAdju} - C1_{ref}$
Imb_{BRP2}	0
Imb_{BRP3}	0

When all the actors act perfectly, the balance responsible party 2 and 3 are perfectly balanced. It means that other actors do not affect them and the first condition is observed for both actors.
This is not the case for the balance responsible party 1. The imbalance is the difference between the forecast made by the supplier, ea_C1_{sAdju} and the reference consumption, $C1_{ref}$. To study how it may affect the balance responsible 1, we discuss the four manners the reference consumption is determined and what it changes for the balance responsible 1.

4.2. Discussion about the reference consumption

There are four methods to calculate the reference consumption. All of them are based on the previous consumption of the site and differ mainly on the period considered to calculate the reference consumption. Each method and the conditions are explained in [9].

4.2.1. Reference consumption by forecast. In that case, the reference consumption is a scheduled consumption sent to the TSO by the demand response operator. There are no restrictions for the power values but it is compared to the historic consumption. If it deviates too much from the reality, the operator loses the possibility to send schedules. Therefore, there are no law-based relations between the reference consumption, $C1_{ref}$, and the forecast made by the supplier, $C1_{SAdju}$.
The difference between both values is considered the responsibility of the supplier even though he does not manage both. If he perfectly performs as he planned but his planning is different from the operator's schedule, he will be imbalanced. Consequently, the first condition is not respected.

4.2.2. By average or median. In that case, the reference consumption is the average or median of previous periods. It could be the same period of the week for the four previous weeks or the same period of the day for the ten previous days.
In this case, the supplier cannot balance his position in the ex-ante period because his forecast is not necessarily equal to the average/median of the previous periods.
The mechanism does not comply with the first condition. To solve it, the supplier must know that a load reduction will happen during the week and must only buy the quantity corresponding to the reference consumption.

4.2.3. Double-corrected reference rectangle method. In that case, the reference consumption is the minimum between the average power over the period preceding the load reduction period and the average power over the period following the load reduction period.

The first condition is respected as long as the reference consumption is typical of the consumption that would have happened without the load reduction.

4.2.4. Algebraic rectangle site-by-site method. This method is applied to demand response aggregators only. The reference consumption is calculated over the period of 10 minutes preceding the period of load reduction and the reference value is the average power. The period of load reduction is either several 10-minute periods separated by a 20-minute period, or a 30-minute period.
Like the previous method, it is based on the consumption close to the period of load reduction. Consequently, it respects the first condition as long as the reference is typical of the consumption.

4.2.5. Analysis of an error made by the demand response operator. The second part of our study is the introduction of an error from the demand response operator and the analysis of its spreading into the system. The demand response operator can make two kinds of error:

1. The operator forecasts a reference consumption different from the one calculated by the TSO. It gives the following equation:
$$C1_{ref} = C1_{ref,DRO} \pm Err \tag{26}$$
2. The operator does not control the reduction of consumption perfectly as planned.
$$ep\_C1_S = ea\_C1_S \pm Err \tag{27}$$

The introduction of an error of prevision results in the relations presented in Table 5.

Table 5. Propagation of the error of prevision

Variable	Value at the end of the invoicing process	Variable	Value at the end of the invoicing process
P1	$ea\_C1_O + ea\_C1_{SAdju}$	P2	ea_P2
F1	$ea\_C1_O + ea\_C1_{ref}$	C2	ea_C2
$C1_O$	ea_C1_O	$C1_{SAdju}$	$C1_{ref}$
$C1_S$	$C1_{ref}$ - ea_C2 + **ep_Err** + ea_P2	$C1_F$	ea_C2 - **ep_Err** - ea_P2
V_{LR}	ea_C2 - **ep_Err** - ea_P2	V_{SM}	ea_C2 - ea_P2
$C1_{ref}$	$C1_{ref}$	$C1_{ref,DRO}$	$ep\_C1_{ref,DRO}$
Imb_{BRPs}	0	Res	0
Imb_{RE1}	$- C1_{ref} + ea\_C1_{sadju,C1}$	Imb_{RE2}	0
Variable	**Value at the end of the invoicing process**		
Imb_{RE3}	- **ep_Err**		
Imb_{BRP123}	$- C1_{ref} + ea\_C1_{sadju,C1}$ - **ep_Err**		
Net	$C1_{ref} - ea\_C1_{sadju,C1}$ + **ep_Err**		

Our results show that the error is contained to the responsibility domain of the demand response operator. Consequently, the NEBEF mechanism complies with the second formal condition we introduce.

5. Conclusion

In the paper, we present a model whose objective is to analytically study how the design of the exchanges made by the TSOs complies with the EU's legislative framework. The model simplifies the management of the electricity system by a TSO from a business point of view.
Only two entities are defined in the model: the actors and the markets. They are represented by one or several equations with the energy-related terms as variables. The variables are symbolic in order to study how they can spread across the system because of the interactions between the actors. The equations are a translation and an interpretation of the rules and must be challenged with other models for the management of the electricity system.

IOP Conf. Series: Earth and Environmental Science **897** (2021) 012017 doi:10.1088/1755-1315/897/1/012017

Another simplification is the discrete representation of the system at two steps of the continuous process only. The first step is the gate closure time and the second one is the end of the invoicing process.
In the paper, we apply our model to the NEBEF mechanism in order to analyse its compliance with the principle of balancing responsibility. The formalisation of the principle translates into two conditions. The first condition is the possibility for a balance responsible party to balance its perimeter at the end of the invoicing process if he perfectly acts according to his schedule. The second condition is the spreading of an error made by an actor. The error should only affect the balance responsible in charge of the actor. We demonstrate that the NEBEF always complies with the second condition. The balancing position of the balance responsible party in charge of the consumer who provide his flexibility cannot be imbalanced due to an error made by the demand response operator.
However, the mechanism does not comply with the first condition because of the calculation method of the reference consumption. If the demand response operator forecasts the reference curve and the balance responsible party does not know the reference consumption, the latter cannot balance his perimeter even if he performs perfectly.
For the other methods, the limit is the adequacy of the reference consumption to be representative of the consumption that would have occured without the reduction of consumption
We conclude that the NEBEF mechanism reproduces the physical consumption as accurately as possible but presents some limits due to the calculation methods of the reference consumption.

6. References

[1] European Parliament and European Council 2019 Regulation (EU) 2019/943 on the internal market for electricity, article 5
[2] R A.C.van der Veen, A Abbasy and R Hakvoort 2012 Agent-based analysis of the impact of the imbalance pricing mechanism on market behavior in electricity balancing markets *Energy Economics*
[3] C Möller, S Rachev and F Fabozzi 2011 Balancing energy strategies in electricity portfolio management, *Energy Economics*
[4] CEN-CENELEC-ETSI 2012 Smart Grid Coordination Group, Smart Grid Reference Architecture
[5] Code de l'Énergie, Article L3 321-10
[6] Code de l'énergie, article L.321-15
[7] RTE 2020 Règles MA-RE, section 2, Règles relatives au dispositif de Responsable d'Équilibre, Chapitres A à D
[8] European Commission 2017 Regulation (EU) 2017/2195 establishing a guideline on electricity balancing, article 2
[9] RTE 2021 Règles pour la valorisation des effacements de consommation sur les marchés de l'énergie, NEBEF 3.3

IOP Conf. Series: Earth and Environmental Science **897** (2021) 012018 doi:10.1088/1755-1315/897/1/012018

Power Conversion Performance of a Single-Phase PV Inverter with Fuzzy Logic Algorithm

G I Giurgi[1], D Petreus[2], D V Giurgi[3] and L Szolga[4]

[1]Faculty of Chemistry and Chemical Engineering, Babes Bolyai University, Chemistry, Organic Chemistry Department, Cluj-Napoca, Romania
[2]Faculty of Electronics, Telecommunications and Information Technology, UTCN, Applied Electronics Department, Cluj-Napoca, Romania
[3]Faculty of Electronics, Telecommunications and Information Technology, UTCN, Communications Department, Cluj-Napoca, Romania
[4]Faculty of Electronics, Telecommunications and Information Technology, UTCN, Basis of Electronics Department, Optoelectronics Group, Cluj-Napoca, Romania

Lorant.Szolga@bel.utcluj.ro

Abstract. The paper presents a low-power conversion system focusing on implementing new solar inverter control techniques implemented with Fuzzy Logic. The power generated by a solar panel requires robust approaches and efficient methods to be used at its maximum. Therefore, a promising strategy is a Fuzzy Logic based on the Maximum Power Point Tracking (MPPT) algorithm. To gather efficient power conversion, our proposed model uses a control loop composed of Fuzzy Proportional Integrative (PI) regulators, Clarke and Park transform, followed by a synchronization grid mechanism Second-order generalized integrator (SOGI) based phase-locked loops (PLLs). The proposed technique examines photovoltaic system (PV) performance with respect to its non-linearities and eventual shaded conditions that can occur in the PV array. The shading effect is tested by varying the irradiance, which determines the variation of the output current and implicitly of the output power. The simulation results show that the inverter control system is very efficient, generating stable and nearly sinusoidal current and voltage characteristics. Thus, the inverter converts over 99 % of the power generated by PV arrays.

1. Introduction

Developing alternative energy resources with high efficiency and low emission has become very important by increasing concerns about fossil fuel deficit, high oil prices, global warming, and environmental and ecosystem damage [1].

The photovoltaic (PV) system has become an important renewable energy source due to its availability and easy access. Since it is a DC power source, an inverter must convert it to AC power, powering up AC loads or transport it to the utility grid. The level of difficulties of the system implementation is associated with invertor topology, switching topology, and system environment or platform. The success of PV power usage is associated with proper control techniques of the inverter. Researchers have focused on various inverter control issues, including self-consumption losses, nonlinearity behaviour, output fluctuation, weather dependence, and low PV efficiency [1].

IOP Conf. Series: Earth and Environmental Science **897** (2021) 012018 doi:10.1088/1755-1315/897/1/012018

There are more than two PI controllers in grid-connected systems, and they usually require high-quality transit performance that is defined. Thus, it is necessary to adjust the value of PI gains during the transit process to achieve better performance. One of the most effective methods for adjusting PI gains is fuzzy logic control (FLC) [2]. In general, the FLC applies to intelligent control systems, nonlinear systems, and other complex applications due to its simplicity and flexibility [2].
MPPT algorithms are essential because PV arrays have a nonlinear voltage-current characteristic with a unique point where the produced power is highest. This point depends on the temperature of the panels and the irradiance conditions. Furthermore, the increase of average PV system size may lead to new strategies like eliminating the DC-DC converter, usually placed between the PV array and the inverter, and moving the MPPT to the inverter, resulting in increased simplicity, overall efficiency, and a cost reduction.
The goal is to implement control techniques based on Fuzzy logic for single-phase inverter in low-power photovoltaic systems [3].

2. PV System Design

The configuration of the grid-connected PV inverter system considered in this paper is illustrated in figure 1. The system contains a control system based on Fuzzy Logic, solar panel, inverter, and RL-load. The control system consists of several sub-control modules: MPPT and current-control Fuzzy logic functions, Park and Inverse Park transformations, SOGI-PLL grid synchronization and pulse width modulated (PWM) signal generator.
The PV array is composed of two 75W Suntech Power STP075S-12/Bp panels, series-connected. The solar panels block has two inputs for both irradiance (varying) and temperature (constant).
The PV's output fits into a Fuzzy Logic MPPT block which provides a voltage reference for the inverter. The single-phase inverter is formed of a DC link (two 10mF capacitors), a power stage (inverter IGBT), and filters. Its high capacitance helps stabilize the input voltage for the invertor. Therefore, the obtained signal goes to the Fuzzy Logic PI controller, giving the current in regulated form. This current, Id_ref, represents a Park transform parameter, which controls the system's active power.
The method generates the feedback currents (Id and Iq) using the inverter's current Iinv. Iq_ref is initialized with 0 to avoid reactive power in the PV system. The Fuzzy logic PI controllers amplify and integrate the signals and adjust the error to fulfil the reference currents' equality conditions. The obtained control currents are transformed back to inverse Park transformation and fed to the PWM generator to send the inverter's command pulses. The inverter feeds current into the grid through an interconnecting load RL. Finally, a phase-loop (SOGI-PLL) synchronizes the reference current injected by the inverter with the grid voltage. Consequently, a phase-frequency alignment takes place between the inverter's sinusoidal current and the grid's voltage. Given this, the inverter converts and delivers maximum power to the grid.

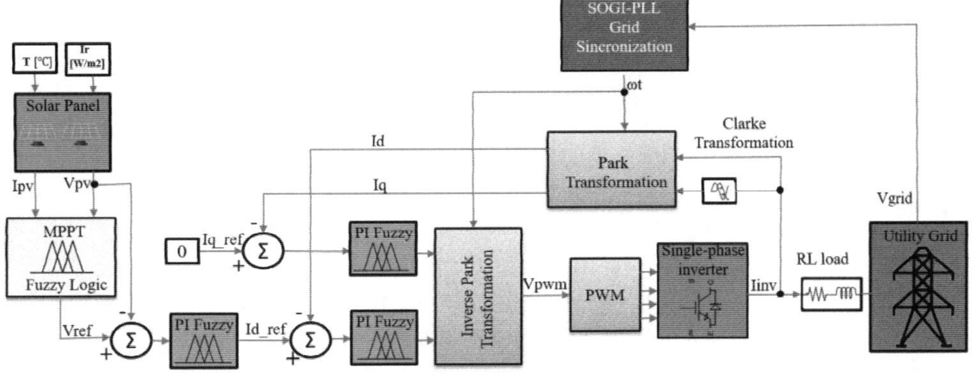

Figure 1. The block diagram of the PV system.

2.1. Maximum power point tracking algorithm (MPPT)

In the past decades, the most used algorithms for PV inverters in commercials are Perturb & Observe and Incremental Conductance. The results obtained for these two algorithms are similar, so it is difficult to say which one is the best. Lately, Fuzzy Logic Control (FLC) proved to perform more rapidly than the Perturb and Observe or Incremental Conductance algorithm. In figure 2 is presented the flow chart of the Fuzzy MPPT method. The main elements that make up an FLC system are the fuzzifier unit at the input terminal, knowledge-based (rule base) and the inference engine, and defuzzifier at the output terminal [4].

The Fuzzy MPPT block's inputs are the voltage (VPV) and the current (IPV) provided by the solar panel, while the output terminal is the voltage reference for the inverter (Vmppt) [4].

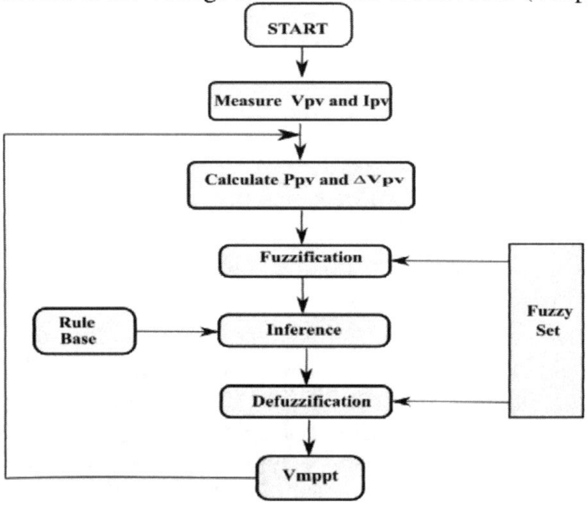

Figure 2. FLC MPPT flow chart.

Inside the Fuzzy MPPT block, the calculated power and voltage variation (ΔV) to the fuzzy logic controller. The following equations determine the parameters of the Fuzzy MMPT:

$$P_{PV}=I_{PV} * V_{PV} \tag{1}$$

$$\Delta Vpv= V(k) - V(k\text{-}1) \tag{2}$$

Figure 3 illustrates the implementation of the algorithm using the elements available in Simulink.

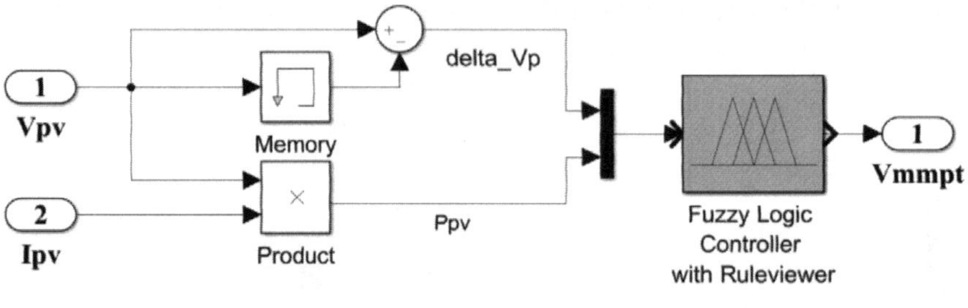

Figure 3. Simulink implementation of the MPPT fuzzy algorithm.

To implement the MPPT algorithm, we use the Fuzzy Logic Designer function, and the fuzzy inference method used is the Mamdani method, which uses Max-min composition. The fuzzy set includes 49 statements (if/then rules), representing the membership functions of input/output variables, ΔVpv, Ppv, and Vmppt. Each variable has seven fuzzy subsets. The input variables are normalized to match certain intervals, ΔVpv between [-1,1], respectively Ppv between [0, 150]. The output variable Vmppt belongs to the interval [34,35]. Any input value that does not belong to the domain is considered too large and generates signal errors. For simplicity, we use triangular and trapezoidal membership functions, through which the Fuzzy controller allows the rapid reduction of signal errors, which improves the transient response of the system [5]

2.2. Grid synchronization

Because of its advantages and wide practical usage, SOGI-based PLL circuits will be implemented in the simulation model. SOGI-based PLL is a highly efficient method for synchronizing PV inverter systems into single-phase, easy-to-design, and implement networks capable of running close to a filter without delay with the required bandwidth [6].

Second-Order Generalized Integrator (SOGI) is used to generate Quadrature signals. These quadrature signals are given as input to the Parks Transform block. Parks transform is used to transform stationary frame of reference to rotation frame of reference. PI controller behaves as a low pass to cancel our higher frequency parts. The PI controller parameters are crucial as altering either of the parameters will lead the system to lose synchronism [6].

2.3. Current control loop

The control provides two nested loops: one slower outer and one inner faster. Its output is the reference for the outer one. Generally, in the control of the inverter, PI controllers are used.

The main issues for these classical regulators are related to the transient regime of the controlled signals. In our model, the traditional controllers were replaced with PI controllers based on Fuzzy logic thanks to their efficiency. In the first step, we configure the fuzzy inference system (FIS) to produce a linear control surface from the E (error) and CE (error change) inputs to the out output the peak positions corresponding to the input sets [7].

The rule base includes 9 rules expressed in terms: If ...Then, using the logical AND operator. Membership functions are the input variables E and CE, each with three fuzzy subsets. The membership functions corresponding to the output variable are of singleton type and are expressed by linguistic terms.

We calculate the gains kp, ki, corresponding to the conventional PI regulators. The controller's gains, PI in the voltage loop were obtained using the trial and error method, reaching the values kp = -5 and ki = -30. The calculation PI gains corresponding to the current loop involves the choice of a constant time, τ _i, of 50 μ s, a value of the load inductance of 200 μ H and a value of the load resistance of 20m Ω . The value of the inductance is sufficient because the work involves modelling a low-power photovoltaic system. The parameters kp and ki are estimated as below (3), (4):

$$k_p = \frac{L}{\tau_i} = \frac{200 \cdot 10^{-6}}{50 \cdot 10^{-6}} = 4 \tag{3}$$

$$k_i = \frac{R + r_{on}}{\tau_i} = \frac{20 \cdot 10^{-3}}{50 \cdot 10^{-6}} = 400 \tag{4}$$

In order to be able to model the PI Fuzzy controller, we need the scaling factors GE (error gain), GCE (error change gain), and GCU output (output gain). These scaling factors are calculated based on the gains kp, ki, assuming the maximum reference step 1, the maximum error E is 1. Since the input range of the error E is [-10,10], the error gain is chosen GE = 10 and the rest scaling factors are calculated with the following formulas [8]:

$$GCE = GE \cdot \left(\frac{kp - \sqrt{kp^2}}{2 \cdot ki} \right) \tag{5}$$

$$GCU = \frac{ki}{GE} \tag{6}$$

After calculations for the first PI Fuzzy controller corresponding to the voltage control loop, the following values are obtained for the scaling factors: GE = 10, GCE = 1.66, and GCU = -3. In figure 4, we have the Simulink implementation of the PI Fuzzy controller

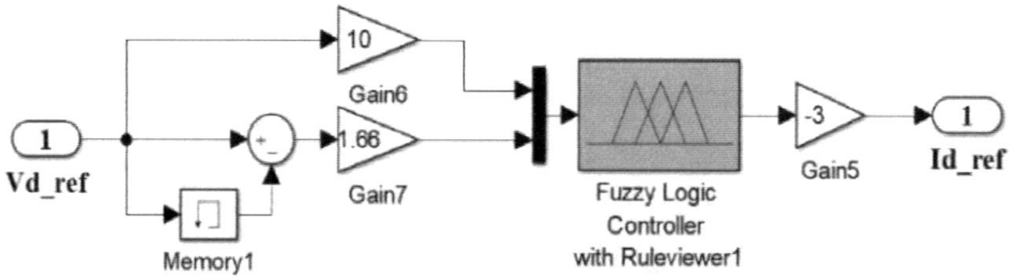

Figure 4. The PI fuzzy controller corresponding to the voltage control loop.

3. Experimental setup

The Matlab model of the Fuzzy Logic Control implementation on PV modules is shown in figure 5. The temperature is constant at 25 ℃, together with the irradiance represents inputs of the PV system. The solar power (PPV) and active power (POUT) is used to show the power conversion of the PV system [9]. We use a varying irradiance input to test the effectiveness of the proposed FLC algorithm. The irradiance varies from 1000 W/m2 to 750 W/m2 and 500 W/m2. The parameters used for the PV panel are shown in table 1 [10-12].

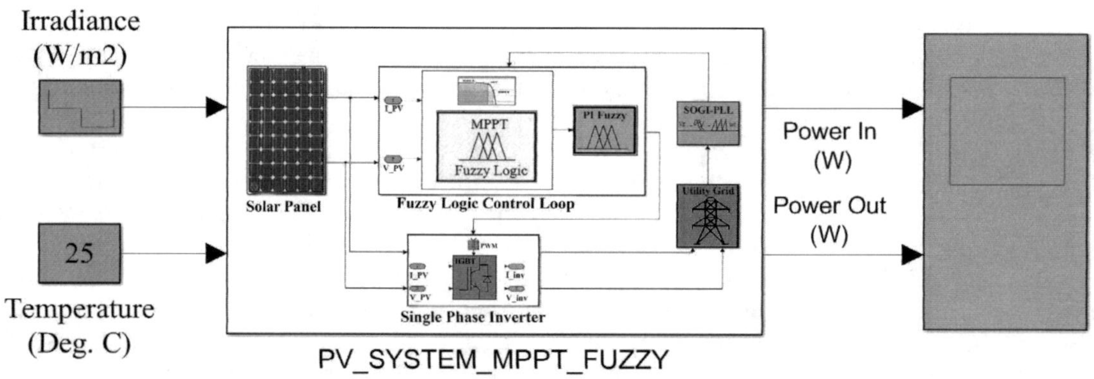

Figure 5. Matlab model of PV system with Fuzzy Logic Control.

Table 1. The main features of solar panel

Parameter	Value
Open circuit voltage (Voc)	21.7 V
Voltage at maximum power point (Vmp)	17.3 V
Short-circuit current (Isc)	4.72 A
Current at maximum power (Imp)	4.35 A
Maximum Power (Pmax)	75 W

4. Simulation results

The following figures illustrate the simulation results of the system presented in figure 5. Figure 6 shows the variation of irradiance to produce the partial shading effect. Figure 7 and figure 8 show current and power characteristics from the solar panel at different values of irradiance and fixed values for temperature. The MPPT controller was under the same operating conditions as illustrated in figure 9. At the initial moment, for the maximum irradiance value (1000 W/m2), the Vmppt voltage is at 34.8 V. If the irradiance drops to 500 W/ m2, there is a slight decrease in the voltage to 34.5 V, corresponding to the new maximum power point. So, it can be stated that the inverter reference voltage (VMPPT) is constant at constant irradiance and varies very small with irradiation changes.
This shows that the Fuzzy-based MPPT algorithm offers a very good precision in establishing the maximum power point.

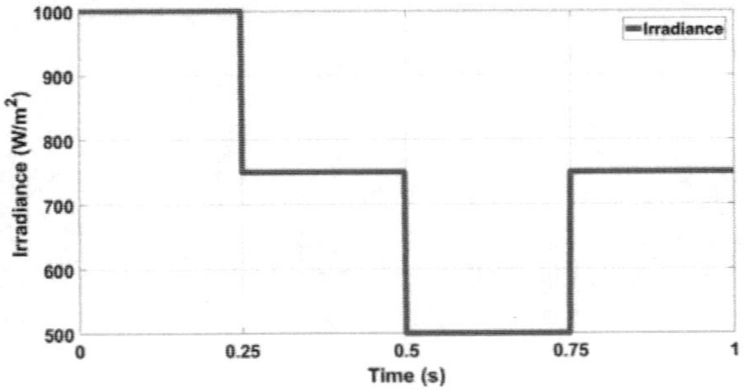

Figure 6. Variation of iradiance.

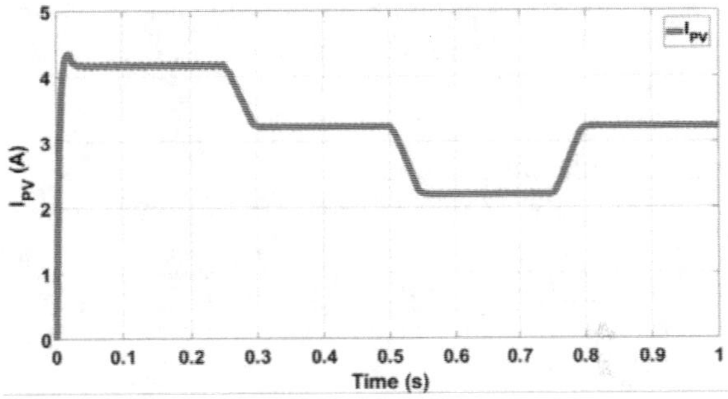

Figure 7. Irradiance effect in I_{PV} characteristics from the solar panel

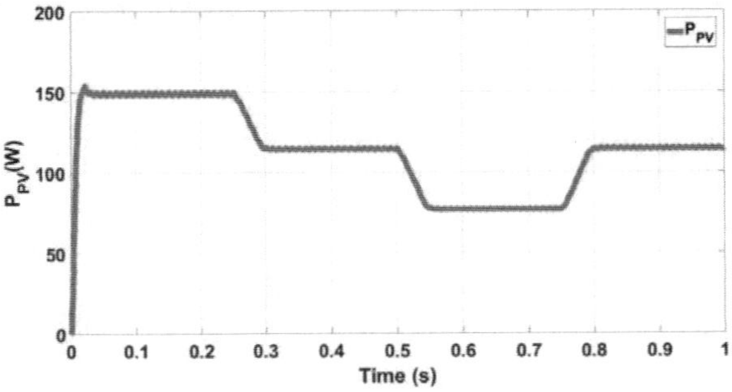

Figure 8. Irradiance effect in P_{PV} characteristics from the solar panel.

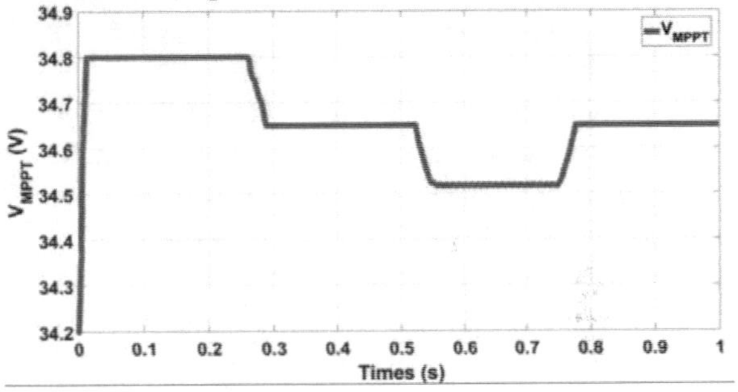

Figure 9. Inverter reference voltage generated by the Fuzzy MPPT controller.

The current and power extracted from the panel are the parameters that vary the most with irradiation change, and the voltage remains nearly constant. The output current and power of the inverter are also proportional to the irradiance. The current characteristics of the inverter are used in the current control loop to obtain maximum power (figure 10).
Figure 11 illustrates the power transfer performance at different values of irradiance and fixed values for temperature. It can be seen that the power output of the solar panel (PPV), varies between 149.8 W and 150.2 W and the active power (POUT) varies between 149.5 W and 149.9 W in the case of 1000 W/m2. That means that the systems have an efficiency greater than 99.5 %, due to the Fuzzy MPPT algorithm and the control loop based on PI Fuzzy controllers.

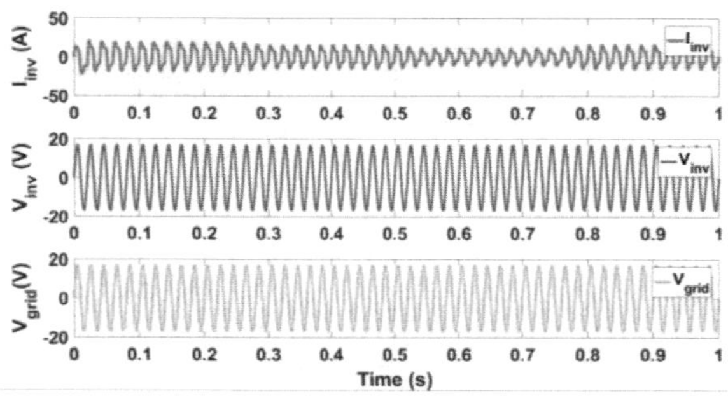

Figure 10. Output inverter current (I_{inv}), inverter voltage (V_{inv}) and grid voltage (V_{grid}).

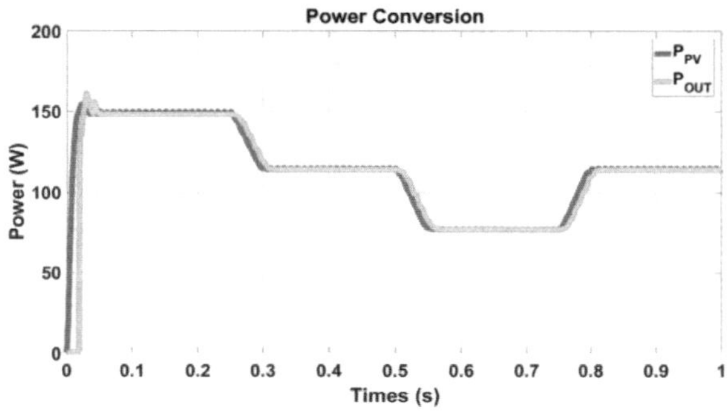

Figure 11. Power conversion performance.

5. Conclusions

The modeling and real-time testing of a single-phase PV inverter control based on Fuzzy logic have been presented in this paper. The simulation results showed that the inverter control system employing Fuzzy MPPT and Fuzzy PI controller effectively produces stable and nearly sinusoidal waveforms of both voltage and current. Moreover, the inverter can transfer the available excess PV power to the utility grid.

6. References

[1] Chel A, Tiwari G N and Chandra A 2009 *Energ. and Buildings* **41** 1172
[2] Hannan M A, Ghani Z A, Mohamed A and Nasir Uddin M 2015 {IEEE} *Trans. Ind. Appl*, **51** 6 4775
[3] Venkatesan M, Rajeswari R and Deverajan N 2015 *IJRER* **5** 739
[4] Lazarov V, Zarkov Z, Stoyanov L, Kanchev H and François B 2013 *IEEE* **63** 357
[5] Teekaraman Y, Kuppusamy R, Baghaee H R, Vukobratović M, Balkić Z and Nikolovski S 2020 *Energies* **13** 3974
[6] Urhekar R and Kulkarni S U 2016 *IJSRSET* **2** 741
[7] Hariri M H M, Desa M K M, Masri S and Zainuri M A A M 2020 *Energies* **13** 4279
[8] Azharuddin S M, Babu T S, Bilakanti N and Rajasekar N 2016 *Electr. Pow. Compo. Sys.* **44** 774
[9] Babu T S, Azharuddin S M, Bilakanti N and Rajasekar N 2016 *Renew. Sust. Energ. Rev.* **8** 015503
[10] Roop M E and Gonzalez S 2009 *{IEEE} Trans. Energy Convers.* **24** 195
[11] Blaabjerg F, Chen Z and Kjaer S B 2004 *IEEE T. Power Electr.* **19** 1184
[12] Chowdhury S, Das D K and Hossain S 2021 *JEA* **2** 7

Acknowledgments
The present work was accomplished by the support of the grant "REGRENPOS" PN-III-P2-2.1-PED2019-2601, founded by the Romanian Ministry of Education and Research

The Refractive Periscope – a Novel Concept

A Haltovsky[1]**, A Shemar**[2]**, A Schwarz**[2] **and E Gur**[2,*]

[1]Jerusalem College of Technology, Department of Electro-Optics Engineering, Israel.
[2]Azrieli College of Engineering, Department of Electrical Engineering, Israel.

* Corresponding author: erangu@jce.ac.il

Abstract. A periscope is an optical device whose function is to allow sighting of objects that are not in direct line of sight with the observer Eye/ Screen/ Detector. The concept behind the periscope is tilting the optical axis by breaking the rays of light with a mirror or prism so that the image is obtained in the desired location. The periscope has a variety of uses, many of which belong to the military realm. For example, observing from submarines above the sea level surface, looking at the war zone from inside of a tank, directing light into hidden places and more. In this article, the authors will review a new development based on wave guide concept which composes a tube with light reflective walls. This allows us to build a significantly Small-dimensional periscope with respect to a standard periscope that does not use lenses. The authors will also review a number of applications suitable for the reflection periscope.

1. Scientific Background

The simplest periscope is composed (as shown in Fig. 1) from two 45 ° rotated reflective surfaces in relation to both the observed object and the viewer's eye.

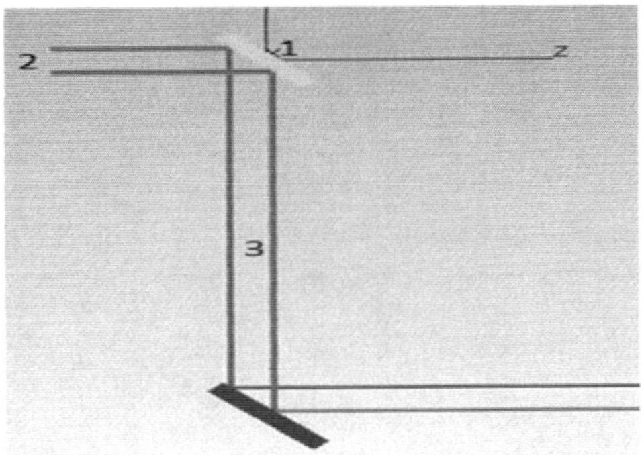

Figure 1. Schematic Periscope.

The first documented periscope was invented in the 15th century in order to watch over the heads of people in a crowd. In the 19th century, the periscope began to be used for military purposes, which received tremendous momentum in World Wars I and II, in the early 20th century [1-2].
To describe the operation of the periscope we define the following (as in Fig.1): The periscope axis (No. 1) – The horizontal line that exits from the center of the upper Mirror The Sign-in window (No.

2) – The aperture through which the light enters the periscope. The periscope tube (No. 3) – The periscope body, in which the rays propagate after refracting.

The above structure has two major drawbacks:

1. The tube diameter must be at least the size of the incoming beam diameter, which makes the periscope a large and cumbersome tool.
2. As the length of the tube (the z-axis) increases in relation to the length and width (Plane X-Y) of the entry window, the angle in relation to the periscope axis in which the object is observed becomes smaller, because the rays are reflected from the walls of the tube.

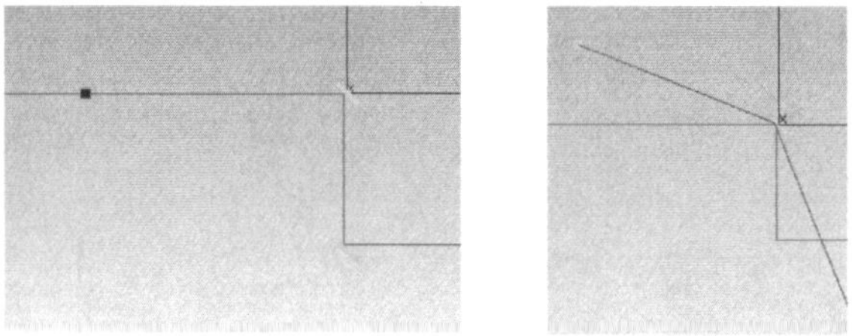

a b

Figure 2. (a) Object on Periscope axis (b) An object outside the periscope axis..

In modern periscopes [3-4], as the one demonstrated in Fig.3, a telescope and a field lens increase the field-of-view of the periscope. This increases the system size, adds geometric distortions, and creates constraints on the length of the system and the periscope's tube diameter.

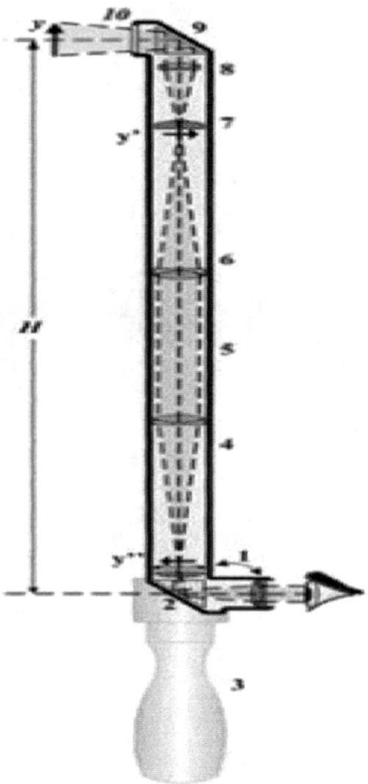

Figure 3. A modern periscope.

IOP Conf. Series: Earth and Environmental Science **897** (2021) 012019 doi:10.1088/1755-1315/897/1/012019

In the present paper, the authors will demonstrate a new development, based on the reflective walls of the periscope. The reflective walls allow the light to propagate along the periscope in additional routes rather than the direct downward path. That suggested approach allows us to: A. use a mirror positioned at an angle different than 45 degrees in relation to the periscope axis. B. significantly reduce the periscope Dimensions. C. convey the image of an object located far above the viewer or slightly below the optical axis of the periscope.
In the following paragraph the authors will review the four different models suggested for this periscope:
1. The basic model – A Mirrors Tube. 2. Adding a prism supplement at the edge of the periscope. 3. Adding spherical mirrors at the entrance and exit of the periscope. 4. A periscope tube based on a total internal reflection.

2. Different Models of The Reflective Periscope

2.1 Reflection periscope - The basic model

The most basic development is a periscope whose upper mirror is tilted at an angle greater than 45 degrees with respect to the periscope's axis. The periscope consists of a rectangular tube where mirrors serve as front and back walls and where the side walls do not reflect light. At the lower edge of the periscope, there is another mirror, parallel to the upper mirror. In this structure, except for the periscope entrance and exit, the periscope thickness can be significantly reduced, as shown in Fig.4 and Fig.5.

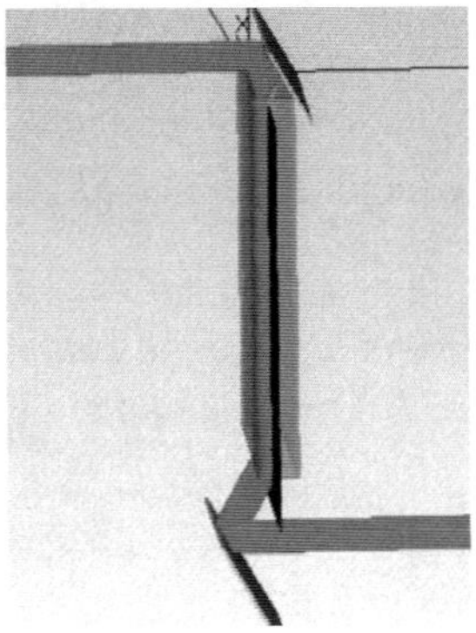

Figure 4. Mirrors tilted in 60 degrees; a beam is delivered in a tube that is half the diameter as the original beam.

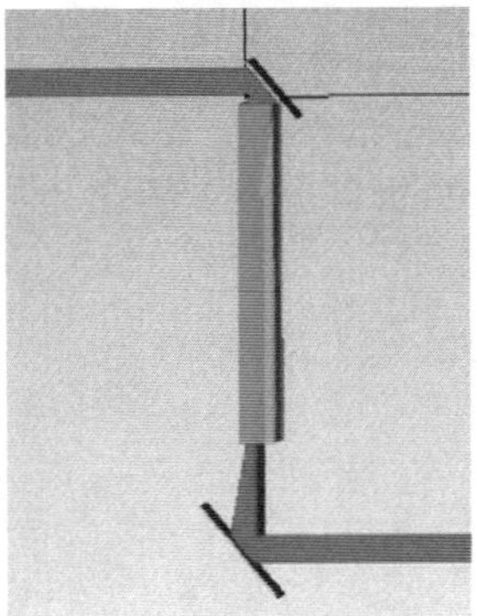

Figure 5. Mirrors tilted in 50 degrees; a beam transmitted by a tube that is 44% of the original beam diameter.

Figures 6 through 8 demonstrate an additional advantage of the system. Objects that are far above the periscope axis (the red line depicts a ray propagating from the object through the periscope.) may also be imaged by this periscope (Fig.6). In figures 7 and 8 one can see rays of light entering at different angles and the way they exit the Periscope.
Despite the great advantage of this system, there are also some structural problems that require attention:
1. In the meridional (horizontal) plane, only a small range of angles will pass through the periscope (depending on the relation between its length and width).
2. A major disadvantage of this system is the need for a relatively large structure at the entrance and exit of the periscope (The wall opposite to the tilting mirror should be at such a distance that a ray hitting the top of the mirror will be reflected to the wall and not above it).

IOP Conf. Series: Earth and Environmental Science **897** (2021) 012019 doi:10.1088/1755-1315/897/1/012019

3. Another problem to consider is a geometrical problem; In a long periscope there will be a large gap along the periscope between rays at different angles, this may result in some rays not hitting the lower mirror at all or hitting it at the opposite angle with respect to the vertical axis. Possible solution to these problems can be given by adding a prism at the lower end of the periscope. Below are several applications and suggested models.

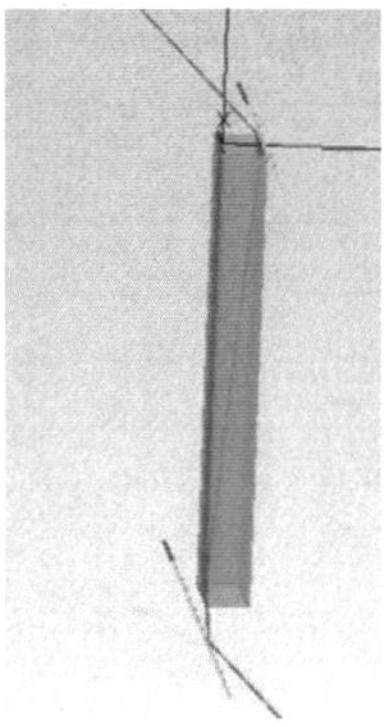

Figure 6. An object that is at 45 degrees in relation to the optical axis exit the periscope at the angle at which it enters.

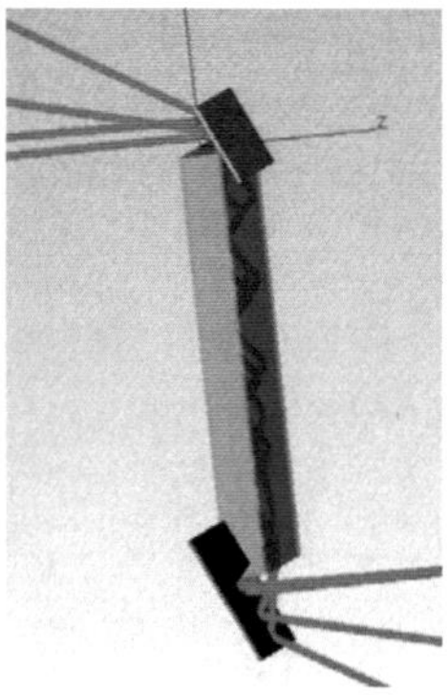

Figure 7. Objects at different angles and how they exit from the Periscope.

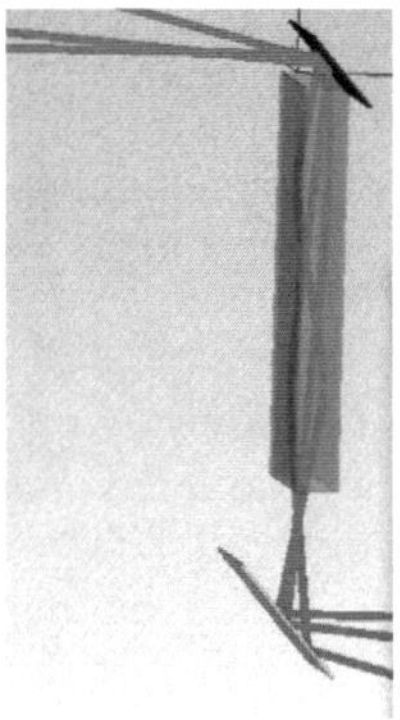

Figure 8. Objects at different angles and how they exit from the Periscope.

2.2 Applications

Through the years many applications for the standard periscope have been suggested [5-14], in this section we introduce application for the novel refractive periscope.

2.2.1 Transparent sealed materials. Fig. 9 shows that it is possible to create curves along the path of the rays, which can have significance in different applications of the periscope. In the setup shown in Fig. 9, the object looks transparent.

Figure 9. The transparent basic model plus the curve.

Figure 10. Periscope glass tube. Sun light to building is without imaging.

2.2.2 Transferring daylight into a windowless room. Another important application that can be improved with the model we built is daylight transfer into rooms without a window or adding it to a room and enhancing the light quality as seen in Figure 10. Although this can also be done using a standard periscope, our model is much more suitable both due to the fact it can deliver the actual image as in a real window and due to the large range of angles of light collected there's no need for lenses, making the periscope cheaper. In such a system, where precise imaging of a specific object is not required, mirrors can be added to the sides of the periscope, allowing the propagation of a large range of incoming angles through the periscope, and by that reduce its width.

2.2.3 Reflection periscope – Prism addition. In this model, the authors add a prism (or a beam splitter) to the original model, placed at the lower end. This causes rays with large incoming angles to reflect and hit the bottom mirror again, thus that all the rays will exit the periscope at the same angle at which they entered. This model is built on the principle of total internal reflection from the prism walls. This principle and the conditions for its existence are elaborated in following models.

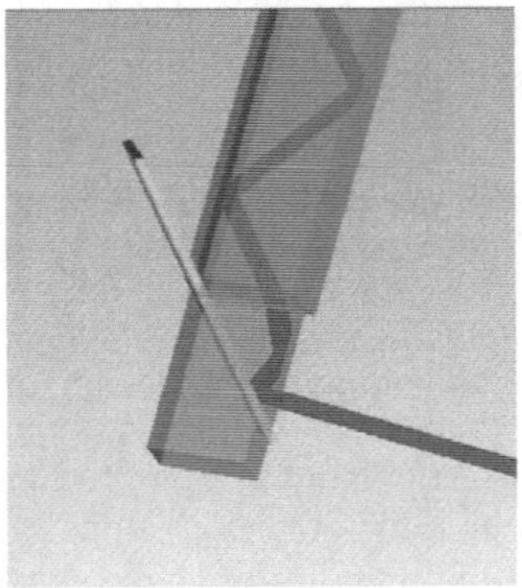

Figure 11. The advantage of adding a prism to the periscopes edge.

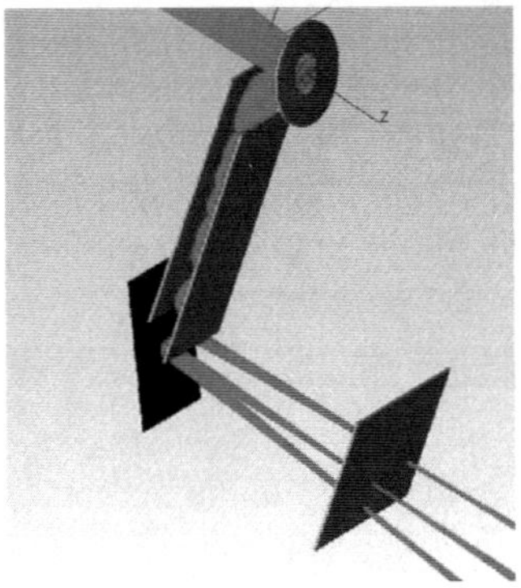

Figure 12. Focusing the image on the screen with a single circular mirror.

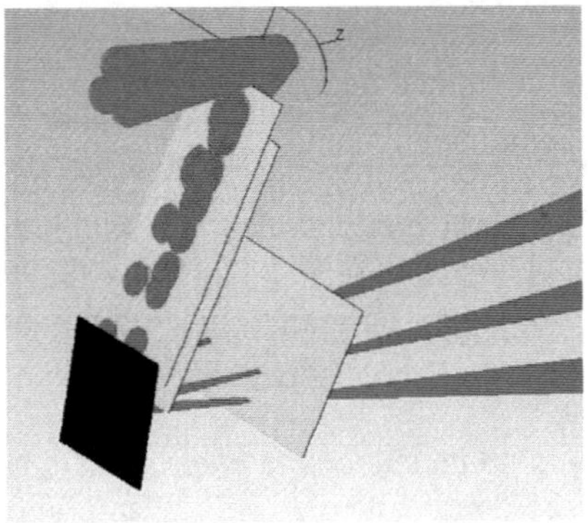

Figure 13. Focuses the image on the screen with a single circular mirror.

2.2.4 Reflection Periscope - Spherical or cylindrical mirrors. In this model, the upper mirror or the two mirrors, the upper and lower, are spherical mirrors. As a function of the mirror's curvature, one may gain several benefits in this case:

1. The ability to focus the image on a screen without the need for a lens add-on (Figs. 12 and13).
2. Expansion or narrowing of the laser beam (Fig. 14).
3. Reducing the beam so that the periscope tube can be further reduced both in depth and width (Fig. 14).

IOP Conf. Series: Earth and Environmental Science **897** (2021) 012019 doi:10.1088/1755-1315/897/1/012019

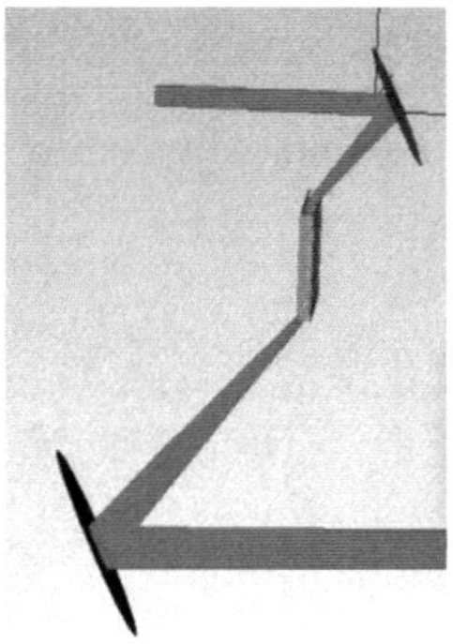

Figure 14. Expansion of the beam using two circular mirrors.

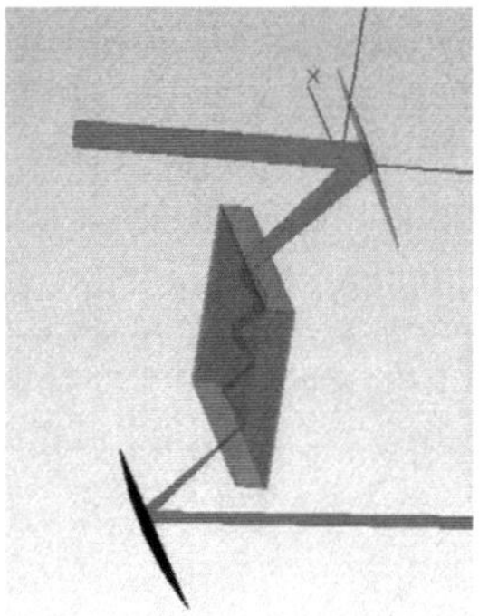

Figure 15. Narrowing the beam with two circular mirrors.

2.2.5 Reflection Periscope – Transparent material. In this model, the Periscope tube is not filled by air but by a transparent material with a higher refractive index. The principle of this model is based on the total internal reflection principle based on Snell's law:

$$n_1 \sin \theta_1 = n_2 \sin \theta_2 \tag{1}$$

According to this principle, in the transition from a high refractive index medium to a low refractive index medium, there is a critical angle in which the ray will not penetrate the lower medium but would rather be reflected (as in a mirror) to the medium of the higher refractive index. To find the critical angle, place a 90 ° angle at the exit angle. To simplify the calculation, suppose that the low refractive element is air. We accept:

$$\theta_c = \sin^{-1}\left(\frac{1}{n_1}\right) \tag{2}$$

One can obtain, for example, that for a refractive index of 1.5, total internal reflection requires an incident angle of 41.81 degrees. The mirror can be tilted up to 69.1 degrees (45 + 41.81/2) above the periscope axis in order to enable us to make the incoming parallel rays to be reflect within the periscope, (Attach a ray diagram). Increasing the refractive index, will result in the increase of the range of angles in which the mirror can be tilted.

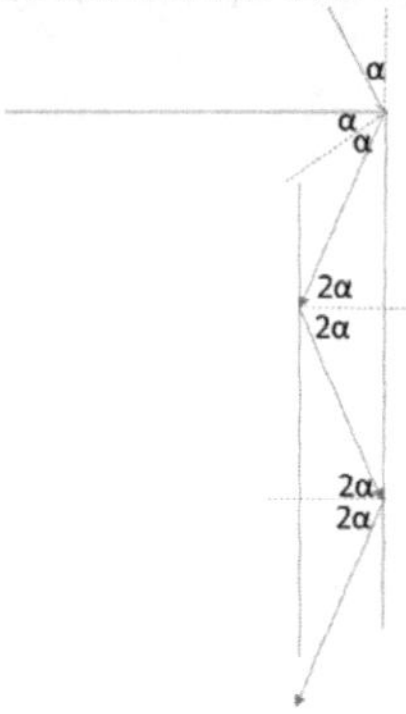

Figure 16. Rays diagram with incident and reflect angles.

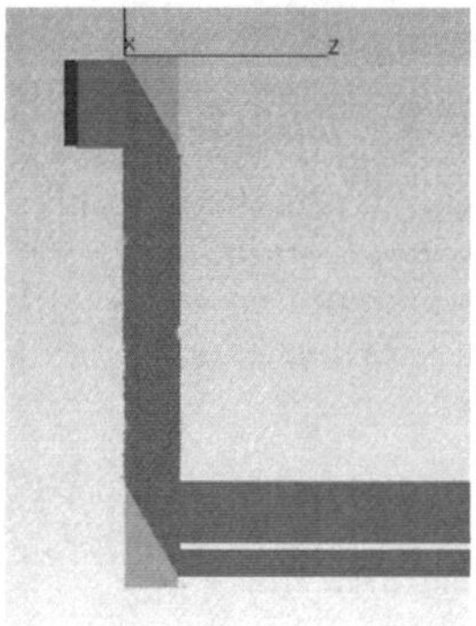

Figure 17. Narrow periscope, glass tube. Ratio of source's radius/tube's radius = 16/10. The image splits with the top of the object at the bottom of the image.

There are several advantages to this model. First, the mirror-reflecting rays can hit the periscope's entrance aperture yet stay inside the periscope, allowing a much narrower structure for the periscope. This comes in addition to the advantages mentioned above for Model 2, at the output of the periscope. Second, this model enables the incoming rays to have larger angles on the horizontal plane that can propagate through the periscope, since according to Snell's law, the beam within the periscope will propagate at a lower horizontal angle.

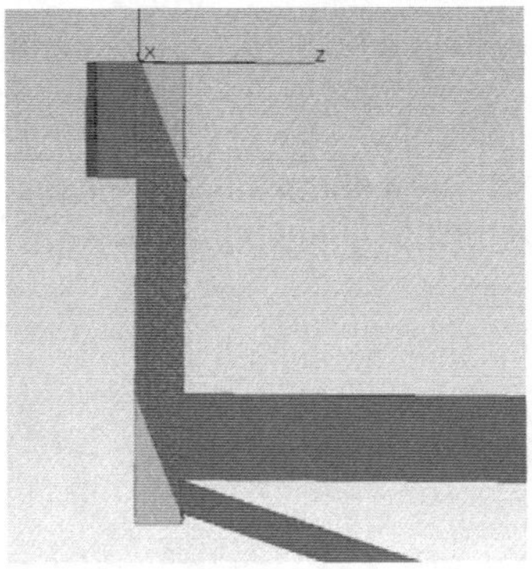

Figure 18. Narrow periscope, glass tube. Ratio of source's radius/tube's radius = 25/10. 8 upper mm exit with a vertical angle (seemingly should have another internal reflection)

Figure 19. Periscope glass tube. Horizontal entrance angle of the blue ray: 30 degrees.

2.3 The mirror tilting angle as a function of the refraction index

At this point we calculate the length-to-width ratio allowed in the basic model. We examine an incident ray of light that has a zero vertical angle with respect to the optical axis, and a five degrees horizontal angle. The top mirror is tilted at an angle of 65 degrees with respect to the axis. The light hits the right side of the entrance aperture.
The angle of the ray in the vertical axis after hitting the mirror will be 50 degrees ((90-65)+(90-65)=50). We expect the length of the tube to be 8.79 times bigger than its width (sin 50/sin 5). On the other hand, if the ray passes through glass with a refractive index of 1.8, according to Snell's law the new angle in which the ray propagates will be 2.77 degrees which allows the length of the periscope to be 15.85 (sin 50/sin 2.77) times bigger than its width.

3. Conclusions and summary:

In this work the authors examined a variety of periscope models and discussed the advantage and disadvantages of using different materials and elements. This research brings a new approach for the known periscope concept. The standard periscope transfer image using mirrors in 45 degrees and pipe that blocks stray light from the outside. Moreover, the now days periscopes are based on lenses or 45 degrees prisms. In solar energy field some companies use light pipes to transfer sun light into closed space rooms but needs to be close to ceiling and don't have the ability to transfer image. Our novel approach of transferring image is based on mirror waveguide with suitable angle which reduce the periscope size but preserve the image same as in standard periscopes. Future tests will examine the light efficiency of the suggested periscope to ensure an improvement compared to standard periscopes.
The authors would like to acknowledge the Jerusalem College of Technology – Lev Academic Centre, for enabling us to use the TracePro© Software for design and analysis of illumination and optical systems.

IOP Conf. Series: Earth and Environmental Science **897** (2021) 012019 doi:10.1088/1755-1315/897/1/012019

4. References:

[1] https://maritime.org/doc/fleetsub/pscope/index.htm, 1946, Maritime Park Association.
[2] Lt. Cmdr. Irving R, 1960, US Navy, Navigation, 7, 2, 140-144.
[3] Walker B H, 2000, SPIE Press, Chapter 9, Washington.
[4] Bellis M, 2020, https://www.thoughtco.com/history-of-the-periscope-4072717.
[5] Castigliano S G, 1950, The American Journal of Roentgenology and Radium Therapy, 64, 3, 492-494.
[6] Gregory P J, 1979, Journal of Experimental Botany, 30, 1, 205–214.
[7] Tonkin S P and Dolman D L, 1990, IEEE Proceed. 137, 2, 149-156.
[8] Trainer F W and Eddy J E, 1964, Geological Survey Research. Paper 501-D, 203-207.
[9] Hay J G, and Gerot J T, 1991, International Journal of Sport Biomechanics, 7, 4, 392-399.
[10] Stremplewski P, Komar K, Palczewski K, Wojtkowski M and Palczewska G, 2015, Biomedical Optics Express, 6, 9, 3352-3361.
[11] Yoder P R, 1986, Contemporary Optical Instrument Design, Fabrication, and Testing 656, 225-230.
[12] Heddle D W O, 1971, J. Phys. E: Sci. Instrum. 4, 12, 981.
[13] Wakabayashi F, 2008, J. Chem. Educ. 85, 6, 849-853.
[14] Lidwell M O, 1995, SPIE Proc. 2539, Zoom Lenses.

IOP Conf. Series: Earth and Environmental Science **897** (2021) 012020 doi:10.1088/1755-1315/897/1/012020

Adsorptive removal of chromium (VI) from synthetic waters using magnetic lignocellulosic composites

Tanya Medina-Espinosa[1], Christopher Asimbaya[1], Salomé Galeas[1], Nelly M. Rosas-Laverde[1], Alexis Debut[2] and Víctor H. Guerrero[1,*]

[1]Department of Materials, Escuela Politécnica Nacional, Quito 170525, Ecuador
[2]Centro de Nanociencia y Nanotecnología, Universidad de las Fuerzas Armadas (ESPE), Sangolquí 171103, Ecuador

[*]Corresponding author: victor.guerrero@epn.edu.ec

Abstract. The removal of heavy metals from water is one of the major challenges that humanity must address to avoid negative potential impacts on the environment and human health. During the last few years, several adsorbents have been examined, in a search for highly efficient and cost-effective materials. In this work, we investigated the use of laurel, canelo and eucalyptus lignocellulosic sawdust residues (LRs) impregnated with magnetite nanoparticles (MNP), to remove Cr^{6+} ions. Each LR was added to an aqueous solution in which MNP were being synthesized by coprecipitation. Two composite adsorbents were obtained, with LR:MNP ratios of 1:1 and 3:1. The materials obtained were characterized by X-ray diffraction, scanning and transmission electron microscopy, and infrared and Raman spectroscopy. The results obtained showed that the laurel composite was the best adsorbent, reaching a maximum removal efficiency and capacity of 99.8 % and 30.5 mg/g, respectively. The optimal contact time was 30 min and the process fitted the Langmuir isotherm model, showing small effects of the fraction of sawdust residues used to support the MNP. Further studies will be performed to optimize the composition of the composites aiming to reduce the amount of costly MNP used while ensuring a high removal performance.

1. Introduction

Wastewater from industries is one of the main sources of environmental pollution. Among water pollutants, heavy metals have gained great importance due to their persistence and toxicity [1]. The industries with this type of effluent include those that operate with metallic coatings, mining, automotive industry, tanneries, foundry operations, and alloy industries [2].

The chromate anion (CrO_4^{2-}) is very common in industrial wastewater, polluting and affecting human health. Hexavalent chromium, even at relatively low concentrations (from 0.10 mg/L), is a mutagen and a potential carcinogen [3]. That is why it is necessary to treat effluents, avoid contamination and reduce the risk generated by the presence of heavy metals. Several studies have been conducted on the removal of heavy metal ions such as Cr^{6+}. Activated carbon is widely used in adsorption, in addition to chemical precipitation, electrochemical reduction, ion exchange, biological methods, and electrocoagulation [4,5]. However, these methods have certain limitations in terms of efficiency, costs, and the generation of secondary products requiring subsequent treatment [6].

Adsorption is an alternative treatment for the removal of heavy metals using lignocellulosic residues such as sawdust, wood chips, fruit peels, etc. These residues are widely available, have a low cost, and a relatively high specific surface area. The adsorptive removal is directly influenced by the amount of lignin present in the material [7,8]. The adsorption is promoted by reactions between the oxygens of the ionizable functional groups of lignin, and the metal ions in the solution [2].

Adsorption is directly related to the nature of the adsorbent material, and sometimes the kinetics of heavy metal adsorption is not optimal. In recent years there has been a great development of nanotechnology, which has helped to increase the adsorption efficiency of metal ions due to the increase in active sites [9]. One of the nanoparticles that has been most studied for this purpose is magnetite (Fe_3O_4) due to its magnetic properties that allows an easy recovery, and because it is a relatively simple and economically convenient method [10]. Additionally, this type of iron oxide nanoparticles can be obtained through a variety of environmentally friendly methods [11,12].

Magnetite nanoparticles (MNP) increase the performance of adsorbents by being incorporated into their surface. Among the adsorbents that have been studied together with MNP are sawdust, residues of tea, orange peels and almonds, and polymers such as chitosan. The applications of these composite materials are focused on the removal of colorants and heavy metals, especially zinc, chromium (VI), cobalt, lead, strontium and arsenic [13–15]. Using MNP modified with an anionic surfactant, zinc removals of 89.3% have been obtained from 20 mL solutions with initial concentration of 10 mg/L, in 5 min and with 10 mg of the adsorbent [16]. Orange peels impregnated with MNP remove up to 85% of cadmium after 40 min [17]. Removals of 95% of chromium (VI) have been obtained from 100 mL solutions with an initial concentration of 50 mg/L, in 100 min and with 0.5 g of MNP [18].

In the above-mentioned investigations, the influence of the contact time and the amount of the composite material in the adsorption tests is mainly considered. However, the relative amounts of the MNP and the lignocellulosic sawdust residues (LR) used to obtain composite adsorbents has received relatively little attention. In this work, we studied the effect of the contact time and the fraction of the sawdust residues in composite adsorbents used for chromium (VI) removal, considering for this purpose two LR:MNP ratios, and three types of residues. Since the MNP are costly, compared to the LR, it would be beneficial to reduce the amount of magnetite required to obtain an efficient composite adsorbent.

2. Materials and methods

All the reagents used were analytical grade and the solutions were prepared using type 2 ultra-purified water. The reagents included ferric chloride hexahydrate ($FeCl_3.6H_2O$), ferrous chloride tetrahydrate ($FeCl_2.4H_2O$), and hydrochloric acid (HCl) from Panreac, and sodium hydroxide (NaOH) from Merck. Argon was used as an inert atmosphere for the synthesis of magnetite and the composite material. Potassium dichromate ($K_2Cr_2O_7$) from Lobachemie was used for the removal tests.

Three types of wood residues were used: Ecuadorian laurel (Cordia alliodora), canelo (Ocotea javitensis) and eucalyptus (Eucalyptus globulus), with a particle size between 74 and 150 µm. The residues were washed several times with ultra-purified water to remove impurities. Subsequently, each residue was dried at 80 °C for 12 h [13].

MNP were synthesized by the co-precipitation method, according to the chemical reaction:

$$FeCl_2+2FeCl_3+8NaOH \rightarrow 2Fe_3O_4+4H_2O+8NaCl \quad (1)$$

A 25 mL iron solution was prepared dissolving 8.649 g of $FeCl_3.6H_2O$, 3.181 g of $FeCl_2.4H_2O$, and 1 mL of 0.4 M HCl in distilled water; 250 mL of a 0.9 M NaOH solution was also prepared. Argon was bubbled into each solution for 15 min. The iron solution was added to the NaOH solution dropwise, and stirred at 1000 rpm for 45 min at 50 °C, with constant argon bubbling [19]. To obtain the composite adsorbent, each LR was added to the previous solution and the stirring was maintained for 45 min at 50 °C under argon atmosphere. The LR:MNP ratios used were 1:1 and 3:1. The resulting suspension was washed with distilled water until the pH of the solution was neutralized; the precipitate was magnetically separated and dried at 70 °C for 12 h. Finally, the composite obtained was disintegrated.

The lignin composition was determined according to ASTM D1106-96(2013). The functional groups in each LR and composite adsorbent were identified by a Perkin Elmer Spectrum 100/Spotlight 200 FTIR spectrometer [20]. Additionally, the magnetite and the composites were analyzed using Raman

spectroscopy (Horiba LabRAM HR Evolution) and X-ray diffraction (XRD) (Panalytical Empyrean). To determine the particle size of the magnetite, transmission electron microscopy (TEM) was performed with a FEI Tecnai G2 Spirit Twin microscope. The surface morphology of the composite materials was studied by scanning electron microscopy (SEM), using an Aspex PSEM Express equipment [13].
The conditions for the removal tests were established based on the work of previous authors. The amount of the composite adsorbent material used was 0.9 g, added to 100 mL of a 20 mg/L chromium (VI) solution, at pH 2.5 and 25 °C at 200 rpm. The removal using the MNP was studied under the same conditions. Each removal test was carried out considering different contact times, from 15 min to 160 min [18,21]. Filtered solutions of each test were analyzed using a HACH DR 1900 spectrophotometer according to the APHA 3500-Cr established method. Finally, adsorption isotherms were fitted for the chromium (VI) ion removal. The tests were carried out with 10, 20, 50, 100, 150 and 200 mg/L solutions. Freundlich and Langmuir models were used according to the corresponding linearized equations:

$$\log q_e = \log K_F + (1/n) \log C_e \tag{2}$$

$$1/q_e = 1/q_{max} + 1/(K_L q_{max})\,(1/C_e) \tag{3}$$

where q_e is the ion adsorption capacity per unit mass of adsorbent (mg/g); C_e is the equilibrium concentration (mg/L); K_F and n are Freundlich isotherm constants that relate adsorption capacity and intensity, respectively; q_{max} is the maximum ion adsorption capacity as monolayer (mg/g); and K_L is a constant related to the adsorption energy for the Langmuir isotherm [7,10].

3. Results and discussion

3.1. Materials characterization

The amount of lignin (dry basis) was 31.84 % in laurel, 28.28 % in canelo and 18.28 % in eucalyptus. According to Lee and Rowell (2004), the material with the greatest lignin content, has a greater number of phenolic groups and therefore a greater capacity to remove metal ions [20]. Therefore, laurel was expected to adsorb higher amounts of chromium (VI), compared to the other two residues.
As illustrated in figure 1, TEM showed that the average diameter of the MNP synthesized was 11.3 nm. This figure also shows the SEM micrographs of the composites obtained with 1:1 and 3:1 canelo:MNP ratios, where the surface of the lignocellulosic residue coated with the magnetite nanoparticles is observed. Magnetite agglomerates with different shapes and sizes are formed due to magnetic attraction. In the composite with a 3:1 ratio, a smaller amount of nanoparticles is observed coating the canelo fibers. Similar results were observed for the laurel and eucalyptus composites.
XRD patterns of the synthesized magnetite and the composite materials in a 1:1 ratio are shown in Figure 2.a. Magnetite exhibits characteristic peaks at indices 220, 311, 400, 422, 511, and 440 between 30° and 63°. These results correspond to the cubic spinel crystal structure of magnetite [22]. Every composite obtained, with the three LR and the two LR:MNP ratios studied, showed the magnetite peaks mentioned. This allowed us to verify the presence of the nanoparticles on the adsorbent surface. Figure 2.b shows the Raman spectrum obtained during the characterization of the MNP. The most representative peak is found around 670 cm^{-1}. The appearance of the band around 350 - 385 cm^{-1} in the magnetite spectrum may be due to the bond between the OH- groups and the Fe^{3+} ions. Furthermore, by having nanoparticulated material, the width of the lines of the Raman spectra can be broadened [23].
Figure 3 shows the FTIR spectra of the lignocellulosic canelo residue, and the composites with 1:1 and 3:1 LR:MNP ratios, where the characteristic peak of lignin is in the band around 1500 cm^{-1}, which corresponds to the vibration of the aromatic skeleton of the benzene ring. The stretching of the OH bonds of the phenol and alcohol groups is showed in the bands around 3300 cm^{-1} [13,24,25]. The presence of magnetite nanoparticles on the lignocellulosic residues is evidenced in the spectra in the band around 600 cm^{-1}, which corresponds to the vibrational stretching of the Fe-O bond [13,24,25].

Figure 1. TEM images of the MNP and SEM images of the composites of canelo impregnated with MNP in (a) 1:1, and (b) 3:1 ratios.

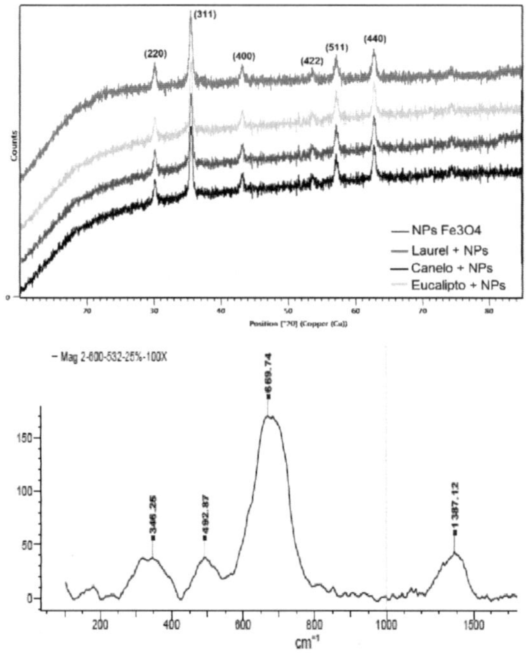

Figure 2. (a) X-ray diffraction patterns of the MNP and the composites with a 1:1 LR to MNP ratio. (b) Raman spectra of the MNP synthesized.

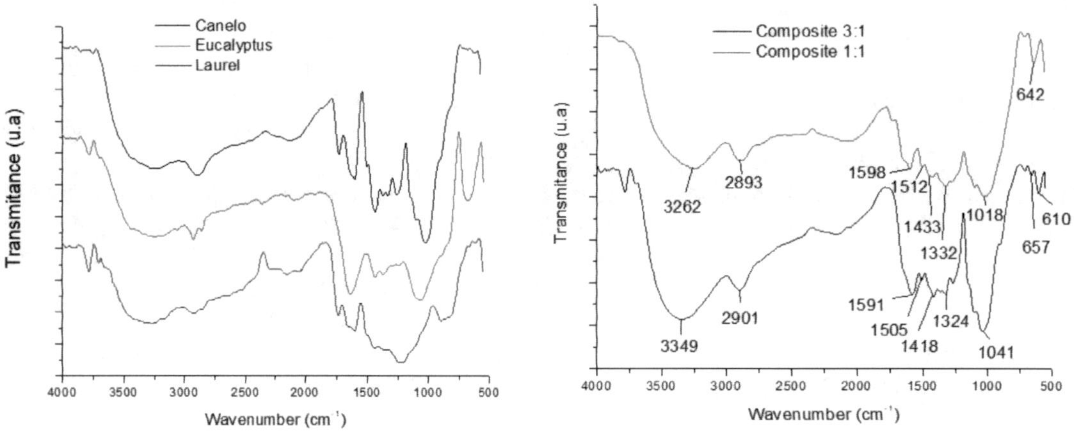

Figure 3. FTIR spectra for (a) canelo, eucalyptus and laurel residues, and (b) composites with 1:1 and 3:1 canelo residue:magnetite nanoparticle ratios.

3.2. Effect of the adsorbent dose

Figure 4 shows that the chromium (VI) removal efficiencies using the composite adsorbents with 1:1 and 3:1 LR:MNP ratios are practically the same. This illustrates the direct influence of magnetite nanoparticles on the adsorption of heavy metal ions, which lies in the presence of Fe and O atoms that, by adsorbing OH- radicals from the solution, increase the OH- groups on the magnetite surface. The interactions between these groups and the metal ions improve the removal performance [26]. Table 1 summarizes the results obtained in the adsorption tests with each composite material.

Table 1. Chromium (VI) adsorptive removal efficiency using magnetic lignocellulosic composites and a 30 min contact time.

LR : MNP ratio	Laurel - MNP	Canelo - MNP	Eucalyptus - MNP
1:1	99.78 %	99.60 %	98.77 %
3:1	99.64 %	99.59 %	98.42 %

3.3. Effect of the contact time
Figure 4 shows a steep slope for the first 30 min during the removal of chromium (VI) ions. From this contact time on, the increase in the removal percentage is minimal, approximately 1%, until reaching equilibrium. In the first minutes, there is a greater amount of free active sites in the composites, which after being occupied, decrease the adsorption of the metal ion. The laurel composites show higher removals; however, the difference with the removals obtained with canelo is 0.3 %, and with eucalyptus 1 %. This dependence on the removal by lignocellulosic composites is related to the amount of lignin in each residue [20]. In the chromium (VI) removal tests with laurel composites, percentages of around 99% were reached. As expected, the lower the initial concentration, the shorter will be the contact time and the higher the removal percentage [7]. It should be noted that, by adsorbing the heavy metal using only magnetite nanoparticles, a 96.95% chromium (VI) removal efficiency was obtained, with a 100 min contact time and using 0.9 g of adsorbent in a 100 mL solution with initial concentration of 20 mg/L of the contaminant. Similar removal efficiencies were obtained using the lignocellulosic residues impregnated with MNP, but in a much shorter time (approximately 30 min) under the same conditions.

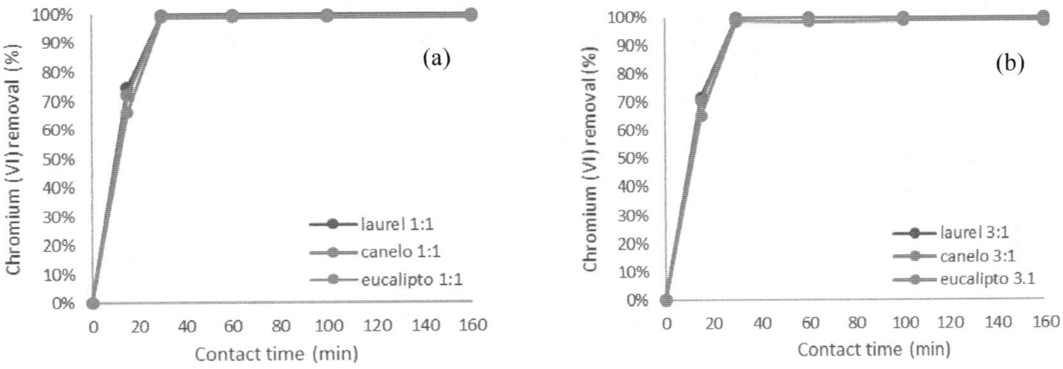

Figure 4. Chromium (VI) removal using composites with LR:MNP ratios of (a) 1:1 and (b) 3:1.

3.4. Adsorption isotherms
Table 2 shows the parameters for the fittings of the Langmuir and Freundlich adsorption isotherms to the data obtained when using the laurel composite adsorbents with LR:MNP ratios of 1:1 and 3:1. The model that best fits the chromium (VI) adsorption is Langmuir's. However, the difference between the correlation coefficients for the two models is not significant. This can be attributed to the surface heterogeneity of the lignocellulosic residues, which mostly generates multilayer adsorption, as well as the nanoparticles agglomerated on the surface of the residue that give rise to active sites with different adsorption energy capacity [7].

Table 2 also shows that the maximum adsorption capacity of the chromium (VI) ions on the laurel composite with MNP is around 30 mg/g. This demonstrates that there is a higher adsorption of the heavy metal with this type of adsorbent. As an example, Yuan et al. (2009) obtained a maximum chromium adsorption capacity with magnetite nanoparticles of 9.86 mg/g. With the composite material, the surface area was increased and therefore the maximum adsorption capacity of the heavy metal [18].

Table 2. Isotherm model parameters derived for the chromium (VI) adsorptive removal using laurel impregnated with MNP.

Model	Parameters	1:1 Laurel : MNP ratio	3:1 Laurel : MNP ratio
Langmuir isotherm	R^2	0.9989	0.9974
	K_L (L/mg)	0.809	0.724
	q_{max} (mg/g)	31.44	30.48
Freundlich isotherm	R^2	0.9920	0.9860
	K_F	15.097	13.883
	n	1.202	1.172

4. Conclusion

In this work we studied the removal of chromium (VI) ions from synthetic wastewaters, giving particular attention to the fraction of lignocellulosic residues used to obtain magnetic composite adsorbents. The results showed that the maximum contaminant removal is achieved in a shorter time using the composites, in comparison to the use of magnetite nanoparticles alone. The magnetic lignocellulosic composite that allowed the highest metal ion adsorption was the one made using laurel. Its high performance can be attributed to the higher content of lignin in its composition, with resulted in a removal of around 99.8 % for chromium (VI), with a 30 min contact time.

On the other hand, it was observed that there are no significant differences between the results of the adsorption tests performed using the lignocellulosic composites with 1:1 and 3:1 LR:MNP ratios. The adsorption efficiencies and capacities reached are very close. However, the role of the residue is very important since it provides a support for the highly effective magnetite nanoparticles, preventing them from excessive agglomeration while providing adsorption sites. As a result, the magnetite nanoparticles could be incorporated in relatively low fractions in magnetic lignocellulosic adsorbents, leading to more economical treatment methods. On this basis, further studies will be performed aiming to optimize the fraction of nanoparticles required to obtain highly efficient heavy metal removal.

5. References

[1] Vareda J P, Valente A J M and Durães L 2019 Assessment of heavy metal pollution from anthropogenic activities and remediation strategies: A review *J. Environ. Manage.* **246** 101–18

[2] Shukla A, Zhang Y-H, Dubey P, Margrave J . and Shukla S S 2002 The role of sawdust in the removal of unwanted materials from water *J. Hazard. Mater.* **95** 137–52

[3] Zhang J, Li Y, Zhou J, Chen D and Qian G 2012 Chromium (VI) and zinc (II) waste water co-treatment by forming layered double hydroxides: Mechanism discussion via two different processes and application in real plating water *J. Hazard. Mater.* **205–206** 111–7

[4] Stoica-Guzun A, Stroescu M, Jinga S I, Mihalache N, Botez A, Matei C, Berger D, Damian C M and Ionita V 2016 Box-Behnken experimental design for chromium(VI) ions removal by bacterial cellulose-magnetite composites *Int. J. Biol. Macromol.* **91** 1062–72

[5] Zamora-Ledezma C, Negrete-Bolagay D, Figueroa F, Zamora-Ledezma E, Ni M, Alexis F and Guerrero V H 2021 Heavy metal water pollution: A fresh look about hazards, novel and conventional remediation methods *Environ. Technol. Innov.* **22** 101504

[6] Aftabtalab A, Sadabadi H, Chakra C, Rao K and Mahofa E 2014 Magnetite nanoparticles (Fe3O4) synthesis for removal of Chromium (VI) from waste water *Int. J. Sci. Eng. Res.* **5** 1419–23

[7] Kapur M and Mondal M K 2016 Magnetized sawdust for removal of Cu (II) and Ni (II) from aqueous solutions *Desalin. Water Treat.* **57** 12620–31

[8] Castro D, Rosas-Laverde N M, Aldás M B, Almeida-Naranjo C E, Guerrero V H and Pruna A I 2021 Chemical Modification of Agro-Industrial Waste-Based Bioadsorbents for Enhanced Removal of Zn(II) Ions from Aqueous Solutions *Materials (Basel).* **14** 2134

[9] Lasheen M R, El-Sherif I Y, Sabry D Y, El-Wakeel S T and El-Shahat M F 2014 Removal and

recovery of Cr (VI) by magnetite nanoparticles *Desalin. Water Treat.* **52** 6464–73
[10] Padmavathy K S, Madhu G and Haseena P V 2016 A study on Effects of pH, Adsorbent Dosage, Time, Initial Concentration and Adsorption Isotherm Study for the Removal of Hexavalent Chromium (Cr (VI)) from Wastewater by Magnetite Nanoparticles *Procedia Technol.* **24** 585–94
[11] Kumar B, Smita K, Galeas S, Sharma V, Guerrero V H, Debut A and Cumbal L 2020 Characterization and application of biosynthesized iron oxide nanoparticles using Citrus paradisi peel: A sustainable approach *Inorg. Chem. Commun.* **119** 108116
[12] Kumar B, Smita K, Cumbal L, Debut A, Galeas S and Guerrero V H 2016 Phytosynthesis and photocatalytic activity of magnetite (Fe3O4) nanoparticles using the Andean blackberry leaf *Mater. Chem. Phys.* **179** 310–5
[13] Cheng Z, Gao Z, Ma W, Sun Q, Wang B and Wang X 2012 Preparation of magnetic Fe3O4 particles modified sawdust as the adsorbent to remove strontium ions *Chem. Eng. J.* **209** 451–7
[14] Abas S N A, Ismail M H S, Kamal M L and Izhar S 2013 Adsorption process of heavy metals by low-cost adsorbent: A review *World Appl. Sci. J.* **28** 1518–30
[15] Afkhami A, Saber-Tehrani M and Bagheri H 2010 Modified maghemite nanoparticles as an efficient adsorbent for removing some cationic dyes from aqueous solution *Desalination* **263** 240–8
[16] Adeli M, Yamini Y and Faraji M 2017 Removal of copper, nickel and zinc by sodium dodecyl sulphate coated magnetite nanoparticles from water and wastewater samples *Arab. J. Chem.* **10** S514–21
[17] Gupta V K and Nayak A 2012 Cadmium removal and recovery from aqueous solutions by novel adsorbents prepared from orange peel and Fe2O3 nanoparticles *Chem. Eng. J.* **180** 81–90
[18] Yuan P, Fan M, Yang D, He H, Liu D, Yuan A, Zhu J X and Chen T H 2009 Montmorillonite-supported magnetite nanoparticles for the removal of hexavalent chromium [Cr(VI)] from aqueous solutions *J. Hazard. Mater.* **166** 821–9
[19] Devaraj N K, Ong B H and Matsumoto M 2008 Yield control of chemically-synthesized magnetite nanoparticles *Synth. React. Inorganic, Met. Nano-Metal Chem.* **38** 208–11
[20] Lee B G and Rowell R M 2004 Removal of heavy metal ions from aqueous solutions using lignocellulosic fibers *J. Nat. Fibers* **1** 97–108
[21] Yasemin B and Zeki T 2007 Removal of heavy metals from aqueous solution by sawdust adsorption *J. Environ. Sci.* **19** 160–6
[22] Petcharoen K and Sirivat A 2012 Synthesis and characterization of magnetite nanoparticles via the chemical co-precipitation method *Mater. Sci. Eng. B Solid-State Mater. Adv. Technol.* **177** 421–7
[23] Slavov L, Abrashev M V., Merodiiska T, Gelev C, Vandenberghe R E, Markova-Deneva I and Nedkov I 2010 Raman spectroscopy investigation of magnetite nanoparticles in ferrofluids *J. Magn. Magn. Mater.* **322** 1904–11
[24] Panneerselvam P, Morad N and Tan K A 2011 Magnetic nanoparticle (F3O4) impregnated onto tea waste for the removal of nickel(II) from aqueous solution *J. Hazard. Mater.* **186** 160–8
[25] Sundrarajan M, Ramalakshmi M and Nadu T 2012 Novel Cubic Magnetite Nanoparticle Synthesis Using Room Temperature Ionic Liquid **9** 1070–6
[26] Nethaji S, Sivasamy A and Mandal A B 2013 Preparation and characterization of corn cob activated carbon coated with nano-sized magnetite particles for the removal of Cr(VI) *Bioresour. Technol.* **134** 94–100

Acknowledgments
The authors acknowledge the support provided by the project grant PIGR-19-05, from the Escuela Politécnica Nacional.

IOP Conf. Series: Earth and Environmental Science **897** (2021) 012021 doi:10.1088/1755-1315/897/1/012021

Optimal Planning Method of Integrated Energy System Considering Carbon Cost from the Perspective of the Whole Life Cycle

Bin Bian[1,*], Zhihuan Du[1], Kui Zhou[2], Tao Huang[2], and Fengbo Lv[2]
[1]State Grid Beijing Electric Power Company Economic and Technical Research Institute, 102209, Beijing, China
[2]Tsinghua Sichuan Energy Internet Research Institute, Chengdu 610213, Sichuan, China

* Corresponding author: 1570126408@qq.com.

Abstract. China commits its goal of peak carbon dioxide emissions before 2030 and achieving carbon neutrality before 2060. The integrated energy system (IES) is one of the critical approaches to achieving the commitments. While the prevailing evaluation method for calculating the carbon emissions of IES neglected parts of factors influence, the result could not reflect the carbon emissions comprehensively. Considering the insufficiency above, in this paper, the evaluation method of carbon emission based on the whole life cycle of IES is proposed. First, based on the IES energy hub model, a typical park's carbon emission model has been established. Then, the carbon emission coefficients of energy and equipment in production, transportation and operation are analysed, respectively. Hence, a low-carbon operation optimisation model of the IES is proposed. Later, with the lowest annual carbon emission of the integrated energy system as the optimisation target, the IES's optimal carbon emission allocation and operation plan are proposed, based on the balance between energy supply and demand in the process of energy and equipment use and operation. As a result, the carbon emission of the IES of the park reduces effectively.

1. Introduction

China has repeatedly stated to the world on international occasions its goal of peak carbon dioxide emissions before 2030 and achieving carbon neutrality before 2060, emphasising its commitment to green and low-carbon development [1]. The 13th Five-Year Plan for Electricity Development states that China's power development should gradually focus on the development of environmentally friendly, safe and highly productive renewable energy systems, optimizing the capacity allocation of units and absorbing overcapacity [2]. China's power industry accounts for about 50% of the country's total CO_2 emissions [3]. Therefore, increasing the proportion of low-carbon renewable energy in the power industry, improving energy use efficiency, optimising unit capacity allocation, as well as absorbing overcapacity, is conducive to the development of China's low-carbon economy [2].

The integrated energy system is based on the complementary characteristics of electricity, heat, gas and other energy types as well as the principle of energy gradient utilization [4], and provides unified planning and coordinated and optimized operation of multi-energy systems, so that the production, transmission, distribution, transformation, storage and consumption of different energy sources can be organically coordinated under a multi-temporal scale, which helps to reduce the cost of energy use and improve energy utilization efficiency [5]. In recent years, China has been paying more attention to energy conservation and environmental protection, and the proportion of total clean energy consumption to total energy consumption has been increasing in recent years, which has formed a trend of clean energy gradually replacing high pollution and high energy consumption traditional energy, IES can forecast energy supply and demand, improve the utilization rate of clean energy, and promote the large-scale consumption of clean energy [6]. IES uses a variety of heterogeneous energy supply, can accept a variety of clean energy, can form a single or multiple integrated energy system supply system according to the area of individual use area, so that IES has flexibility, reliability, low carbon and scalability [7]. In the context of low carbon development, along with the continuous research on IES, how to develop low carbon IES has become a hot issue in current research. With the goal of carbon peaking and carbon neutrality, the requirements for low-carbon IES should also be further improved. In order to achieve low carbon operation, the integrated energy system of the park needs to consume as much low carbon renewable energy as possible while meeting normal operation. IES can greatly enhance the flexibility of the system, and the cooling, heating and electricity demand of end-users can be met by various forms. When the theoretical output of renewable energy is greater than the space for consumption, it can be stored directly through electric storage, or it can be used to supply heat to users through electric-to-heat conversion, and in the case of surplus heat supply, electricity can be stored indirectly through electric-to-heat conversion + storage, which in turn enhances the flexibility of the whole system in terms of electricity consumption and the level of renewable energy consumption, thereby reducing overall carbon emissions.

Domestic and foreign scholars have started researches on the low-carbon operation of IES, but issues including carbon emission transfer in the process of energy coupling and transformation, and the promotion of the optimal operation of IES from the perspective of low carbon, remained to be further studied. Presently, the carbon trading market is being gradually promoted throughout the country. By establishing legal carbon emission rights, carbon trading is able to trade such rights, then costs are followed with carbon emission and benefits are followed with carbon emission reduction. Such mode is a trading mechanism that can effectively reduce carbon emission [8]. However, there are still few studies on the impact of carbon trading price on IES.

Life Cycle Assessment (LCA), which covers the whole life cycle of products, originated from the European and American research on industrial products in the 1970s [9]. Based on the law of conservation of energy and the law of conservation of mass, the life cycle assessment could be used to calculate the carbon emissions generated by the energy and equipment used by IES during the production, transportation and applied stages, leading to a more accurately determination on the total carbon emission of the system [10]. Such a method therefore is significant for comprehensively assessing the actual IES carbon emissions.

2. Carbon emission calculation method of IES based on LCA

The essence of an integrated energy system is to apply multiple energy technologies to realize the close connection between multiple energy supply and diversified energy demand, as shown in Figure 1. These technologies cover multiple links of energy production, energy conversion, energy consumption. The traditional calculation method takes electricity and gas purchased as the two major carbon sources, calculated the overall carbon emission of the system as the consumption of corresponding fuel (electricity/natural gas) multiplied by the corresponding emission factor.

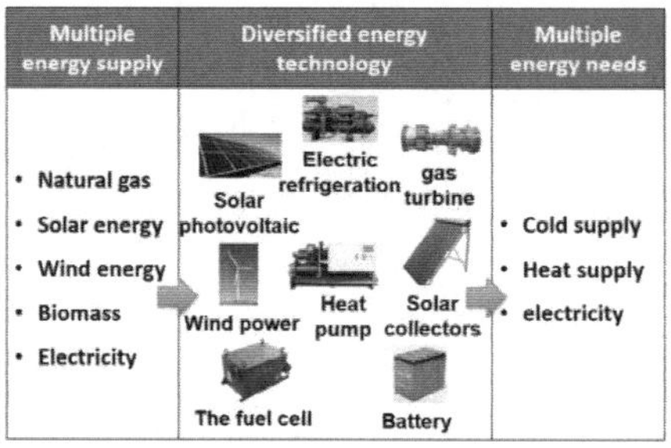

Fig. 1. The connection between the supply and demand of the integrated energy system.

However, the total carbon emissions of the IES cannot be adequately characterised if only the carbon emissions from fossil energy generation are considered. For this reason, it is necessary to introduce the LCA energy chain carbon analysis method, which aims to analyse the system not only in terms of the direct greenhouse gas emissions generated by the system activity itself or by the input materials, but also in terms of the accompanying effects of the operation of the system or the material-related transformations, i.e. the total carbon emissions generated per unit of energy provided by the IES.

Firstly, the greenhouse gases emitted by IES were determined as metering pollutants, and every energy source in IES was divided into energy chains according to the order of energy activities. Then, the transmission links are summarized and simplified. Finally, the carbon emission coefficient generated by energy activities in every link is measured. The main measurement sources on the power generation side are coal power, new energy power generation and natural gas, while carbon emissions in the secondary energy's (such as electricity) transmission and distribution links are not taken into consideration since the energy storage is mainly focused on the electricity consumption side.

The life-cycle carbon emission accounting formula is based on the calculation of conventional carbon emission, and its value is equal to the sum of the carbon emissions in every stage, which can be expressed as:

$$C1 + C2 + ... + Cn \tag{1}$$

where, C represents the total carbon emission in the whole life cycle; C1, C2,... and Cn represent the corresponding carbon emission value in each emission process.

(1) Life cycle carbon emissions of coal power

LCA greenhouse gas emissions from coal power calculating scope involves three parts: coal production, coal transportation and coal-fired power generation [13]. Emissions from mining are mainly coal bed methane emissions, coal spontaneous combustion and energy consumption emissions, emissions from transportation are mainly energy consumption emissions from coal transportation, and emissions from power generation are mainly emissions from coal-fired power generation and related auxiliary activities.

(2) Life cycle carbon emissions of wind power and photovoltaics

The greenhouse gas emissions of wind power, photovoltaic and other new energy LCA can be divided into two segments: production and construction, and transportation. The greenhouse gas emissions of the operation segment mainly come from the energy consumption of transportation and equipment maintenance during the production and replacement of consumables, and this part of emissions is negligible compared to the emissions of the production and construction phase. There have been many studies on the life cycle carbon emissions of wind power and photovoltaic [14]-[17], and this study cites Guo [16], who concluded that the life cycle carbon emission factors of wind power and photovoltaic are 9.5 gCO_2/kWh and 86 gCO_2/kWh respectively.

(3) Carbon emissions during the life cycle of natural gas

Natural gas by source includes oilfield gas and coal gas. Oilfield natural gas LCA greenhouse gas emissions include the extraction, transportation and use of natural gas, while the life cycle of coal gas

IOP Conf. Series: Earth and Environmental Science **897** (2021) 012021 doi:10.1088/1755-1315/897/1/012021

includes coal mining, transportation, conversion, natural gas transportation and use, and there are differences in their carbon emission factors, which are 0.48 kg CO_2/kWh [18] and 0.54 kgCO_2/kWh respectively [18]. To simplify the calculation, the life-cycle carbon emission factor for natural gas is 522 gCO_2/kWh, weighted according to the composition of natural gas market in China at this stage [19].

(4) Life cycle carbon emissions of energy storage equipment

Since the indirect use process of electrical energy storage equipment has a large carbon footprint throughout its life cycle, and waste disposal is difficult, this part of the carbon footprint should also be measured in the total carbon footprint for IES. Therefore, the life-cycle boundary of energy storage equipment is defined as the cradle-to-gate process, which takes into account the production, construction, transportation and waste recycling [20].The actual operating life of the energy storage system is related to its selected battery model, frequency modulation participation, etc., which affects the determination of the replacement coefficient. This article uses typical lithium iron phosphate battery parameters to model the carbon emissions of energy storage equipment in the park. In this scenario, depth of discharge (DOD)=80%, charge-discharge efficiency 90%, estimated cycle life n=2500 times, daily charge-discharge times m= 1. The replacement coefficient is approximately 0.15, so that the indirect emission of the electric energy storage equipment during use is 13.7 gCO_2/kWh. Energy storage equipment waste is generally entrusted to relevant companies for professional treatment, which is generally divided into centralized landfill and metal recycling. In order to simplify the calculation, it is assumed that the energy storage equipment waste is uniformly commissioned by a site 200km outside the park. It is further obtained that the electrical energy storage equipment is being discarded. The carbon emissions produced during the process are 7 gCO_2/kWh. Finally, the life cycle carbon emission coefficient of energy storage equipment is 112 gCO_2/kWh.

The LCA emission coefficients of each carbon source are shown in Table 1:

Table 1. LCA emission coefficients of various carbon sources in the integrated energy system

Carbon source	Carbon emission factor		sum	unit
	Construction and transportation links	Operation/ generation link		
Coal power	119.2	856.0	975.2	gCO_2/kWh
Wind power	9.5	0	9.5	gCO_2/kWh
Photovoltaic	86	0	86	gCO_2/kWh
natural gas	164	357	522	gCO_2/kWh
Energy storage	91.3	20.7	112	gCO_2/kWh

3. Optimal model of IES Considering Carbon Cost

3.1 The mathematical model of IES

The model of IES is described as Energy Hub (EH) [11] shown in Figure 2. This architecture can accurately describe the energy flow and transformation relations in different links of the integrated energy system, but the structure is relatively complex, and the optimization solution needs to be determined both among different links and among different technical paths in the same link, resulting in a difficult implantation of modelling and solving. Therefore, global optimal is tough to realize.

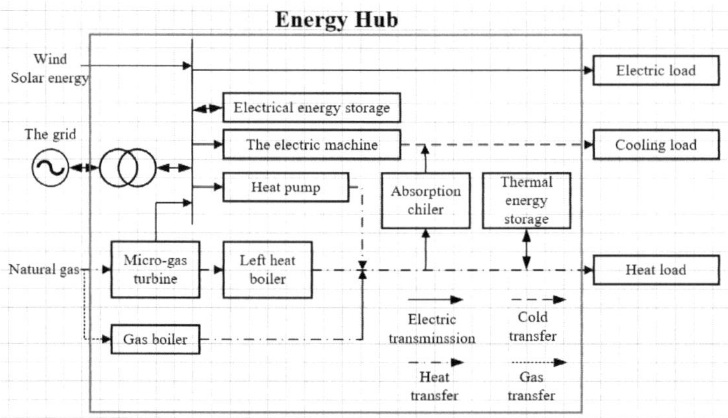

Fig. 2. Energy concentrator model of the park's IES

This paper solves the problem by assuming that the input energy of the EH has M kinds and the output energy has N types, the energy conversion matrix can be used to describe the energy conversion relationship, as follows:

$$\begin{pmatrix} E_1^{out} \\ E_2^{out} \\ \mathrm{M} \\ E_N^{out} \end{pmatrix} = T \begin{pmatrix} E_1^{in} \\ E_2^{in} \\ \mathrm{M} \\ E_M^{in} \end{pmatrix} \tag{2}$$

Where, E_i^{out} represents the energy output of the EH; E_i^{in} represents the energy input of an EH; T is an N by M matrix.

According to the direction of energy flow, the integrated energy system can generally be divided into energy production, energy conversion, energy storage, energy transmission, and energy use. It is necessary to carry out mathematical modelling for the key equipment of each link, and clarify the constraints, mainly including System energy balance constraints, energy storage equipment constraints, interaction power constraints with the main network, controllable unit constraints, climbing constraints, and gas source point constraints.

(1) Energy production

The energy production link of the integrated energy system is mainly the process of converting primary energy such as wind energy and solar energy into electric energy. This process can be expressed as::

$$e_{f,m,hr}^{g,in} = \sum_{egt} e_{egt,f,m,hr}^{g,in} \tag{3}$$

Among them, $e_f^{g,in}$ refers to the primary energy consumption; $e_f^{\max}$ is the available primary energy; f refers to different types of primary energy; m refers to different months, and hr refers to each hour of the day.

The consumption of solar and wind energy is limited by the available amount of local resources, and cannot exceed the maximum available amount of resources at that moment. The above relationship can be expressed as:

$$e_{f,m,hr}^{g,in} \le e_{f,m,hr}^{\max} \tag{4}$$

(2) Energy conversion

Energy conversion equipment includes gas turbines, heat pumps, electric refrigerators, etc. There is a certain efficiency in the conversion between different energies, and usually the range of change is not large when the equipment is running stably. Therefore, the mathematical model of the energy conversion unit can be simply expressed as:

$$q_t^{out} = q_t^{in}\eta(P) \tag{5}$$

Among them, q_t^{out} and q_t^{in} respectively represent the output and input of a certain energy conversion equipment, $\eta(P)$ represents the energy conversion efficiency, and its value is related to the load factor P.

Certain energy conversion equipment, such as heat pumps/refrigerators, usually have different operating modes in winter and summer, namely heating conditions and cooling conditions. The coefficient of performance (COP) of the heat pump/refrigerator is different under heating and cooling conditions. Therefore, the input and output relationship is as follows:

$$e_{ter,m,hr}^{c,out} = \sum_{ect} e_{ect,sec,m,hr}^{c,in} \cdot (j_{m,hr}^{heating} \cdot \eta_{ect,sec,ter}^{heating} + j_{m,hr}^{cooling} \cdot \eta_{ect,sec,ter}^{cooling}) \tag{6}$$

Among them, ter refers to different tertiary energy carriers, namely heat, electricity and cold, $ter \in \{p,h,c\}$, c refers to cold; $e_{ter}^{c,out}$ denotes the output of tertiary energy carriers produced in the energy conversion process; $\eta_{ect,sec,ter}^{heating}$ and $\eta_{ect,sec,ter}^{cooling}$ are energy conversion technologies in heating mode And the coefficient of performance in the cooling mode; $j^{heating}$ and $j^{cooling}$ are 0-1 variables, which indicate whether the heat pump/refrigerator is working under what conditions.

(3) Energy transmission

The integrated energy system of the park mainly has two kinds of energy transmission: electricity and heat. The power transmission loss of the integrated energy system of the park is almost negligible, so the energy transmission link is mainly considered Hot net. As a network system for heat energy transmission and exchange, the heating network reflects the changes between the temperature and the heat medium in the node. At present, the transmission medium of domestic heating system is mainly hot water [12].The heating network model is mainly to model the heat energy loss. According to the basic principle of thermal steady-state heat transfer, the thermal energy loss is:

$$\Delta Q = 2\pi \frac{T_H - T_e}{\sum R} l \tag{7}$$

Among them, ΔQ is the pipeline heat loss during transmission, T_H is the water supply temperature, T_e is the average temperature of the medium around the pipeline, R is the average thermal resistance per unit length of the pipeline, and l is the length of the pipeline.

(4) Energy storage

The entrance energy of the energy storage link is equal to the export energy of the energy conversion link. Energy storage equipment can increase the freedom of operation of energy production technology and energy conversion technology. The energy generated in the energy conversion link is equal to the sum of the export energy of the energy storage equipment and the net energy storage of the corresponding energy storage equipment. Among them, the net energy storage refers to the difference between the energy released by the energy storage device and the energy stored. The energy balance of energy storage equipment can be expressed as:

$$e_{ter,m,hr}^{s,out} = e_{ter,m,hr}^{s,in} + e_{ter,m,hr}^{-} - e_{ter,m,hr}^{+} \tag{8}$$

IOP Conf. Series: Earth and Environmental Science **897** (2021) 012021 doi:10.1088/1755-1315/897/1/012021

Among them, $e_{ter}^{s,out}$ refers to the outlet energy of the energy storage device; e_{ter}^{-} refers to the energy released by the energy storage device; e_{ter}^{+} refers to the energy stored in the energy storage device.
At each moment, the total stored energy in the energy storage device changes with the net stored energy, which can be expressed by the following formula:

$$E_{ter,m,hr+1} = E_{ter,m,hr} + e_{ter,m,hr}^{+} - e_{ter,m,hr}^{-} / \eta_{es,ter} \geq 0 \tag{9}$$

Among them, E_{ter} represents the energy contained in the energy storage device; $\eta_{es,ter}$ represents the efficiency of the energy storage device. The energy exported from the energy storage link must be greater than the terminal demand at each moment.

3.2 The objective function of the model

In previous studies on the optimal operation of integrated energy systems, most of the goals have been to minimize the total annual cost of the system. The total annual cost includes investment costs, operation and maintenance (O&M) costs, and fuel costs. The investment cost is determined every year from the depreciation of the total equipment investment to its operating period, as shown in the following formula:

$$Min\, f = C_{capital} + C_{O\&M} + C_{fuel} \tag{10}$$

Investment cost, operation and maintenance (O&M) cost and fuel cost are respectively determined by the following three formulas:

$$C_{capital} = \sum_{tech} Inv_{tech} \cdot Cap_{tech} \cdot \frac{I}{1-(1+I)^{-Y_{tech}}} + \sum_{es} Inv_{es} \cdot Cap_{es} \cdot \frac{I}{1-(1+I)^{-Y_{es}}} \tag{11}$$

$$C_{O\&M} = \sum_{tech} (OM_{tech} \cdot \sum_{m} \sum_{hr} e_{tech,m,hr}^{g} \cdot d_m) + \sum_{es} OM_{es} \cdot Cap_{es} \tag{12}$$

$$C_{fuel} = \sum_{f} \sum_{m} \sum_{hr} (e_{f,m,hr}^{g,in} \cdot P_{f,hr} + e_{p,m,hr}^{gp} \cdot P_{gp,hr}) \cdot d_m \tag{13}$$

Among them, Inv represents the unit investment cost of the equipment; Y represents the life of the equipment; I represents the depreciation rate, which is set to 10% in this study; OM represents the unit operating cost; dm represents the number of days per month; Pf, hr and Pgp , hr respectively represent the hourly fuel price and grid electricity price.
In previous studies, operation and maintenance costs mainly considered fuel/electricity costs, as well as system maintenance, operation and maintenance. With the introduction of carbon peak and carbon neutral goals, as China proposes a carbon peak and carbon neutral vision, the use of market means to reduce carbon emissions has attracted much attention, and the carbon emissions cost of the integrated energy system should also be integrated into the objective function for optimization.
The carbon trading mechanism is a mechanism that treats carbon emissions as a commodity. By assigning a certain carbon emission credit to each carbon emission source, if the actual carbon emission of the carbon emission source is greater than the allowance, you need to go to the carbon trading market to purchase the excess credit; Conversely, when the actual carbon emissions of the carbon emission source are less than the allocated amount, the remaining carbon emissions can be sold as commodities on the market. Under the carbon emission trading mechanism, carbon emission companies will spontaneously look for production or operation methods with lower carbon emissions in order to maximize their own interests. It not only promotes the energy saving and emission reduction of enterprises, but also makes carbon emissions have economic value.

IOP Conf. Series: Earth and Environmental Science **897** (2021) 012021 doi:10.1088/1755-1315/897/1/012021

This study combines the carbon trading mechanism to introduce the cost of carbon emissions into the total cost of the integrated energy system, and provides innovative ideas for the optimization of the integrated energy system that considers economy and low-carbon behavior.
The carbon transaction cost model of the integrated energy system is as follows:

$$C_{co_2} = S_{co_2,t}\left(E_{total} - E_{rate}\right) \tag{14}$$

Among them, $S_{co_2,t}$ is the trading price of carbon emission rights in period t, E_{total} and E_{rate} are the total emissions of the system and the given carbon emission allowances, respectively.

The objective function of the integrated energy system optimization taking into account the carbon cost is:

$$Min\, f = C_{capital} + C_{O\&M} + C_{fuel} + C_{co_2} \tag{15}$$

4. Model solving method

The carbon emission model constructed in this paper can be solved by using LINGO for mixed certificate integer programming. The steps are as follows:

- Step 1: Determine the type of input energy and output energy;
- Step 2: When calculating unit time *t*, the output energy of *i* input energy in each stage of production, conversion, transmission and storage;
- Step 3: Calculate the total output energy S of each energy link for 8760 hours a year, and output the optimal result when the adjacent results meet the error value.

The specific flow chart is shown in Figure 3:

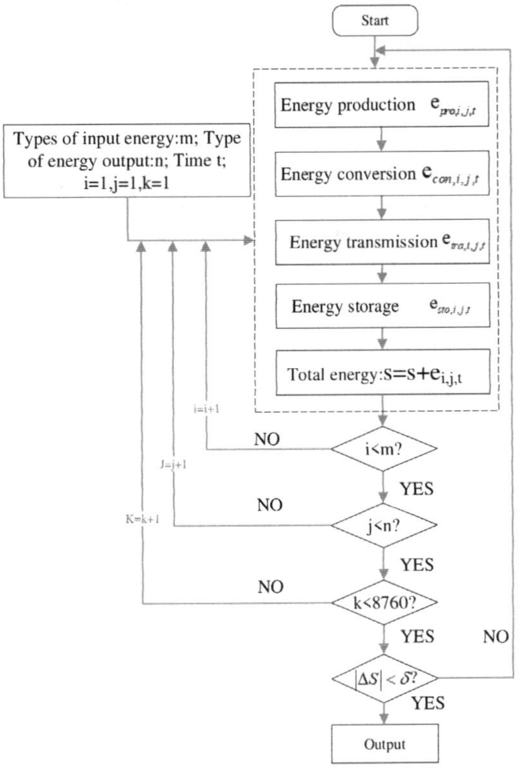

Fig. 3. Flow chart of solving carbon emission model

IOP Conf. Series: Earth and Environmental Science **897** (2021) 012021 doi:10.1088/1755-1315/897/1/012021

5. Case analysis

5.1 Case Introduction

The example in this paper is based on a typical park with a design life of 25 years and energy demand for cooling, heat and electricity. The energy production units in the park's integrated energy system are photovoltaic power generation units, fans and gas boilers, the energy conversion units are electric compression chillers, the energy storage equipment is storage batteries, the equipment parameters are shown in Table 2, and the topology of the park's integrated energy system is shown in Figure 4.

Table 2. Design equipment parameters of the park's integrated energy system

Equipment name	Investment costs(Yuan/kW)	Maintenance costs(Yuan/kW)	Effectiveness
Photovoltaic power generation	4000	0.005	—
Wind turbine generator	3000	0.005	—
Gas boiler	130	0.05	0.8
Electric refrigeration	1023	0.01	4.0
Battery	1259	0.01	0.9

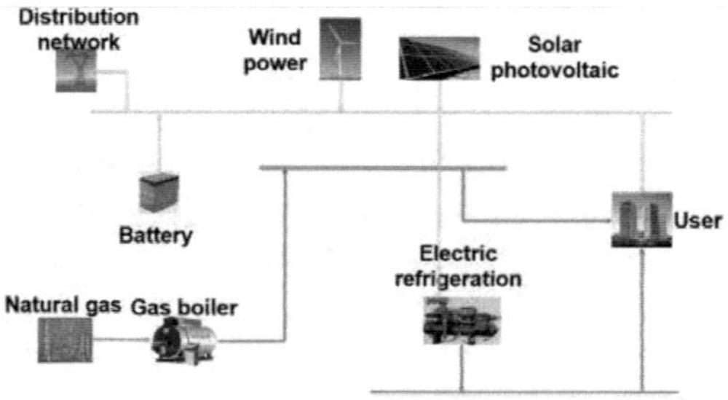

Fig. 4. Topology diagram of the park's IES

The calorific value of natural gas is 36MJ/Nm³, and the purchase cost is 3.3 Yuan/Nm³, and its CO_2 emission is 1.8kg/Nm³. The electricity purchase price of the system adopts the local time-of-use electricity price. The specific price is shown in Table 3. The energy used for power generation from the external grid is a combination of coal, nuclear, biomass, natural gas, renewable energy, etc. , The unit calorific value is 29.270MJ, and the unit comprehensive energy is used for power generation to produce 3.411kg CO_2.

Table 3. Time-of-use electricity price

Time period	Electricity price（Yuan）	time
Peak price	1.4	10:00-14:00;18:00-20:00
Parity price	0.87	07:00-09:00;15:00-17:00 21:00-22:00
Valley price	0.37	00:00-06:00;23:00-00:00

The typical daily power load curve of the park is shown in Figure 5. The load increases slightly from 0:00 to 6:00 There was a slight decrease, and the load rebounded slightly from 18:00 to 19:00, and then from 19:00 to 23:00, the load showed an overall downward trend.

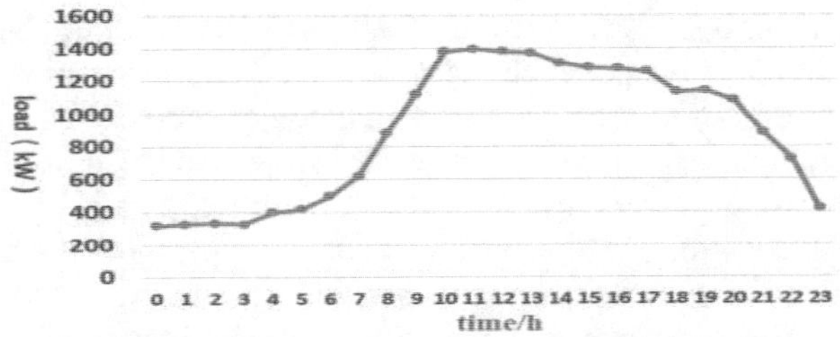

Fig. 5. Typical daily power load curve in the park

The heating load curve of the heating season in the park is shown in Figure 6. The load increases hourly from 0:00 to 13:00, reaches the maximum around 13:00, and then the load starts to decrease slowly, and the heating load from 13:00 to 23:00 shows a downward trend.

Fig. 6. Typical daily heating load curve in the park

The cooling load curve in the cooling season of the park is shown in Figure 7. The load decreases slightly from 0:00 to 4: 00, the load increases gradually from 4: 00 to 15: 00, reaches the maximum at 15: 00, and the load shows a downward trend from 15: 00 to 23: 00.

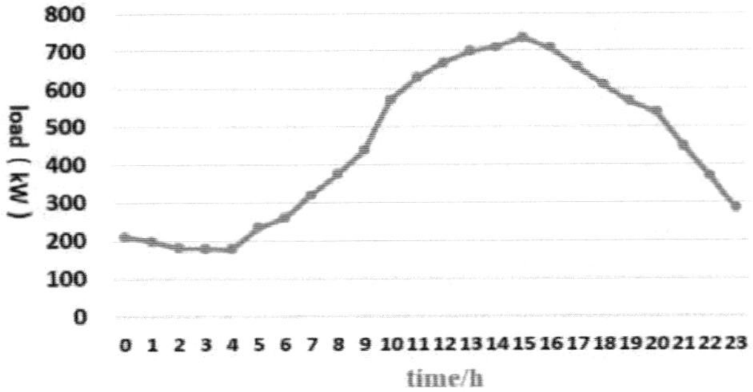

Fig. 7. Typical daily cooling load curve in the park

5.2 Scene setting

In order to deeply analyse the impact of the full life cycle carbon cost on the planning and design of the park's integrated energy system, combined with the objective function in Section 3.2, this study constructs three scenarios for analysis by introducing different objective functions:

- Scenario 1: In the integrated energy system optimization design model, the objective function only considers the main equipment investment, fuel consumption, power purchase costs, and operation and maintenance costs, and does not consider carbon emissions/carbon transaction costs.
- Scenario 2: Incorporate carbon emissions/carbon transaction costs into the objective function of the integrated energy system optimization planning and design, and consider the corresponding costs caused by direct fuel emissions and indirect emissions from external power sources. Among them, the carbon emission coefficient of fuel emissions comes from

the "Provincial Greenhouse Gas Inventory Compilation Guidelines", and the indirect emission coefficient of external power sources is determined by the power structure of the regional power grid and the emission coefficients of each power source type.

- Scenario 3: Incorporate carbon emissions/carbon transaction costs into the objective function of the optimal planning and design of the integrated energy system, where the carbon emissions of the system are considered from the perspective of the entire life cycle. Considering the full life cycle carbon emissions of energy and energy storage equipment during production, transportation and operation, it is also converted into carbon emission costs and calculated into the system cost. The relevant carbon emission coefficients are shown in Table 1.

5.3 Optimization Results

Taking economy and environmental protection (carbon emissions) as the target constraints, combined with local meteorological data, the integrated energy system constructed this time was simulated for 8760 hours, with a design life of 25 years, and the optimal configuration plan for the park's integrated energy system was obtained. The simulation results are as follows: the main equipment investment cost is 4,792,500 yuan, the annual maintenance cost is 148,900 yuan, the annual natural gas consumption is 1,028,700 yuan, the annual electricity consumption is 10.073 million kWh, and the annual power purchase cost is 6,389,600 yuan. The prediction curve is shown in Figure 8. The equipment design and installation capacity of the optimized plan is shown in Table 4.

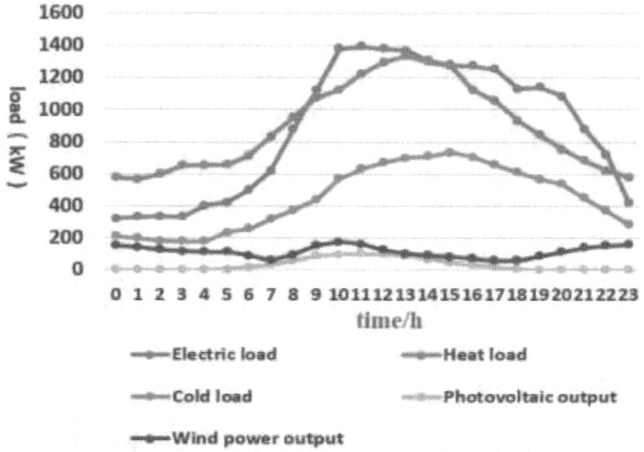

Fig. 8. Forecast curve of park daily load and wind output

Table 4. Design and installation capacity of equipment

Equipment name	Design capacity(kW)
Photovoltaic power generation	179.2
Wind turbine generator	420
Gas boiler	1800
Electric refrigeration	750
Battery	958.1

5.4 Analysis of the results of the program

This article analyzes the cost of the optimal configuration plan according to three scenarios, and the specific results are shown below.

1) Scenario 1 only considers the investment cost of the main equipment of 4.792 million yuan, the annual natural gas consumption of 1.0287 million yuan, the annual electricity purchase cost of 6.3896

million yuan, and the operation and maintenance cost of 148,900 yuan. Obtained the comprehensive cost of the park's integrated energy system scenario 1 is 7.7589 million yuan。

2) Scenario 2 calculates the carbon emissions from direct fuel emissions and indirect emissions from external power sources on the basis of comprehensive cost considerations, and uses 10.0723 million kilowatt-hours of purchased electricity annually. External power sources consider the energy structure of local power generation. Coal power generation accounted for 65.97%, natural gas power generation accounted for 13.24%, wind power generation accounted for 8.53%, photovoltaic power generation accounted for 12.25%, and its operating power generation carbon emission coefficient refers to Table 1, and the cost of each carbon source under scenario 2 is shown in Table 5. Combining the carbon emission coefficients of the operation/power generation link in Table 1, the carbon emission price combined with the local transaction average price is 28.49 yuan/ton, and the annual carbon emission price of the system operation/power generation link under Scenario 2 is 201,300 yuan, and the comprehensive cost is 7.9602 million yuan.

Table 5. Carbon emissions of each carbon source operating power generation link under Scenario 2

Carbon source	External power purchase（kWh）	System internal load（kWh）	Carbon emissions（t）
Coal power	6746200	-	5687.86
wind power	872300	804000	0
Photovoltaic	1252700	292000	0
natural gas	1353900	2493800	1366.37
battery	-	562000	11.63

3) Scenario 3, on the basis of scenario 2, considers the carbon emissions of each carbon source in the integrated energy system during the production and transportation process, including coal power, photovoltaic, wind power, consumed natural gas and energy storage equipment, according to the carbon emission coefficient in Table 1. Converted to the cost of the integrated energy system, it is obtained that under Scenario 3, the annual carbon emissions of each carbon source in the construction and transportation link total 1618.07 tons, as shown in Table 6. Combined with the comprehensive price of carbon trading, the annual carbon emission cost of each carbon source construction and transportation link under scenario 3 is 46,100 yuan.

Table 6. Emissions from carbon construction and transportation links of various carbon sources under Scenario 3

Carbon source	Carbon emissions（t）
Coal power	792.05
wind power	15.80
Photovoltaic	131.22
natural gas	627.69
battery	51.31
total	1618.07

The comprehensive cost of the park's integrated energy system under the three scenarios is shown in Figure 9:

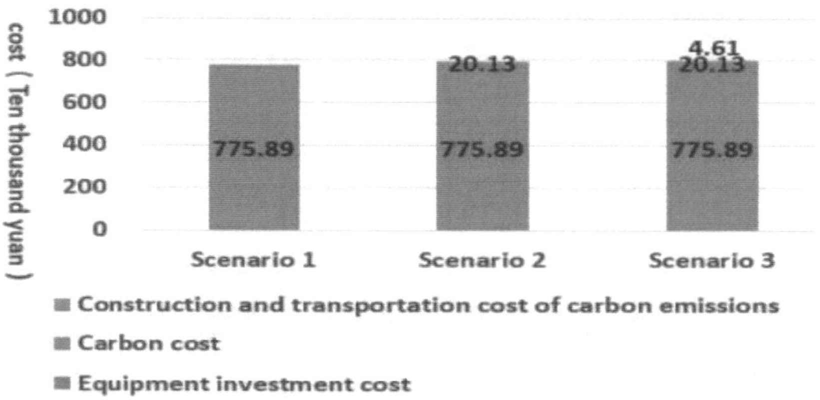

Fig. 9. The comprehensive cost of the park's IES under the three scenarios

In Scenario 3, the comprehensive annual cost of the system is RMB 8.0063 million. The cost of carbon emissions during the entire life cycle is RMB 247,400, accounting for 3.1% of the total cost. This cost ratio will be compared with the cost of carbon emissions during the production, transportation and operation of power generation. Changes in carbon emissions prices are positively correlated.

6. Conclusion

This paper establishes a low-carbon operation optimization model of the integrated energy system. On the basis of satisfying the energy balance of supply and demand, proposes the optimal configuration and operation plan for the carbon emission of the integrated energy system of the park, and uses the carbon emission analysis method of the entire life cycle energy chain to calculate this time. The research park contains the comprehensive cost of carbon emissions throughout the life cycle.

1) The optimization scheme of the integrated energy system of the park proposed in this paper proposes a low-carbon configuration and operation scheme under the premise of meeting the balance of energy supply and demand in different seasons and periods. On the basis of ensuring the safe and stable operation of the system, the integrated energy system achieves high environmental benefits.

2) This article fully considers the full life cycle carbon emissions of the energy and equipment used in the park's integrated energy system during the production, transportation, and operation of power generation. The established low-carbon operation optimization model of the integrated energy system has strong applicability .

3) The minimum time scale of the model built in this paper is 1h, and the response capacity of the integrated energy system equipment can reach the minute or second level. How to optimize the park's integrated energy system in a smaller time scale will be a further step.

7. Reference

[1] GAO Hu. Efforts to promote high-quality energy development and accelerate the transformation of energy green, low-carbon and intelligent [J]. China Energy,2020,42(12):4-5.

[2] YUAN Jiahai，ZHANG Kai. Research on my country's Coal Power Development Transition and Future Forecast in the Middle and Late Period of the "13th Five-Year Plan" ［J］. China Coal，2019，45(8) : 13-19.

[3] MEI Tianhua，BIAN Qiaoyan，XIE Jun， et al. Low-carbon power dispatch and revenue sharing considering carbon emission rights ［J］. Automation of Electric Power Systems，2016，40(22) : 49-55.

[4] LI jinghua， ZHU Mengshu， LU Yuejiang， et al. Summary of Optimal Dispatch of Integrated Energy System . Power grid technology. https://doi.org/10.13335/j.1000-3673.pst.2021.0020.

[5] PAN Hua, LIANG Zuofang, XIAO Yuhan, et al. Optimal operation of regional integrated energy system in multiple scenarios [J]. Acta Solar Energy,2021,42(01):484-485.
[6] ZHU Chenguang, TIAN Yuan, GUAN Shaofeng, et al. Regional integrated energy system supply and demand forecasting and optimized operation technology [J]. China High-tech. 2020,(13):109-110
[7] SONG Zehui, LV Chaogang, DENG Pengyuan, et al. Integrated Energy System Overview [J]. Chemical Design Newsletter,2019,45(10):209]
[8] ZHANG Xiaohui, YAN Keke, LU Zhigang, et al. Low-carbon economic dispatch of wind power system based on carbon trading [J]. Power grid technology. 2013,37(10):2698
[9] HUI Jingxuan, WAN Liyang. Life cycle assessment method and its application in the field of renewable energy research in my country [J]. China Energy, 2020,42(03):42-43.
[10] SUN Lili. Research on Energy Cost of Power Generation—Analysis Based on Life Cycle Assessment Theory [J]. Price theory and practice. 2018,(05):152.
[11] Geidl M, Andersson G. Optimal Power Flow of Multiple Energy Carriers[J]. IEEE Transactions on Power Systems, 2007, 22(1): 145-155.
[12] LI Xiaolu, SHAN Fuzhou, SONG Yanmin, et al. Optimal dis-patch of multi-region integrated energy systems consideringheating network constraints and carbon trading [J]. Automationof Electric Power Systems, 2019, 43 (19): 52-59.
[13] XIA Dejian,REN Yulong,SHI Yuefeng. Life Cycle Carbon Emission Coefficient Measurement of China's Coal Power Energy Chain [J]. Statistical Research,2010,27(08):82-89.
[14] YANG Dong,LIU Jingru,YANG Jianxin, et al. Analysis of Carbon Footprint of Wind Turbine Based on Life Cycle Assessment [J]. Journal of Environmental Science, 2015,35(03):927-934.
[15] JIA Yalei,WANG Jixuan,HAN Zhonghe, et al. Environmental load analysis of wind power, photovoltaic power generation and coal-fired power generation based on LCA [J]. Journal of Power Engineering,2016,36(12):1000-1009.
[16] GUO Minxiao. Life cycle CO_2 emission accounting for wind, photovoltaic and biomass power generation [D]. Tsinghua University,2012.
[17] WENG Lin,CHEN Jianbo. Analysis of the environmental and economic benefits of photovoltaic systems based on the calculation of carbon emissions throughout the life cycle [J]. Journal of University of Shanghai for Science and Technology,2017,39(03):282-288.
[18] FU Zihang. Life Cycle Analysis and Horizontal Comparison of Carbon Emissions from Coal-to-Natural Gas [J]. Natural gas industry,2010,30(09):100-104+130.
[19] FAN Jinlu,WU Lixin,REN Shihua. Comparative Research on the Cost of Coal-fired Power Generation and Natural Gas Power Generation under the Constraint of Carbon Emission Reduction [J]. China Coal,2016,42(12):14-17+23.
[20] TAN Yanqiu. Life cycle greenhouse gas impact analysis of battery energy storage systems in power system applications [D]. Nanjing University,2017.

Acknowledgments

Supported by the Science and Technology Project of State Grid Beijing Electric Power Company.

IOP Conf. Series: Earth and Environmental Science **897** (2021) 012022 doi:10.1088/1755-1315/897/1/012022

Assessment of Energy Conservation Resource considering the Lighting service in Academic buildings seeking sustainable Energy planning

André Sotero[1], Miguel Udaeta[1], André Gimenes[1] and Luiz Fernando Kurahassi[2]

[1]Energy Group of Department of Energy and Electrical Automation Engineering (GEPEA). Polytechnic School, University of São Paulo, Brazil.
[2]Energy Group of Department of Energy and Electrical Automation Engineering (GEPEA). Federal Institute of São Paulo, Brazilype the author addresses here

Abstract. The objective of this work is to evaluate the full potential, since include technic-economic & environmental & social & policy dimensions, of the benefits of applying a demand side energy resource for considering for implementation of an energy efficiency project. Methodologically, this assessment was made by establishing values for energy resources in attributes and sub-attributes in the four different dimensions of the Integrated Resources Planning, which are: environmental dimension, social dimension, political dimension, and economic-technical dimension. The analysis of the energy efficiency project in four different perspectives makes this work innovative, most of the papers found in the academy does not consider the four dimensions studied here. The accounting of the full potential is also considered within this methodology, which makes it possible to evaluate its benefits in traditional and economic-technical aspects, as well as the dimensions of sustainable development. A case study was also developed on the replacement of traditional lamps for LED lamps in higher education institution. Results showed a decrease in CO2 emission in the atmosphere of 11.94 tons, the creation of 21 temporally job positions, reduction in the release of 52g of mercury in the environment and an injection of nearly 2 million dollars into the economy. Therefore, it was systematically proved that the benefits of the energy efficiency evaluated in the four dimensions increase sustainable development to all of society. With this work it is possible to concluded that all the society is impacted in different aspects after the implementation of an energy efficiency project in an energy consumer company e not only these companies

1. Introduction

Energy efficiency has had an impact to society, governments, and companies since the 1970s. It can be explained as a way of ensuring safety in energy supply [1], a result of the world oil crisis in 1973 [2]. In addition to this important factor for the implementation of energy efficiency projects, it also started to be considered as a good opportunity to reduce costs for consumers with the implementation of energy efficiency [3]. In other words, it was also considered to use less energy to produce the same planned volume. However, since the 1990's, energy efficiency has become more relevant to companies and society, forcing governments to create public policies geared towards such projects [4]. Until then, energy efficiency consisted of creating educational materials and programs for a more efficient consumption of energy, and adequacy to the hours when sunlight can be better used during the day [3].

IOP Conf. Series: Earth and Environmental Science **897** (2021) 012022 doi:10.1088/1755-1315/897/1/012022

Later, the implementation of more efficient equipment became an alternative for energy efficiency projects [5].

Since the beginning of the 21st century, energy efficiency has become a way for companies to reduce costs [6], as well as a fundamental way to reduce global warming, a critical subject in debates on sustainability [7]. Energy efficiency now incorporates all the previous concepts; it is considered a group of actions that, when combined, can provide options of sustainability, economic development, cost reduction to companies, and a guarantee of energy supply to society [8].

Due to increasing competition, companies are seeking cheaper and more efficient processes. In this context, the energy bill has become one of the most relevant costs that need to be worked on. Energy efficiency measures tend to have positive economic impacts in addition to their environmental appeal. Light retrofit in builds is a very common energy efficiency project also with efficient results, it offers significant savings in the energy account and potential reduction in the CO2 emissions [9].

This study aims to evaluate the full potential, since include not only for its financial aspect, what is done in the economic-technical dimension, but also, systematically, for all the aspects that regard sustainable development, to provide a complete validation of the benefits of energy efficiency measures, that are made in the environmental, social and politic dimension. To this end, it was conceptually and methodologically based on the Integrated Energy Resources Planning (IRP), which is described as a formation of a set of strategies to optimize the use of energy resources [10], a combination of energy management actions on the supply side and the demand side [11], or even as a planning that incorporates actions and strategies of Demand Side Management (DSM) to structure energy supply [12], with the possibility to provide the same level of importance to the demand and supply resources [13].

Full potential is achieved by the IRP methodology, which provides the basis for energy resources demand side management and has as focus of study: a higher education institution of a large company in the Brazilian educational sector. Specifically, this work analyses a retrofit of the lighting system in the studied company.

The fact that an actual energy efficiency project was studied as it was applied in a consuming company supports and validates its replication in other companies or other energy efficiency projects

2. State of Knowledge

Analysing a project feasibility, since its technical and financial situation to its result monitoring phase is fundamental in any type of company. From this importance arise different models and methods for projects evaluation.

An example is the Balanced Scorecard (BSC), considered a strategic management tool to analyse internal and external scenarios of the company, also, its purpose is supporting the company's managers in the development and execution of strategic actions in the long term [14]. Another method for project evaluation is Goal Question Metrics (GQM), this is an approach that starts with the goals and from them are outlined strategies [15]. It is also cited a very famous method in the financial field, the Monte Carlo Simulation, which uses probability distribution to simulate financial return on a given investment [16]. Finally, a last example is the Life Cycle Assessment (LCA), which analyses environmental impacts associated with a given product, process or activity, the application of this method has as a sequence: to define the project's goal and scope: list analyses that should be made and the outputs of each of them; expected impact assessment [17].

Meanwhile, to evaluate energy efficiency projects, specifics methods can be used. The most common and simple way to measure the energy efficiency obtained by implementing a certain resource is to compare the consumption required for the operation of a consumer with the actual energy supplied to this consumer; the lower the volume of required energy and the closer the delivered energy is, the more efficient will the consumer's energy management be [18].

Another way to measure energy efficiency projects or structure energy management is by creating four different modules: developing a database of the studied company; structuring the company's demand regarding energy efficiency; structuring the company's consumption profile; and estimating future energy consumption. With these data, it is possible to make decisions on energy resources implementation [19].

As mentioned before, the IRP methodology used as reference in this work considers the characterization of energy resources on supply and demand sides in four dimensions of analysis: environmental, social, economic-technical, and political. This characterization is given by the Full Potential Accounting and Validation (FPAV) methodology.

The goal of FPAV is to estimate the impacts of the use of an energy resource and incorporate the externalities of each of the dimensions into the analysis of these resources. More specifically, FPAV is a method in which both the Full Costs Analysis (FCA) and the Analytic Hierarchy Process (AHP) are applied [20]. The FCA can also be defined as a method in which political, social, environmental, and economic-technical subjects can be connected to traditional models [21]. The AHP is a method created in 1987 by Saaty that requires hierarchy or relationship structure in a problem to create measures between the analysed groups or objects [22].

The relationship among the methodologies cited above is best shown on Figure 1. Where AHP and FCA are the inputs for FPVA, on the other hand, Full Validation of the four dimensions is the output for FPVA.

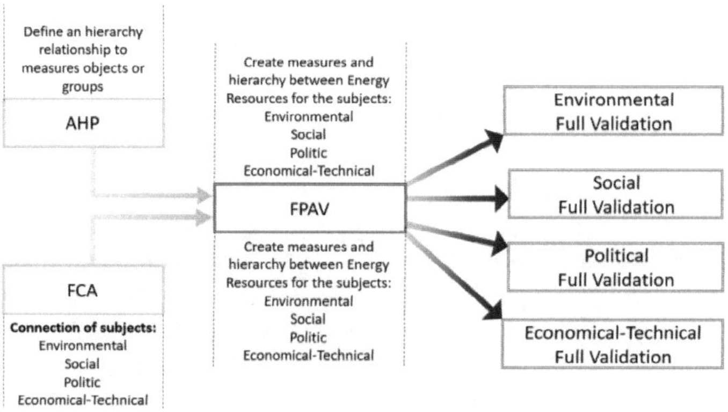

Figure 1. Relationship among AHP, FCA and FPAV

In this study, the starting point of the analysis is the lighting system, an important Energy Resource on the Demand Side (ERDS). Is evaluated the impact of replacing the inefficient fluorescent, incandescent and halogen lamps lighting system with an efficient LED lamps lighting system. To this end, the analysis of all IRP dimensions, and their respective attributes and sub-attributes, was based on the FPAV.

Before the analysis of each dimension, the attributes and sub-attributes that were considered or disregarded in the analysis are presented. The ones indicated with the ✓ symbol was included in the analysis, and the ones that were not analysed are marked with the ✕ symbol.

Analyses that show currency were made in Brazilian real (R$), but were converted to U.S. dollar (USD), considering the exchange rates on December 31th, 2019, when USD 1.00 equalled to R$ 4.031.

It is already known that the lighting system retrofit can show significant results in different aspects, electricity consumption and demand reduction, indoor comfort and environmental benefits [22].

3. Description of the case

The higher education institution studied in this work is in the city of São Paulo and belongs to a large private company in the education market of Brazil and the world. With more than 4.000 attending students and a little more than 30,000 m² of constructed area, including the areas that most consume energy, as classrooms, Health and Engineering laboratories, recording studios for online classes, an auditorium, administrative areas and a large common area for students, and the areas that are not energy intensive consumers, as outdoor garden, staircases, corridors and car parking. For all year 2019 its total energy consumption was 1,132,815.10 KWh and the monthly contracted demand was 200KW.

Due to the pandemic crisis of 2020, this work analyses energy consumption in 2019, when classes were in person and not online, as it had been in 2020. Such consumption curves in which both years are compared are shown in figure 2, in KWh.

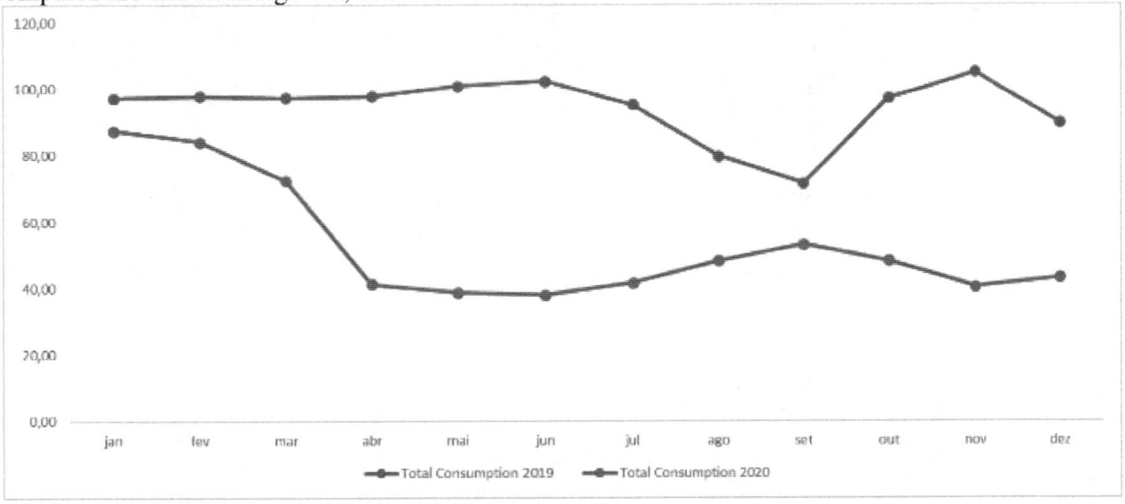

Figure 2. Energy consumption in 2019 and 2020

In overall there are about 40 different energy consumers equipment, for instance, televisions, computers, printers, fans, etc. However, the main ones that are always used in efficiency energy programs by the studied company are the lamps and air conditioner, due to them intensity energy consumption.

One of the essential steps for the elaboration of the FPAV is the listing and selection of resources, where all the possible types of energy resources in the evaluated company must be identified [23], as presented in Table 1, where the first column refers to the energy resources, that are installed in the education building consume or generate energy, on the other hand, the second column refers to the equipment of usage and related to the energy resource.

Table 1. Inventory of the energy resources.

Energy Resource	Related Equipment
Lighting	Lamps
Cooling system	Air conditioner
Sound system	Microphones and speakers
Video system	Video cameras

The implementation of ERDS consists in combining energy efficiency measures with energy use management measures to optimize the end use of energy [21]. Thus, the analysis with the lighting resource was chosen for its representativity in the company's energy bill and for being easily replicable to other facilities of the same company.

Led lamps are less damaging to the environment and consume 85%, 65% and 50% less energy than incandescent, compact fluorescent and sodium-vapor fluorescent lamps, respectively [24].

Table 2 presents the relation of the current lighting system installed in the education institution, referred to as the inefficient lighting system, which was collected by the own researchers, with its respective models and power (in watts). Table 3 presents the replacement system, referred to as efficient system, with LED lamps compatible with the inefficient system's lamps which was proposed by a specialist company of the sector.

Table 2. Inefficient lighting system

Lamp	Power W	Quantity	Total Power in KW

T8 36W Fluorescent	36	3,068	110.45
60W CFL	60	3	0.18
Fluorescent	28	576	16.13
18W CFL	18	59	1.06
60W Incandescent	60	34	2.04
25W CFL	25	62	1.55
20W Fluorescent	20	11	0.22
T5 Fluorescent	14	138	1.93
50W Halogen	50	16	0.80
110W Fluorescent	110	34	3.74
20W CFL	20	28	0.56
32W Fluorescent	32	44	1.41
11W CFL	11	25	0.28
16W CFL	16	1	0.02
Rectangular mercury vapor floodlight	250	8	2.00
100W Incandescent	100	3	0.30
40W Fluorescent	40	39	1.56
	Total	**4,149**	**144.22**

Table 3. Efficient lighting system

Lamp	Power W	Quantity	Total Power in KW
T8 18W LED	18	3,151	56.72
30W LED bulb	30	3	0.09
15W LED tube	15	576	8.64
18W Downlight	18	43	0.77
15W LED bulb	15	93	1.40
T8 8W LED	8	149	1.19
7W LED PAR	7	16	0.11
44W LED HO	44	34	1.50
5W LED bulb	5	25	0.13
10W LED bulb	10	51	0.51
120W LED reflector	120	8	0.96
	Total	**4,149**	**72.01**

4. Development and evaluation methodology

The presented results consist in the validation of the energy resource studied in the IRP dimensions. To make the validation possible, it is necessary not only to apply the FPAV method, but also to determine which attributes and sub-attributes can, and must, be validated in the resource in question.

For some sub-attributes, it is necessary to establish premises that can be applied to all energy resources that may be analysed in a consuming company.

Once the validation in the analysed dimensions is carried out, it is possible to structure the result of the analysed resources, which becomes the output required for decision making by the consuming

IOP Conf. Series: Earth and Environmental Science **897** (2021) 012022 doi:10.1088/1755-1315/897/1/012022

company's managers. The results obtained from the analysis in one facility can be expanded to the other facilities, with a proper comparison parameter, such as the size of the constructed area.

4.1. Environmental dimension

Validation of the environmental dimension consists in analysing three attributes: Airborne environment, Land environment and Aquatic environment, which are presented in Table 4 below, with their respective sub-attributes [20].

Table 4. Environmental dimension attributes and sub-attributes

Attributes	Sub-attributes	Analysis
Airborne environment	Atmospheric pollutants	✕
	Greenhouse gases	✓
	Ozone-depleting gases	✕
Terrestrial environment	Waste	✓
	Land occupation	✕
Aquatic environment	Water quality	✓
	Water demand, consumption, and flow	✕

The environmental characterization of demand side energy resources must be made considering the environmental impacts caused by the implantation of the resource and the impacts avoided by the energy savings brought by such resource [20].

The replacement of the inefficient lighting system lamps for LED lamps reduced total power used, the consumption in KWh of the education institution with the lighting resource can be determined. As informed by the Utility Department Manager of the Company, all the numerical analyses must be done considering 1,485 hours per year of operation. Thus, to find the saving in energy consumption, must be considered that:

IS: Full power of the inefficient system

ES: Full power of the efficient system

H: Hours of operation in one year

S: Savings in KWh

Reduction in energy consumption (in KWh) can be obtained by the following formula:

$$S = (IS - ES) \text{ x } H \quad (1)$$

$$S = (144.22KW - 72.01KW \text{ x } 1{,}485$$

$$S = 114{,}380.84KWh/year$$

Reduction in energy consumption also reduces greenhouse gas emission, including carbon dioxide (CO2). According to the Brazilian National Energy Balance, in 2018, carbon intensity in energy generation in Brazil was 104.4kg CO2/MWh [25]. Considering Brazilian matrix as parameter, the savings in KWh presented above result in a yearly reduction of 11.9 tons of CO2

Since LED lamps are mercury free and do not contaminate land and water [24], the environmental damages to land that can be caused by the disposal of fluorescent lights must be analysed within the scope of the Terrestrial Environment attribute, considering the frequency in which an LED lamp is disposed of, having as basis for calculation its lifetime in hours in relation to the frequency in which fluorescent lamps are disposed of.

This attribute would not suffer any impact if Brazil's National Solid Waste Policy (PNRS) of 2010 were correctly adopted [27]. In that case, the removed lamps would be correctly disposed of, which would not result in any environmental impacts on the terrestrial environment.

As determined by PNRS of 2010, lamps must be disposed of through a reverse logistic system, which should be structured by the manufacturers, importers, distributers, and dealers. However, even though PNRS is considered an environmental landmark in Brazil, it was not capable of changing the population

IOP Conf. Series: Earth and Environmental Science **897** (2021) 012022 doi:10.1088/1755-1315/897/1/012022

behaviour regarding waste disposal. More than that, according to the policy, the cities are responsible for developing local solid waste plans; however, in many of them, the plans were not developed, which led to failure in solving the problem of solid waste disposal.

If we also add to this condition the fact that the education institution does not have any specific solid waste disposal policy, as informed by the managers of the studied company, the analysis of this attribute becomes highly important, once it makes it clear that the lamps removed for the installation of the LED lamps will not have the proper destination for disposal.

Since the T8 18W LED lamp is the direct substitute for T8 36W tube fluorescent lamp, 32W fluorescent lamp and 40W fluorescent lamp, this specific analysis about the environmental impacts involves a comparison between these three models of fluorescent lamps and the T8 18W LED lamps. The fluorescent lamps have a total lifetime of 8,000 hours, while T8 tube LED lamps have a total lifetime of 40,000 hours [26]. Therefore, considering the same premises regarding hours of use, the 3,151 T8 tube LED lamps would be replaced after 26 years; on the other hand, in the same period, 15,755 T8 tube fluorescent lamps would be replaced. Complementing the technical data presented by Cepel (the Electrical Energy Research Centre), the three fluorescent lamps models cited above have 3.5mg of mercury per unit [27]. Thus, by avoiding the disposal of 15,755 T8 tube fluorescent lamps, the disposal of 73.5g of mercury in the environment would also be prevented, a volume that is way above the determination of Resolution No. 420 of 2009, which determines the maximum volume for prevention of 0.5mg per 1kg of dry land [28].

As it is true for the terrestrial environment, the fact that Brazil has not yet successfully implemented PNRS, along with the fact that the studied company also does not have any type of specific policy for lamps disposal, it can be concluded that the modernization of the lighting system also prevents environmental damages to the aquatic environment. Regarding the Water Quality sub-attribute, a single fluorescent lamp has enough mercury to pollute 20 thousand litters of water [31]; therefore, the 15,755 T8 tube fluorescent lamps could pollute more than 317 million litters of water. Still on the aquatic environment, the maximum accepted mercury level in non-predatory fish is 0.5mg per Kg, while in predatory fish, the maximum accepted mercury level is 1.0mg per Kg [32]. Therefore, the 73.5g that were prevented from disposal would be enough to contaminate 111 tons of non-predatory fish and 55.5 tons of predatory fish.

4.2. Political dimension

Within the IRP perspective, in the political dimension, factors that can lead and assess the development and implementation of a given energy resource were analysed [19]. This dimension approaches the Government's actions towards the development and implementation of the energy system, which can be made by creating public policies that encourage the development of projects connected to energy efficiency, such as ICMS Convention 16, published on April 22nd, 2015, that allows the States to exempt from the Brazilian value-added tax (ICMS) the energy obtained from micro generation, when consumed by the generator agent themselves. Table 5 below presents the attributes and sub-attributes of the political dimension.

Table 5. Political dimension attributes and sub-attributes

Attributes	Sub-attributes	Analysis
Political support	Political instruments	✓
	Legal aspects	✓
Possession and energy integration of resources	Possession	✓
	Ownership	✓
	Energy integration	✓
Conjunction and encounter of interests (stakeholders)	Residential, commercial, and industrial consumers	✓
	Generators, distributers, and producers	✓
	Federal, state and city government	✓

	Civil society, NGOs, and associations	✔
	Residential, commercial, and industrial consumers	✔
Level of motivation of the agents (stakeholders)	Generators, distributers, and producers	✔
	Federal, state and city government	✔
	Civil society, NGOs, and associations	✔
	Residential, commercial, and industrial consumers	✔
Level of acceptance of the resource (stakeholders)	Generators, distributers, and producers	✔
	Federal, state and city government	✔
	Civil society, NGOs, and associations	✔

Overall, it can be said that the political support deems fundamental to have actions in a global level, such as the Paris Agreement, whose purpose is to limit the increase of Earth's temperature in 1.5°C, regarding pre-industrial levels [30]. Therefore, Brazil is compromised to the UN, as it is a signatory of the aforementioned resolution, and must establish policies and goals that contribute to the reduction of Earth's temperature, which can be achieved through the encouragement of new energy efficiency projects.

Another form of political encouragement in Brazil is Law No. 9,991, published in July 2000, which determines that holders of concessions or licenses destine 1% of their operational income to research and development in the energy sector, and to energy efficiency programs in end use, where 0.75% should go to the former and 0.25% should go to the latter. These energy efficiency projects include several actions, including the creation of incentives to the replacement of fluorescent or incandescent lamps with LED lamps. In this case, the Public Call for Energy Efficiency Projects 01/2017 of Eletropaulo, the energy company of the state of Sao Paulo (currently Enel), can be used as an example, in which the concession holder destined USD 1,240,479.32 (considering the exchange rate of December 31th, 2019) to energy efficiency projects, including improvements in lighting facilities, meaning, in this case, the replacement of fluorescent, incandescent and halogen lamps with LED lamps.

Regarding to the Ownership sub-attribute, considering that the company buy the lamps, it has total control over the equipment, however, remains to the company to elaborate a maintenance agreement to avoid the normal equipment's operation.

Regarding the Possession and Energy Integration of resources attribute, its sub-attributes can be analysed together, when it refers to ERSS (Energy Resources on the Supply Side). The first one means refers to the LED technology, which is not a Brazilian property, once the LED lamps sold in the country are imported from China market. The second one, when applied to the ERDSs analysis, consists of exchanging ERDS technology between different parts [19]. This specific resource is produced in China, however, there is no effective exchange of technology between China and Brazil.

As for the three other attributes of the political dimension – Level of acceptance of the resource (stakeholders), Level of motivation of the agents (stakeholders), Conjunction and encounter of interests (stakeholders) – they are directly related to the stakeholders involved and interested in the project and consist not only of those in the education institution, but also employees of the company, public and private agents who are part of, and are directly involved in, the Brazilian energy sector, as well as other institutions that are not part of this sector .

Provided that energy efficiency works tend to benefit society in general, not only on environmental issues, but also financial matters, it is correct to infer that stakeholders inside and outside the organization present converging strategies and goals among themselves. It is worth highlighting that the energy providers are not necessarily benefited by the energy efficiency projects, once the implementation of these projects can reduce energy consumption and, consequently, the provider's recipe. However, they still show an interest in the process, specifically because their financial planning is impacted after the implantation of an energy efficiency project.

4.3. Social dimension

With IRP, social dimension represents the possible impacts caused to population, society and the inhabitants of a certain area that is suffering direct or indirect interference from an energy project [16]. This dimension directly or indirectly, positively or negatively, impacts the life of the individual who has some sort of relation to the area affected by the energy resource project [33]. The attributes and sub-attributes of this dimension are presented in Table 6.

Table 6. Social dimension attributes and sub-attributes

Attributes	Sub-attributes	Analysis
Environmental imbalance in the social environment	Impact on Agriculture	✗
	Impact on Health	✗
Perception of comfort	Odor pollution	✗
	Thermal pollution	✓
	Visual pollution	✓
	Sound pollution	✗
Influence on development	Human development	✗
	Economic / Infrastructure	✗
Impact on projects space occupation		✗
Job creation	Direct jobs	✓
	Quality and safety	✗

About the Comfort Perception attribute, which measures the comfort a person experiences after the implantation of the energy project, the first thing to emphasize is that the education institutions must follow lighting rules as defined by the Brazilian Association of Technical Standards (ABNT), as described in NBR 8995. This standard explains that compliance to the illuminance specifications described therein leads to a better visual comfort, as well as improves the performance of those who study or work in the environment, which is entirely related to the Impact on Health, Visual Pollution and Thermal Pollution sub-attributes.

Regarding visual pollution, as it is presented in detail in the economic-technical dimension, the inefficient system has low capability to enlighten a classroom than the efficient system, despite the lower colour temperature emitted by LED lamps when compared with fluorescent lamps, 3,500 Kelvins versus 6,000 Kelvins, respectively [26]. Even more, there is also a benefit in the Thermal Pollution, whereas LED lamps emit less heat than fluorescent lamps, there is also an improvement in the thermal attribute [29]. Depending on the characteristic of the local where the lamps are installed, the heat reduction can be felt by the people present, or even, less power from the cooling system may be required to maintain the temperature in the local. Therefore, the application of the resource positively impacts the Thermal Pollution and Visual Pollution attributes.

In the Job Creation attribute, more specifically the Direct Jobs sub-attribute, and based on information obtained from employees of the studied company, we can cite the departments involved in this project, with their respective functional positions, as described in Table 7. In this case, it does not consist in direct job creation, however it still means that workforce is allocated in the project

Table 7. Employed workforce in the lamp replacement project

Department	Position	Action
Supply	Director; Manager; Responsible buyer	Negotiation and purchase with the responsible supplier
Infrastructure	Director; Manager; Responsible engineer	Demand survey and installations monitoring
Legal	Lawyer	Development of the purchase agreement
Financial	Analyst	Invoice payment and tax collection

Sustainability	Analyst	Measurement of the benefits and disclosure of results
Education institution	Director; Manager; Operations assistants	Following and approving the work of lamp replacement

In the lamps vendor it is also possible to quantitative the workforce employed during their replacement operation. According to a documental research made in the studied company, 21 employees of the vendor was allocated in this project, as presented in Table 8

Table 8. Work team in the lamp replacement

Project engineer	1
Administrative manager	17
Operations supervisor	1
Electrician	7
Assistant	11
Total employees	21

4.4. Economic-technical dimension

The economic-technical dimension, in comparison to the other ones, includes most of the measurable parameters, since the focus of the analysis of this dimension is to calculate the cost of energy generation, when considered on the supply side, or the cost of the energy saved, in the case of demand resources [19]. Moreover, the validation of this dimension is made by calculating the different technical and economic indexes [23]. It must also be considered that this dimension consists in searching for the lowest possible cost and the highest expected financial return [33].

The attributes and sub-attributes of this dimension are presented in Table 9.

Table 9: Economic-technical dimension attributes and sub-attributes

Attributes	Sub-attributes	Analysis
Technology domain	Project and logistics	✓
	Technology and equipment	✓
Technical ease	Distance from the consumption centre	✕
	Implantation time	✓
Energy quality		✕
Energy potential	Power range	✓
	Energy volume	✓
Reliability	Availability	✕
	Intermittence	✕
Generation cost	Internal return rate	✕
	Net present value	✓
	O&M Cost	✕
	Implantation cost	✓
	Useful life	✓

Once there is an alternative to reduce energy consumption through the implantation of new energy efficiency projects, they start to be evaluated in a pragmatic manner. The evaluation must consider two different paths: the technical analysis, which must report whether the project will reach the goal of energy consumption reduction; and the validation of the financial viability of the project implementation, determining whether the final financial result will be the one expected by the company [34].

IOP Conf. Series: Earth and Environmental Science **897** (2021) 012022 doi:10.1088/1755-1315/897/1/012022

The analysis of this dimension is made based on the technical differences between fluorescent and LED lamps, which makes it possible to carry out financial assessments that can prove the highest economic viability of either lighting technologies.
Regarding the technical analysis, a comparison between the T8 tube fluorescent lamps and the T8 tube LED lamps, is presented to the reader, once both models of lamps were the most numerous in the studied build. These specifications are presented in the table below, which was developed based on the report obtained from Cepel, and on the report issued by the company Eaton [35].

Table 10. Technical profile of the LED and fluorescent lamps

Characteristics	T8 tube fluorescent	T8 tube LED
Size	120cm	120cm
Power (Watts)	36	18
Kelvin	3000	6500
Lumens / Watts	72.22	102.78
Hg Content	1.7mg	0
Light emission	21%	35%

The first attribute analysed is the Technology Domain. It represents the nationalization index of the technology used, which can be considered: national technology, if the entire production process is made in Brazil; foreign technology, in case the entire production process is made abroad; or mixed, when part of the production process is made abroad, and part of it is made in Brazilian territory [21].
More specifically, this attribute includes the type of technology, equipment and logistics used in the execution of the project, which are explained by the Technology and Equipment and Project and Logistics sub-attributes, respectively. The final sub-attribute cited refers not only to the place of origin of the analysed resource, but also to the logistic transportation until its final destination. The first sub-attribute cited refers to the technology availability on the site where the project will be carried out. The LED lamps are characterized for having a foreign and mixed technology, since part of the product is manufactured in China, and part of it is made in Brazil [35].
The Technical Ease attribute, composed by the Implantation Time and Distance from the Consumption Centre sub-attributes is analysed in this work only through the first sub-attribute, which represents the period of execution of the project, from the development of the technical project to the beginning of operations. Moreover, the same sub-attribute varies according to the technology used, project location, financial resources, and available workforce [20]. Finally, according to the project scope provided by the company, the conclusion of the project, including the removal and disposal of the inefficient system lamps and the installation of the LED lamps takes three working days.
The Energy Potential attribute is characterized as the capacity of each technology to generate the energy required to attend a given region, considering the available physical potential [19]. This attribute, when analysed on demand side, represents the savings potential obtained from power, in KWh, of the consumed energy [19].
This difference in power is presented in tables 2 and 3, and it is possible to observe that LED lamps have less power than the other lighting technology used.
The difference in power between one technology and the other is decisive to identify the financial viability of replacing the inefficient system for an efficient lighting system; the lower the lamp power is, the less will energy consumption be, leading to financial savings in the energy bill. This sub-attribute makes it possible to analyse the last attribute of this dimension, Generation Cost, an attribute that is entirely financial, and is composed by the following sub-attributes: Useful Life; Implantation Cost; O&M Cost; Net Present Value; and Internal Return Rate.
Since it is a lamp replacement project, the financial analysis developed in this dimension approaches the purchase cost of new lamps, considering the material and additional services, removal and disposal of the inefficient system lamps, and installation of the LED lamps. Other important items are also considered for the development of the project's business case, and are presented below:

IOP Conf. Series: Earth and Environmental Science **897** (2021) 012022 doi:10.1088/1755-1315/897/1/012022

- Freight: CIF (Cost In Freight), in which the cost of the lamps sold involves delivery of the material, including transportation insurance until destination.
- WACC (Weighted Average Capital Cost): WACC is 13.5%, as informed directly by the company.
- Conditions for usage: The same ones used in the analysis of the other dimensions.
- Lamps lifetime: Overall, the lifetime for the inefficient lighting system lamps will be considered 5 years, and 25 years for the efficient lighting system lamps.
- The financial values were initially calculated in Brazilian real and were later converted to U.S. dollars, with the exchange rate mentioned in the Methodology section herein.
- The energy fee of the consumption used is the one currently charged by the energy provider where the institution was installed, on the same year of the consumption's analyses, 2019.
- The value consists of the sum of the distribution charge (TUSD – Fee for the Use of the Distribution System) and the energy fee (TE) at peak hours, which were, respectively, R$0.642 + R$ 0.382, resulting in a R$1.024 fee, that is, USD 0.254 per kWh. And sum of TUSD and TE at the out peak hours, respectively R$0.080 + R$0.237, resulting in R$0.317, that is, USD 0.079 per kWh.
- As informed by the company, the energy consumption represents 70% during peak hours and 30% during out peak hours.
- The energy fee of contracted demand used is, R$ 0.110 or USD 0.027 per KW as informed in the company's energy bill.

In a business case, the main goal is to calculate the project's viability, through the calculation of the Net Present Value (NPV).

Table 11: Comparison of consumption between LED and CFL lamps

Lamp	Unitary power W	Quantity	Total power KW	Total investment in USD
18W T8 LED	18	3,151	56.72	$19,801.73
30W bulb LED	30	3	0.09	$122.74
15W tube LED	15	576	8.64	$4,777.25
18W Downlight	18	43	0.77	$2,747.04
15W bulb LED	15	93	1.40	$467.23
8W T8 LED	8	149	1.19	$1,186.62
7W LED	7	16	0.11	$191.85
44W HO LED	44	34	1.50	$792.07
5W bulb LED	5	25	0.13	$33.24
10W LED bulb	10	51	0.51	$159.43
LED Reflector 120W	120	8	0.96	$579.25
Total		**4,149**	**72**	$30,858.45

Following up, in table 12, the discounted cash flow is presented, considering the investment made, the reduction in consumption and in the contracted demand on the energy bill (in Brazilian Real) obtained after the replacement of the lamps, and the eventual financial gain after the application of the revenue resulted from the savings. These analyses are shown on the formulas below.

4.4.1 Financial reduction: First, is calculated the reduction in KW power. As seen as seen in formulas 2 and 3

$$S = IS - ES \quad (2)$$

$$S = 144.22KW\ 72.01KW$$

$$S = 72.21KW$$

IOP Conf. Series: Earth and Environmental Science **897** (2021) 012022 doi:10.1088/1755-1315/897/1/012022

This same reduction means a possible reduction on the contracted demand, therefore, the contracted demand would be:

Demand reduction = actual contracted demand – New contracted demand (3)

Demand reduction = 144.22 KW 72.01KW

Demand reduction = 72.21 KW / month

71.21 KW x 12 = 866.48 KW / year

Financially, considering the contracted demand cost of USD 0.027/KW, the annual addiction saving with this reduction is:

Saving with demand = KW reduction x contracted demand cost (4)

Saving with contracted demand = 866.48 KW x $ 0.027

Saving with contracted demand = $ 23.67 / year

Next, the financial monthly saving is calculated based on the reduction of energy consumption during peak hours and out of peak hours
Consumption saving at peak hours

Consumption saving at peak hours=Total consumption saving x 70% (5)

Consumption saving at peak hours= 107,227.40KWh x 70%=75,059.18 KWh per year

Consumption saving at out of peak hours

Consumption saving at out of peak hours=Total consumption saving x 30% (6)

Consumption saving at peak hours= 107,227.40KWh x 30%=32,168.22 KWh per year

Financial consumption saving per year

(75,059.18 x 0.254) +(32,168.22 x 0.079)=$19,074.20+$2,526.08=$ 21,600.28 (7)

Yearly total financial saving

($ 21,600.28+$ 23.67 = $ 21,623.95 (8)

In this analysis it is assumed that the entire investment is made by the consuming company, i.e., the object of study in this work. In this case, the business transaction is recognized as CAPEX, once an investment in asset is made by the own company.
It assumed that the investment amount of USD 30,858.45 is based on a turnkey agreement, in which is considered not only the cost of the lamps, but also the freight and the payment of the labor force for the additional services.

Table 12: Discounted cash flow

Capex Model	Year 1	Year 2	Year 3	Year 4	Year 5
Investment	$30,858.45				
Consumption profit	$21.623,95	$21.623,95	$21.623,95	$21.623,95	$21.623,95
Balance per period	$21.623,95	$21.623,95	$21.623,95	$21.623,95	$21.623,95
Financial profit	$2.220,78	$2.448,85	$2.472,28	$2.474,68	$2.474,93
Accumulated balance	$23.844,73	$24.072,80	$24.096,23	$24.098,63	$24.098,88

With the discounted cash flow, it is possible to calculate the NPV of the project, using the following formula:

$$NPV = -II + \Sigma \times (CFn / (1+WACC)) \quad (9)$$

NPV = $39,012.51

IOP Conf. Series: Earth and Environmental Science **897** (2021) 012022 doi:10.1088/1755-1315/897/1/012022

Calculate the payback of an investment is also a representative index that must be considered when taking a decision of implement or not an energy project. A financial analysis made about replace inefficient lighting system by LED lamps in an university campus proved that is possible to achieve a payback a short payback [23]. Also, a good and short payback is found in this study, less than one year and half.

$$\text{Payback} = \text{Initial Investment} / \text{Yearly Saving} \qquad (10)$$

$$\text{Payback} = 1.43$$

Another possible financial analysis that does not lead to a direct impact on the company that purchases the lamps is the revenue obtained from such project, which can be done based on the temporary job vacancies creation, which analysis presented in the social dimension and represented in table 8.
Based on the current minimum daily wage on the period when the project was executed – R$34.83 (R$1,045.00 per month), which converts to dollar is $8.64 ($259.26 per month), as established by decree No. 9.255 of December 29th, 2017 – it is possible to calculate the labor force revenue generated on the work period, as presented below.
First, that the daily minimum wage is $7.89 dollars. Second, as already described, it is a 3 workdays project. Finally, there are different kinds of occupation in the project, moreover, each occupation has its own mensal minimum wage quantity, as explained in table 14.

Table 13: Calculation of revenue generated from the project

Occupation	Minimum salary quantity	Number of employees used	Total wage in USD
Operational Assistant	1	11	$285,19
Electrician	1.5	7	$272,22
Operational Supervisor	2	1	$51,85
Administrative Supervisor	3	1	$77,78
Project engineer	4	1	$103,70

Therefore, it is calculated that the total labor force revenue generated is USD 790.74.

5. Results and analyses

After all analyses done, the results are showed in the below tables:

Table 14: Results in environmental dimension

Attributes	Sub-attributes	Results
Airborne environment	Greenhouse gases	Reduction of 11,9 tons per year of CO2
Land environment	Waste	Prevent the disposal of 53g of mercury in the environment
Aquatic environment	Water quality	Avoid the risk to pollute more than 306 million litters of water
Aquatic environment	Water quality	Avoid the risk to contaminate 111 tons of non-predatory fish and 55 tons of predatory fish
Airborne environment	Greenhouse gases	Reduction of 11,9 tons per year of CO2

Table 15: Results in political dimension

Attributes	Sub-attributes	Results
Political support	Political instruments	Publication of the law 9.991 in 2000
Political support	Legal aspects	Signatory country to the Paris Agreement in 2016

IOP Conf. Series: Earth and Environmental Science **897** (2021) 012022 doi:10.1088/1755-1315/897/1/012022

Possession and energy integration of resources	Possession	LED technology does not exist in Brazil and the lamps are imported from China, thus there isn't possession over the ERDS
Possession and energy integration of resources	Ownership	The company has total control over the equipment, once it is bought from the vendor, however, a maintenance agreement must be done
Possession and energy integration of resources	Energy integration	There is no technology exchange between China and Brazil, what beneficial for the Brazilian's buyers
Conjunction and encounter of interests (stakeholders) Level of motivation of the agents (stakeholders) Level of acceptance of the resource (stakeholders)	Residential, commercial, and industrial consumers Generators, distributers, and producers Federal, state and city government Civil society, NGOs, and associations	These Stakeholders present strategies and objectives converging among them.

Table 16: Results in social dimension

Attributes	Sub-attributes	Results
Perception of comfort	Thermal pollution	Better control of a mild temperature.
Perception of comfort	Visual pollution	Increase in the illuminance level, from 248 lux to 523 lux
Job creation	Direct jobs	21 jobs opportunities

Table 17: Results in economical-technical dimension

Attributes	Sub-attributes	Results
Technology domain	Project and logistics	Mixed technology: China and Brazil
Technology domain	Technology and equipment	Increase in the illuminance level, from 248 lux to 523 lux
Technical ease	Implantation time	3 working days
Energy potential	Power range	Reduction of the power from: 36W (Fluorescent T8) to 18W (LED T8)
Generation cost	Net present value	USD 39,012.51
Generation cost	Implantation cost	USD 30.858,45
Generation cost	Useful life	40,000 hours

Reaching the results expected by the consuming companies is only possible through the implantation of energy efficiency projects. There are three different energy efficiency projects that can be implemented in a corporation: replacing or installing equipment with new, or more efficient, technology to reduce energy consumption; monitoring energy consumption together with the client or the company's employees, in a way that makes it possible to develop educational material that can teach them how to reduce energy consumption; and financial incentives for the development of new energy efficiency projects [34].

IOP Conf. Series: Earth and Environmental Science **897** (2021) 012022 doi:10.1088/1755-1315/897/1/012022

In this context, the factor of political uncertainties must also be highlighted. Investments in energy efficiency projects mitigate risks in the company's future strategic planning, once it keeps it from being exposed to variations in energy fees, which is extremely important in countries where the energy fee is susceptible to constant and meaningful variations.
Overall, considering that there is a world trend in the energy field to reduce consumption, especially due to environmental issues, it is assumed that energy efficiency projects represent actions that lead consumers to reduce energy consumption in their facilities.

6. Conclusions
It can be concluded that the application of an energy efficiency project in consuming companies is awarding not only for the company, but also for the entire society, including the Government. Benefits were observed in all analyzed dimensions. It can be considered here that the study project also impacts the government's tax policy, taxes can be collected with the purchase and installation of lamps, as well as a smaller portion of taxes is collected due to the reduction in energy consumption. however, such analysis is not considered in the IRP methodology.
Even more, FPAV showed to be an excellent tool to assess and validate the impacts from implanting energy efficiency projects, which means that if the same method is applied in other energy efficiency projects, it will be possible to obtain, in a structured manner, the results of this energy efficiency project in each of the dimensions. This makes it possible to evaluate the impacts of each resource in a more objective manner, to help decision making. Eventual ERSSs of the company are not excluded, once FPAV can also be used to validate them.
This work further reinforces the importance of energy management in companies. The systematics to define energy resources that must be studied together with the application of the FPAV method for their assessment, as well as the structure of result in such a way that can help decision making for the company's energy bill, can only be made if worked together, in an organized and planned manner, that is, through the implantation of an energy management model pursuant to each company's profile and reality.
Furthermore, the work shows the possibility to connect knowledge developed in the academic field with the necessities and opportunities found in the corporative field. Also, it opens possibilities to analyze the application of others energy resource, as cooling system through the method stated here. And finally, it is also worth highlighting that the results presented herein can be replicated to other education builds of the company. However, to obtain an accurate result, future studies must be done.

7. References

[1] TANAKA, Kanako. Review of policies and measures for energy efficiency in industry sector. Energy policy, v. 39, n. 10, p. 6532-6550, 2011.
[2] JENNE, C. A[1]; CATTELL, R. K. Structural change and energy efficiency in industry. Energy Economics, v. 5, n. 2, p. 114-123, 1983.
[3] FULLER, Robert H.; SULLIVAN, Dallas. Energy efficiency at the Ohio State University. Energy and Buildings, v. 1, n. 4, p. 401-413, 1978.
[4] PATTERSON, Murray G. What is energy efficiency? Concepts, indicators and methodological issues. Energy policy, v. 24, n. 5, p. 377-390, 1996.
[5] REDDY, Amulya KN. Barriers to improvements in energy efficiency. Energy policy, v. 19, n. 10, p. 953-961, 1991.
[6] HERRING, Horace. Energy efficiency—a critical view. Energy, v. 31, n. 1, p. 10-20, 2006.
[7] THOLLANDER, Patrik; DANESTIG, Maria; ROHDIN, Patrik. Energy policies for increased industrial energy efficiency: Evaluation of a local energy programme for manufacturing SMEs. Energy policy, v. 35, n. 11, p. 5774-5783, 2007.
[8] COPIELLO, Sergio. Building energy efficiency: A research branch made of paradoxes. Renewable and Sustainable Energy Reviews, v. 69, p. 1064-1076, 2017.

[9] GAN, Chin Kim et al. Techno-economic analysis of LED lighting: A case study in UTeM's faculty building. Procedia Engineering, v. 53, p. 208-216, 2013.

[10] UDAETA, M.E.M. Novos Instrumentos de Planejamento Energético e o Desenvolvimento Sustentável – Planejamento Integrado de Recursos Energéticos na USP, 2012, Tese (Livre Docência – Área: Energia Elétrica) - Escola Politécnica da Universidade de São Paulo. Departamento de Engenharia de Energia e Automação Elétricas.

[11] YILMAZ, Pelin; HOCAOGLU, M. Hakan; KONUKMAN, Alp Er S. A pre-feasibility case study on integrated resource planning including renewables. Energy Policy, v. 36, n. 3, p. 1223-1232, 2008.

[12] ZHENG, Yanan et al. IRSP (integrated resource strategic planning) with interconnected smart grids in integrating renewable energy and implementing DSM (demand side management) in China. Energy, v. 76, p. 863-874, 2014.

[13] GIMENES, A.L.V. Modelo de integração de recursos como instrumento para um planejamento energético sustentável. 2004. Tese (Doutorado em Engenharia) – Escola Politécnica, Universidade de São Paulo, São Paulo, 2004.

[14] LESÁKOVÁ, Ľubica; DUBCOVÁ, Katarína. Knowledge and use of the balanced scorecard method in the businesses in the Slovak Republic. **Procedia-social and behavioral sciences**, v. 230, p. 39-48, 2016.

[15] BECKER, Shirley A.; BOSTELMAN, Mitchell L. Aligning strategic and project measurement systems. **IEEE software**, v. 16, n. 3, p. 46-51, 1999.

[16] SMITH, David J. Incorporating risk into capital budgeting decisions using simulation. **Management decision**, 1994.

[17] FINNVEDEN, Göran. Valuation methods within the framework of life cycle assessment. 1996.

[18] BAKAR, Nur Najihah Abu et al. Energy efficiency index as an indicator for measuring building energy performance: A review. Renewable and Sustainable Energy Reviews, v. 44, p. 1-11, 2015.

[19] GRIGORAȘ, Gheorghe; NEAGU, Bogdan-Constantin. An Advanced Decision Support Platform in Energy Management to Increase Energy Efficiency for Small and Medium Enterprises. Applied Sciences, v. 10, n. 10, p. 3505, 2020.

[20] BAITELO, Ricardo Lacerda. Modelo de cômputo e valoração de potenciais completos de recursos energéticos para o planejamento integrado de recursos. 2011. Tese de Doutorado. Tese de doutorado apresentada à Escola Politécnica, Universidade de São Paulo, São Paulo.

[21] UDAETA, M. E. M., Galvão, L. C. R., da Costa Rigolin, P. H., & de Oliveira Bernal, J. L. Full assessment energy-sources for inclusive energy-resources planning. Renewable and Sustainable Energy Reviews, 66, 190-206. 2016.

[22] SAATY, Roseanna W. The analytic hierarchy process—what it is and how it is used. Mathematical modelling, v. 9, n. 3-5, p. 161-176, 1987.

[23] FONSECA, Paula et al. Sustainability in university campus: options for achieving nearly zero energy goals. International Journal of Sustainability in Higher Education, 2018.

[24] MURUYAMA, F.M. Arquitetura do Plano Preferencial de Recursos para o Setor Elétrico no Planejamento Integrado de Recursos Energéticos, 2013. Tese (Doutorado em Engenharia) – Escola Politécnica, Universidade de São Paulo, São Paulo, 2013.

[25] EMPRESA DE PESQUISA ENERGÉTICA – EPE. Balanço Energético Nacional, 2019. Acessado em 11 de maio de 2020. Disponível em http://www.epe.gov.br/sites-pt/publicacoes-dados-abertos/publicacoes/PublicacoesArquivos/publicacao-377/topico-494/BEN%202019%20Completo%20WEB.pdf

[26] EPA – United State Environmental Protection Agency. Acessado em 17 de abril de 2019. Disponível em https://www.epa.gov/sites/production/files/2016-03/documents/drum-top_studyandcover_0.pdf

[27] CONSELHO NACIONAL DO MEIO AMBIENTE – CONAMA. Resolução N° 420 de 28 de Dezembro de 2009. Disponível em

http://www2.mma.gov.br/port/conama/legiabre.cfm?codlegi=620, acessado em 17 de abril de 2019

[28] CARVALHO, Estenio Moita. Análise do descarte de lâmpadas fluorescentes e a elaboração de propostas para o manuseio e reciclagem no Bairro de Mussurunga em Salvador, Bahia. 2016.

[29] SCIENCE FOR ENVIRONMENT POLICY (2017) Tackling mercury pollution in the EU and worldwide. In-depth Report 15, produced for the European Commission, DG Environment by the Science Communication Unit, UWE, Bristol. Available at:

[30] GOVERNO BRASILEIRO. Acessado em 15 de outubro de 2018, disponível em http://www.brasil.gov.br/cidadania-e-justica/2015/09/brasil-cumpre-metas-e-assume-novos-compromissos-na-onu

[31] PRESIDÊNCIA DA REPÚBLICA, CASA CIVIL. Lei N° 9.991 de 24 de julho de 2000, disponível em http://www.planalto.gov.br/ccivil_03/Leis/L9991.htm, acessado em 27 de junho de 2018.

[32] ELETROPAULO METROPOLITANA ELETRICIDADE DE SÃO PAULO S.A. Chamada Pública de Projetos de Eficiência Energética (01/2017). Disponível em https://www.aeseletropaulo.com.br/sobre-a-aes-eletropaulo/eficienciaenergetica/Documents/2017/Chamada%20P%C3%BAblica%20Eletropaulo_2017.pdf, acessado em 01 de julho de 2018.

[33] BIAGUE, M. F. Modelagem da Carteira dos Recursos Energéticos no PIR: Validação do Modelo no PIR de Araçatuba. 2010. Tese (Doutorado em Engenharia) – Escola Politécnica, Universidade de São Paulo, São Paulo. 2010.

[34] MCLEAN-CONNER, Penni. Energy efficiency: principles and practices. PennWell Books, 2009.

[35] EATON. LED and Home Energy Savings: The Next Chapter. Disponível em https://www.energystar.gov/sites/default/files/asset/document/LED_and_Home_Energy_Savings_The_Next_Chapter_508.pdf, acessado em 29 de agosto de 2018.

IOP Conf. Series: Earth and Environmental Science **897** (2021) 012023 doi:10.1088/1755-1315/897/1/012023

Measuring Commuters' Behavior and Preference Towards Sustainable Mobility: Case Study of Suburban Context of Pathumthani, Thailand

Pawinee Iamtrakul[1],* and Jirawan Klaylee[2]

[1]Associate Professor, Director, [2]Ph. D. Student, Center for Excellence in Urban Mobility Research and Innovation, Faculty of Architecture and Planning, Thammasat University, Pathumthani, 12120, Thailand.

The corresponding author's e-mail address: iamtrakul@gmail.com

Abstract. Urban mobility has been facing changing circumstances globally including congestion, air and noise pollution, climate change which search for alternatives to mitigate the influence of fossil fuels consumption, urbanization and the impacts of new technology. Cities are currently facing ever greater social challenges in respect of the environment, transport, health and social cohesion. Especially, suburban city context represents the commuting patterns behavior differs from the general populations. This study focuses on an analysis of the behavior and preference of commuters on mobility options in relation to the perspective of the development of energy-efficient systems of the future mobility choice. This study applied factor analysis with principal component method to identify the subjective factors based on travellers' attitude and perception comparative with their current travel choices and recommend suitable measurements. The consideration of suburban context could then be determined together with demand management scheme in term of sustainable transport system in Pathumthani, Thailand as a pilot case for suburban regions.

1. Introduction

Rapid urbanization presents policymakers with new challenges on many fronts which more than 50 percent of people lived in cities with concentration of the majority of economic activities [1, 2, 3, 4]. Over the next four decades, billions of people will continue to migrate to urban areas until more than 75 percent of the global population lives in cities which generate an unprecedented amount of urban traffic both freights and passengers [5]. The urban population has been rapid growing that require housing, transportation, energy, and water systems to be built simultaneously, ideally in an integrated way, to provide the greatest public access while create the least harm to the natural environment as much as possible [6]. Cities already account for 66 percent of total world energy use that may increase to 80 percent [7]. However, these urban areas consume more than 70 percent for transportation, especially regarding the use of energy which energy efficiency includes more than 50 percent of the world's population [5, 6, 7]. It is due to the reason that growing urban populations can overwhelm existing infrastructures and services, especially transport systems. Traveling to work and school, and to access other services, urban dwellers need to be able to easily traverse the city and get wherever they need to go [8]. Thus, efficient transport infrastructure planning and management can positively response to the mobility challenges faced by cities: accessibility, congestion, energy efficiency,

IOP Conf. Series: Earth and Environmental Science **897** (2021) 012023 doi:10.1088/1755-1315/897/1/012023

emissions and safety which having a significant effect on social inequality and global warming [9]. Studies have shown that even small changes in people's individual behavior can lead to significant reductions in carbon emissions with alternative transport choices and the diversification of energy sources to meet climate goals [10].

Common approaches that aim at changing a person's behavior rely on providing generic normative information, i.e., communicating a given behavior is appropriate in a given context [10]. However, considering the fact of most mobility-related activities are shaped by an individual's spatial, temporal, and social constraints. It is not tightly bound by social norms, generic normative information which is often too unspecific. For commuters who has no preferred choice as to use private transport or does not have any information of transport alternatives to perform an activity equally electively (in a sustainable manner), more targeted forms of communication are needed [11]. It is necessary to consider suitable transportation plan in response to changing behavior of choice makers that promotes sustainable mobility behavior. Among all challenges, it is a must to identify the necessary information to allow for behavior change which requires information processes taking into account both the analysis and planning aspects among sustainable activity alternatives [12]. Our proposed approach aims to ask commuters on a personalized level to change their mobility behavior and make decision for more sustainable choices. The available strategies for improving the sustainability mobility vary and span from actions that affect both transportation supply and demand side.

2. Literature review

Travel behaviors change in the context of personal mobility under given urban environments is an active area of research which aim to motivate users towards making more eco-friendly choices (i.e., adopting transportation habits that rely more on the usage of public transportation, non-motorization, and less private vehicles). One of the first attempts which aimed to persuade people to use public transportation instead of their cars is mainly for emission reductions. The approach can be varied such as displaying pollution information on board (the departure time of the next bus as well as the decreasing in emissions achieved by taking the bus instead of the car). On the other hand, the encourage of greener alternatives (including carpooling, public transport and pedestrian modalities) can be recommended by asking perception feedback when users reduced driving [13]. The demand management can be also recommended at the same time for regulatory measures by restrict access to the city centre. All of this highlights the need to couple 'smart' regulations with or replace them where possible 'soft' policy interventions can be implemented. Even where alternatives are offered, the debate on regulation faces the dilemma of balancing people's freedom of choice quintessentially expressed through the personal car and implementing normative regulations to change behavior [14]. The ASI model (Avoid/Reduce, Shift/Maintain, Improve) can help conceptualise and identify the different factors which incentivise the utilization of sustainable transport modes or disincentivise less sustainable ones as described in the following detail [15];

1. *"Avoid/Reduce"* refers to the need to improve the efficiency of the transport system. Through the transit-oriented and compact development, the need for motorized travel and the trip length can be reduced. Transport demand management (TDM) plays an important role into this objective as well. Residential, work, commercial and leisure districts must become more closely connected and intermixed.
2. *"Shift/Maintain"* instruments seek to improve individual trip efficiency. A modal shift from the most energy consuming and polluting urban mobility mode (i.e., cars) towards more environmentally friendly modes addresses all the aforementioned challenges of transport systems. Especially, shifting and maintaining the following transport modes is crucial which are active transport, walking and cycling, public transport (bus, rail, tram, including autonomous electric, etc.).
3. *"Improve"* focuses on vehicle and fuel efficiency as well as on the optimization of the operational efficiency of public transport. This includes the attractiveness of public transport. Additionally, improvement of the energy sources required for their operation is key. Introducing

renewable energy sources into the mobility sector must become a basic principle for motorized transport which can be illustrated in figure 1.

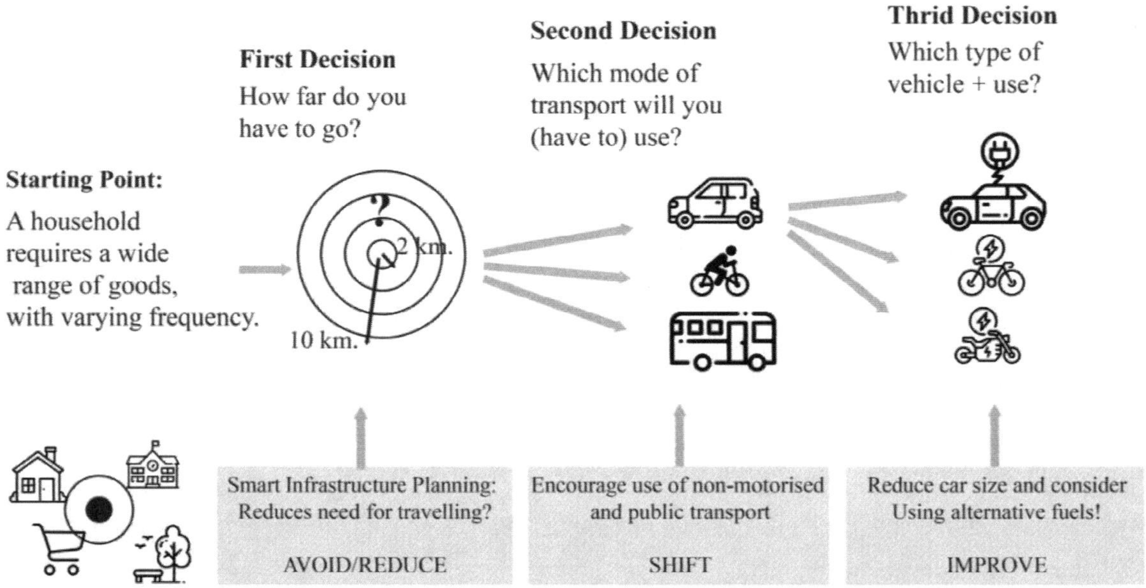

Figure 1. The ASI model
Source: Deutsche Gesellschaft für Internationale Zusammenarbeit (GIZ). (2004)

In this paper, face-to-face interview by using questionnaire survey was designed towards their current travel choices and recommend for the sustainable mobility measurement. More subtle changes in attitudes and perceptions could be explained under sub-urban context together with demand management in term of efficient transport systems. Finally, the recommendation of the policy planning with the explicit goal can be launched for more accessible, safe, efficient, affordable and sustainable infrastructure in the study area of Pathumthani, Thailand.

3. Research methodology and analysis

In this study, 1,000 data sets were collected from Pathumthani residents which focus on measuring behavior and preference changes of sustainable mobility in the sub-urban area. The questionnaire was divided into 3 parts which includes; 1) personal characteristic (social-economic factors), 2) travel behavior and 3) preference on sustainable mobility choices in sub-urban area context. The assessment based on interview was questioned into 6 levels of perception which was ranged from 1-6, from the strongly disagree to totally agree (1-6). The group of factors was analysed based on quantitative approach of factor analysis. In the study, the classification of factor groups applied the explanatory factor analysis method which was used to determine the grouping among different factors. It is an analysis technique by reducing the number of variables with similar statistical characteristics to the same group of factors. To reduce the redundancy of variables, 31 factors were extracted as shown in table 1. To avoid the interference of heteroscedasticity, some variables are processed in logarithms. The descriptive statistical analysis of the main variables is shown in table 2 and analysed by Principal Component Analysis: PCA. From the analysis, it was found that among 31 input factors, 3 groups of factors were extracted. The eigenvalue of that factor was determined when the statistical test exceeded 1.0 and rotation sums of squared loadings were considered. The cumulative percentage (cumulative %) found that the factors obtained after the analysis among all 3 group factors were able to explain the total variability of the variables about 64.40 percent.

Table 1. Input variables for analysis

Variables	X_n	Description
Creating flexibility operational	X_1	Creating flexibility operational of the time, such as working 8 hours a day, able to come to work from 10:00 and finish work by 19.00, etc. will help reduce traffic jams.
Encourages personnel to use public transport	X_2	The encourages of personnel to use public transport as well as provides reward points for lowering prices of travel-related goods and services and help reduce the use of personal cars.
Encourages congestion pricing	X_3	The encourages of congestion fees (Congestion Pricing) only in urban areas with comprehensive public transport, helping to reduce congestion during rush hour.
Encourage people to use clean energy vehicles	X_4	The government sector should establish policies to encourage people with clean energy vehicles.
Private cars in public areas with parking fee	X_5	Private parking for private cars in public areas with service rates (per day and month).
Expansion of a Bus Rapid Transit (BRT) route	X_5	Expansion of a Bus Rapid Transit (BRT) route with special traffic lanes, to promote the use of public transport.
Encourage people to use the carpooling	X_7	The government sector should have the policy to encourage people to use the carpooling by specifying special lanes in conjunction with bus lanes.
More car-sharing in transportation center point	X_8	The service points for a car-sharing to cover the important nodes/locations of the trip.
More bike-sharing in transportation center point	X_9	Service points for bike-sharing to cover the important nodes/locations of the trip.
More use ridesharing	X_{10}	Ride sharing service should be provided to reduce the number of personal vehicles on the road.
Intelligent traffic management (Smart Mobility)	X_{11}	Promote the use of intelligent traffic management (smart mobility) by developing technology to provide travelers with important information when planning their trips.
Seamless public transportation	X_{12}	The connection of public transport should be seamless in order to reduce travel time.
Encourage a work online application system development	X_{13}	Encourage a flexible working online system, in any location which will help better productivity of daily work.
Infrastructure development	X_{14}	Infrastructure development and various facilities to consistence with users' demand.
Bus stops with automated ticketing service	X_{15}	Encourage public transport bus stops to have an automated ticketing service.
Promoting offer selling products in transit stop	X_{16}	Public transport bus stops should be promoted for offering selling products.
Bus stops to provide basic services	X_{17}	Encourage public transport bus stops to provide basic services such as coffee shops, food and beverage vending machines and automatic ATMs, etc.
Standard public transport design	X_{18}	Encourage the standard public transport design.
Applications for specific public transport systems	X_{19}	Provide applications for specific public transport in urban and metropolitan areas as a travel resource and able to propose appropriate routes.
Expanding public transport routes	X_{20}	Expanding public transport network to suburban and metropolitan areas.
Smart cards covering all public transport	X_{21}	The use of smart cards (cards covering all public transport) should be encouraged to ensure a seamless connection for reduce the travel time.
Promoting facility for the elderly, the disabled and children	X_{22}	The promotion should be provided for the elderly, the disabled and children when travelling by public transport.
Connecting feeder system to railway station	X_{23}	Feeder system route should be connected to the Red Line MRT station that connects to Bangkok and its vicinity.
Connecting pedestrian paths and bicycle lanes	X_{24}	Along the canal, housing estate should offer the connection of pedestrian paths and bicycle lanes.
Road safety awareness	X_{25}	Road safety issues should be priority when choosing a mode of transport.
Mitigated road safety problems	X_{26}	Road safety problems can be mitigated by switching from private vehicles to public transportation.
Reduce road accidents	X_{27}	Reducing conflicts such as U-turns and intersections can help reduce road accidents.
Speed-reducing road design	X_{28}	Speed-reduction for road design will help on reducing frequency and severity of road accidents.
Road environment design	X_{29}	Good road environment design affects the choice of traveling.
Provide shading/trees for accommodate active travel choices	X_{30}	Shade of trees influences commuters' decisions on walking and cycling.
Road design allows people for safer travel by personal car	X_{31}	Road design allows people to travel by personal car rather than by public transport or by walking and cycling.

4. Results and discussions

Table 2 presents the most important factor which was built environment design by allow commuters to travel by personal car (X31 = 5.348), followed by transportation choices promoted to the elderly, the disabled and children (X22 = 5.345), connectivity of feeder system to railway station (X23 = 5.332) and shading/trees influences active travel choices (X30 = 5.320). The analysis result showed that people who are Pathumthani resident have a focus on built environment connected to transportation planning options, especially in promoting travel choices for elderly, children and disabled. In addition, there were factors related to the road safety problem which consists of road safety awareness (X25 = 5.249), and speed-reduction for road design (X28) and road environment design (X29), with average value of 4.902, which represent the important level. Factor analysis can be used to classify all factors into 5 main group factors which can be summarized in table 3. The five group factors were generated which were named as; Factor 1: Built Environment, Factor 2: Shared Mobility, Factor 3: Basic services, Factor 4: Infrastructure development and Factor 5: Flexible transportation services. Based on the classification of the five group of factors, the strategy in transportation management can be recommended as depicted in figure 1.

Table 2. Statistical analysis of the main variables

Variables	X_n	Mean	Min	Max	Std. Dev.	Sig. (2-tailed)
Creating flexibility operation	X_1	4.5990	1.00	6.00	1.29687	0.000
Encourages personnel to use public transport	X_2	4.5680	1.00	6.00	1.21196	0.000
Encourages congestion pricing	X_3	4.3760	1.00	6.00	1.24345	0.000
Encourage people to use clean energy vehicles	X_4	4.8640	1.00	6.00	1.11479	0.000
Private cars in public areas with parking fee	X_5	4.3510	1.00	6.00	1.32724	0.000
Expansion of a Bus Rapid Transit (BRT) route	X_5	4.4920	1.00	6.00	1.17700	0.000
Encourage people to use the carpooling	X_7	4.3430	1.00	6.00	1.26133	0.000
More car-sharing in transportation center point	X_8	4.3410	1.00	6.00	1.22891	0.000
More bike-sharing in transportation center point	X_9	4.3290	1.00	6.00	1.24351	0.000
More ridesharing service	X_{10}	4.4110	1.00	6.00	1.17536	0.000
Intelligent traffic management (Smart Mobility)	X_{11}	4.7920	2.00	6.00	1.09359	0.000
Seamless public transportation	X_{12}	4.8370	2.00	6.00	1.13566	0.000
Encourage a work online application system development	X_{13}	4.7830	1.00	6.00	1.14334	0.000
Infrastructure development	X_{14}	5.2170	2.00	6.00	.96272	0.000
Bus stops with automated ticketing service	X_{15}	4.7728	2.00	6.00	1.09719	0.000
Promoting products selling at transit stop	X_{16}	4.8520	2.00	6.00	1.06171	0.000
Bus stops should provide basic services	X_{17}	4.8739	2.00	6.00	1.04225	0.000
Standard public transport design	X_{18}	5.1070	2.00	6.00	.96876	0.000
Applications for specific public transport systems	X_{19}	5.1240	2.00	6.00	.97958	0.000
Expanding public transport routes	X_{20}	5.2492	2.00	6.00	.95118	0.000
Smart cards covering all public transport	X_{21}	4.9629	1.00	6.00	1.15570	0.000
The promotion should be promoted for elderly, the disabled and children	X_{22}	5.3450	2.00	6.00	.87335	0.000
Connectivity of feeder system to railway station	X_{23}	5.3220	2.00	6.00	.85968	0.000
Connectivity of pedestrian paths and bicycle lanes	X_{24}	5.2800	2.00	6.00	.91676	0.000
Road safety awareness	X_{25}	5.2490	2.00	6.00	.91861	0.000
Mitigated road safety problems	X_{26}	5.1630	1.00	6.00	.98661	0.000
Reduce road accidents	X_{27}	5.1900	1.00	6.00	.92777	0.000
Speed-reduction in road design	X_{28}	5.2810	3.00	6.00	.83351	0.000
Road environment design	X_{29}	5.2880	3.00	6.00	.87053	0.000
Shading and trees to support active travel choices	X_{30}	5.3200	1.00	6.00	.87771	0.000
Road design allows people for safer travel by personal car	X_{31}	5.3480	1.00	6.00	.87845	0.000

** Significance (p-value) < 0.05*

IOP Conf. Series: Earth and Environmental Science **897** (2021) 012023 doi:10.1088/1755-1315/897/1/012023

Table 3. Results of factor analysis

No.	Variables	Group of Factors				
		1	2	3	4	5
1	Connected of pedestrian paths and bicycle lanes *(X_{24})*	.738	-.349	.038	-.142	.062
2	Applications for specific public transport systems *(X_{19})*	.719	-.272	.036	.144	.098
3	Road safety awareness *(X_{25})*	.717	-.336	.020	-.135	.020
4	Speed-reducing road design *(X_{28})*	.699	-.382	-.067	-.137	-.248
5	Connected feeder system from rail station *(X_{23})*	.697	-.330	-.032	-.079	.126
6	Reduce road accidents *(X_{27})*	.697	-.302	.009	-.237	-.061
7	Seamless Public Transportation *(X_{12})*	.696	.092	-.223	.387	-.010
8	Standard public transport design *(X_{18})*	.695	-.236	.104	.097	.090
9	Road environment design *(X_{29})*	.694	-.346	-.090	-.154	-.244
10	The promotion should be promoted to the elderly, the disabled and children *(X_{22})*	.674	-.373	-.027	.061	.158
11	Road design allows people for safer travel by personal car *(X_{31})*	.669	-.298	-.139	-.136	-.246
12	Mitigate road safety problems *(X_{26})*	.668	-.256	.074	-.099	-.172
13	Expanding public transport routes *(X_{20})*	.656	-.359	-.031	.200	.146
14	Shading/trees for active travel choices *(X_{30})*	.651	-.296	-.116	-.251	-.279
15	Smart cards covering all public transport *(X_{21})*	.642	-.188	.007	.198	.202
16	Expansion of a Bus Rapid Transit (BRT) route *(X_6)*	.610	.431	-.152	-.095	-.040
17	Encourage people to use clean energy vehicles *(X_4)*	.597	.273	-.259	-.044	.297
18	Encourage a work online system *(X_{13})*	.587	.168	-.035	.503	.018
19	Intelligent traffic management (Smart Mobility) *(X_{11})*	.585	.261	-.195	.294	-.238
20	Expansion of a Bus Rapid Transit (BRT) route *(X_5)*	.552	.362	-.217	-.160	.022
21	Bus stops with automated ticketing service *(X_{15})*	.539	.175	.470	.181	.169
22	Encourage of Congestion Pricing *(X_3)*	.496	.461	-.032	-.336	.221
23	More car-sharing in Transportation center point *(X_8)*	.527	.640	.041	-.116	-.239
24	Encourage people to use the carpooling *(X_7)*	.499	.631	.130	-.061	-.194
25	More bike-sharing in Transportation center point *(X_9)*	.545	.615	.067	-.092	-.183
26	More ridesharing service *(X_{10})*	.543	.594	-.024	.096	-.286
27	Promoting products selling in stop transit *(X_{16})*	.443	.102	.739	.062	-.081
28	Bus stops should provide basic services *(X_{17})*	.457	.013	.732	-.080	.095
29	Infrastructure development *(X_{14})*	.509	-.006	-.133	.494	-.077
30	Creating flexibility operation *(X_1)*	.519	.285	-.111	-.199	.451
31	Encourage of personal to use public transport *(X_2)*	.593	.349	-.150	-.115	.469

The factors of flexible transportation services and infrastructure development is strongly related to the transportation management which provide a variety transportation choices in conforming with land-use planning. The determine settlement can be recommended to avoid traffic congestion or reduce travel distances and travel time, which is similar to the system management efficiency [14, 15, 16, 17]. The factors of shared mobility, built environment and basic services around transit station were related to travel efficiency by promoting modal shift from the most energy consuming to efficient urban transport mode (low-carbon modes); in particular, the shift towards non-motorized transport and public transport [14] & [15]. Especially, in road safety dimension, it can be clearly seen that majority of commuters aware the most for their travel decision.

5. Conclusions

In this paper, we presented our approach for personalized behavioral change support by leveraging travel behavior and personality profiles. Moreover, it can describe the results of a pilot study related to perception on interventions for sustainable urban mobility. The results showed evidence that our proposed approach had some impact on motivating commuters on a personal level to change their mobility behavior and make more sustainable choices. In addition to urban planning and mobility system, in other words, the design of urban infrastructure must be considered together with traffic management. This is very important elements in creating perspectives for alleviating energy efficiency problems while maintaining social inequality which is a consideration of the relevant issues and solutions arise among multidimensional of current urban problems. This study recommended that the result of factor analysis to formulate the set of policy variables is consistence with efficiency traffic management system. Finally, the three pillars of sustainability (the economy, society and the environment) must be treated with equal since suburban attracted migrants with economic prospects, people, activities, interactions and social bring opportunities and threats to sustainable development. Thus, the connectivity with diverse transportation options must be allocated with suitable coverage for time saving, less congestion, producing lower emissions, and promoting efficient land use and improved road safety.

6. References

[1] Giduthuri, V. K. (2015) Sustainable Urban Mobility: Challenges, Initiatives and Planning. Current Urban Studies, 3, 261-265. doi: 10.4236/cus.2015.33022.

[2] Booz & Co. and World Wildlife Fund, Reinventing the City: Three Prerequisites for Greening Urban Infrastructures, 2010, www.wwf.se/source.php/1285816/Reinventing%20the%20City_FINAL_WWF-rapport_2010.pdf.

[3] Arndt, Wulf-Holger & Schäfer, Tanja & Emberger, Günter & Tomaschek, Jan, Lah, Oliver. 2013. Transport in Megacities - development of sustainable transportation systems.

[4] Dur, F., Yigitcanlar, T., 2015. Assessing land-use and transport integration via a spatial composite indexing model. Int. J. Environ. Sci. Technol. 12 (3), 803–816.

[5] Hubacek, K., Guan, D.B., Barrett and J.& Wiedmann, T. 2009. Environmental implications of urbanization and lifestyle change in China: Ecological and Water Footprints. Journal of Cleaner Production (17) 1241-1248.

[6] Steemers, K., (2003). Energy and the City: Density, Buildings and Transport. Energy and Buildings. (35). 3-14. doi - 10.1016/S0378-7788(02)00075-0

[7] Susanne B. B. and Hanna H., (2004). Sustainable Transport: A Sourcebook for Policy-Maker in Developing Cities. Deutsche Gesellschaft für Internationale Zusammenarbeit (GIZ). From https://www.africancityenergy.org/uploads/resource_147.pdf

[8] Florian B., Diana F., Pedro H., Fabrice M., Scott N., Jaana R., Sebastian R., Matt R., Jaeson R., and Jonathan W., (2007) Curbing Global Energy-Demand Growth: The Energy Productivity Opportunity, McKinsey Global Institute, www.mckinsey.com/insights/mgi/research/natural_resources/curbing_global_energy_demand_growth.

[9] D. Hartmann, A. Klein Tank, M. Rusicucci, L. Alexander, B. Broenniman, Y. Charabi, F. Dentener, E. Dlugokencky, D. Easterling, (2013). Observations: Atmosphere and surface.

[10] M. P. Vandenbergh, J. Barkenbus, and J. M. Gilligan., (2008). Individual carbon emissions: The low-hanging fruit. UCLA Law Review, 55:08. 36.

[11] Weiser P., Simon S., Bucher D., Kiefer P. and Martin R., (2015). Towards Sustainable Mobility Behavior: Research Challenges for Location-Aware Information and Communication Technology. GeoInformatica. (1). 1-30.

[12] Anagnostopoulou, E., Urbancic, J., Bothos, E., Magoutas, B., Bradesko, L., Schrammel, J., & Mentzas, G. (2018). From mobility patterns to behavioural change: leveraging travel behaviour and personality profiles to nudge for sustainable transportation. Journal of Intelligent Information Systems, 54, 157-178.

[13] Reitberger, W., Ploderer, B., Obermair, C., Tscheligi, M. (2007). The percues framework and its application for sustainable mobility. Persuasive Technology, LNCS, 4744, 92–95.

[14] Froehlich, J., Dillahunt, T., Klasnja, P., Mankoff, J., Consolvo, S., Harrison, B., Landay, J.A. (2009). UbiGreen: Investigating a mobile tool for tracking and supporting green transportation habits. In Proceedings of the SIGCHI conference on human factors in computing systems (1043–1052). ACM. ISO 690.

[15] Deutsche Gesellschaft für Internationale Zusammenarbeit (GIZ). (2004). Sustainable Urban Transport: Avoid-Shift-Improve (A-S-I). From https://www.transformative-mobility.org/assets/publications/ASI_TUMI_SUTP_iNUA_No-9_April-2019.pdf

[16] Iamtrakul, P., Ruengratanaumporn, I. and Klaylee, J. (2018) The Impact of Urban Development on Social Capital in Urban Fringe Area of Bangkok, Thailand, Lowland Technology International (LTI) journal, 20 (3): 341-350.

[17] Iamtrakul, P. & Chayphong, S. (2021). The Perception of Pathumthani Residents Toward Its Environmental Quality, Suburban Area of Thailand, Geographica Pannonica, 25(2): 136-148.

Acknowledgments

The authors gratefully acknowledge the financial support provided by Thammasat University under the TU Research Scholar of the project entitled "The tackling traffic management by focusing on travel behavior approach towards sustainable campus town development project", contract no.6/2562. The research was conducted by Center of excellence in urban mobility research and innovation, Faculty of Architecture and Planning research fund, Thammasat university.

Environmental Assessment of Recyclable Waste Valorization in United Arab Emirates

A Elfeky[1*], M Abdallah[1], and K Fattah[2]

1 University of Sharjah, Department of Civil and Environmental Engineering, University City, Sharjah, United Arab Emirates

2 American University of Sharjah, Department of Civil Engineering, University City, Sharjah, United Arab Emirates

amalmelfeky@gmail.com, mabdallah@sharjah.ac.ae, kfattah@aus.edu

Abstract. The United Arab Emirates (UAE) has one of the largest per capita carbon footprints and solid waste generation rates in the world. In preparation for the post-oil future, the UAE has planned to diversify its energy mix with renewable sources. Waste valorization, particularly material and energy recovery (MER), presents a promising solution for the waste and energy challenges of the future. It is thus of great importance to ensure that the shift to MER systems lead to a concomitant positive impact on the environment. This research aims to assess the environmental impacts of selected waste management strategies incorporating MER systems in the UAE. The study is focused on the valorization of recyclable wastes through incinerators and/or material recovery facilities (MRFs) compared to existing landfill practices. The mass burn strategy increased the energy produced by 88% compared to the combined MER strategy due to the high energy yield of recyclables. It was found that combining energy and material recovery was the optimum environmental option in terms of global warming and abiotic depletion. The mass burn strategy had a favourable positive impact on freshwater aquatic ecotoxicity and human toxicity, yet it resulted in the highest acidification potential (ACP) and eutrophication potential (ETP). On the other hand, the MRF strategy achieved positive ACP and ETP impacts. Overall, all MER-based strategies achieved better environmental performance compared to existing landfilling practices

1. Introduction

Newly industrialized countries like the United Arab Emirates (UAE) are confronted with numerous waste management problems due to the significant increase in municipal solid waste (MSW) generation in the last few decades. This is largely because of rapid population growth, economic development, and high-consuming society. With a current population of 9.6 million, the waste generation rate was reported to be 1.76 kg/capita/day, which translates to an annual total waste generation of ~6.2 million Mg, most of which is disposed of in landfills [1]. Based on the annual population growth rate of 2.67%, waste generation is estimated to reach ~18 million Mg per year by 2040. Recyclables account for 51% of the total average composition of MSW generated in UAE [2], distributed as 25% paper, 19% plastic, 4% glass, and 3% metal [3]. In response to the challenges associated with traditional waste management methods, the shift to material and energy recovery (MER) systems has become more attractive.

Assessing the environmental effects of waste valorization strategies is vital to strategically prioritize projects and to achieve sustainability [4]. One of the methods to study the environmental impacts is the life cycle assessment (LCA), which is a holistic methodology that considers all environmental burdens throughout the life cycle of products and processes, from cradle to grave [5]. The LCA is suitable for application in waste management due to the complexity of the systems, multiple processes involved, and numerous varying parameters [6].
Recent LCA studies agree that MER systems are more environmentally favorable than landfills over their life cycle. Nevertheless, the performance of such technologies varies depending on specific local conditions, assessment boundaries, input data, and assumptions. For instance, in contradiction to Burnley et al. [7], Buttol et al. [8] and Tunesi [9] found that incineration had a positive environmental impact on acidification and eutrophication compared to landfilling. Likewise, Liamsanguan and Gheewala [10] depicted that incineration was more environmentally superior to landfilling in terms of greenhouse gas (GHG) emissions. Demetrious and Crossin [11] found that treating paper waste in incineration or gasification-pyrolysis produces less GHG emissions than in landfill, while plastic was best managed in landfill. On the other hand, Wang et al. [12] stated that when the total steel and plastic waste were recycled, maximum energy conservation and carbon dioxide (CO_2) reduction were achieved.
Overall, the reviewed literature contains limited research on the environmental performance of MER facilities, locally and regionally. The main purpose of this study was to assess the different environmental impacts of selected waste management strategies incorporating MER systems using an LCA approach. The study was focused on the management of recyclable waste materials through incinerators and/or material recovery facilities (MRFs). The relative performance of the examined individual and combined MER systems was assessed in terms of multiple environmental impacts categories

2. Methodology

The International Organization for Standardization (ISO) provides a set of guidelines (14040-14044) for LCA studies [12]. The standard stages of an LCA study comprise: 1) goal and scope definition, 2) life cycle inventory (LCI), 3) life cycle impact assessment (LCIA), and 4) interpretation, as discussed in the following sections.

2.1. Goal and Scope Definition

The main goal of this study was to evaluate the environmental burdens associated with the management of recyclable materials under specific valorization strategies compared to existing practices. The assessed waste management strategies in this study are based on incineration and MRF compared to the existing landfill practices. The functional unit to which all inputs and outputs of the study are referred to be the management of 1 Mg of recyclable waste. The proposed LCA system boundary includes the separation, treatment, and disposal processes. The transportation and collection processes have been found to have minimal impacts relative to other processes [13] and was therefore excluded from this study. The computations assumed an ideal 100% separation efficiency of recyclables and organics at the source. Currently, a fraction of the recyclable waste in UAE is segregated in MRFs (32% of total waste) and diverted from landfills [14]. Based on the literature, 90% of the proposed waste in an incinerator facility is assumed to be recovered as heat and gas, whereas the remaining 10% is converted into ash and disposed of at a landfill. The heat produced by incinerators is used to generate electricity through steam turbines.
The current waste management scheme was set as the business as usual (BAU) strategy compared to the proposed alternative waste management strategies in Figure 1, as follows:

- Strategy ER: The strategy constitutes single-bin collection of commingled waste that was processed at an incineration facility. The recovered heat was used to generate electricity, whereas the produced ash was disposed of in an inert landfill.
- Strategy MR: The strategy constitutes separate collection in two bins (recyclables and commingled waste). The separated recyclables are sent to an MRF, where the recovered

materials are baled and marketed. The non-recovered materials at the MRF are disposed of in a landfill.

– Strategy MER: The strategy constitutes separate collection in two bins (recyclables and commingled waste). Recyclables are sent to an MRF, whereas MRF rejects are processed at an incineration facility. The produced ash was disposed of in a landfill.

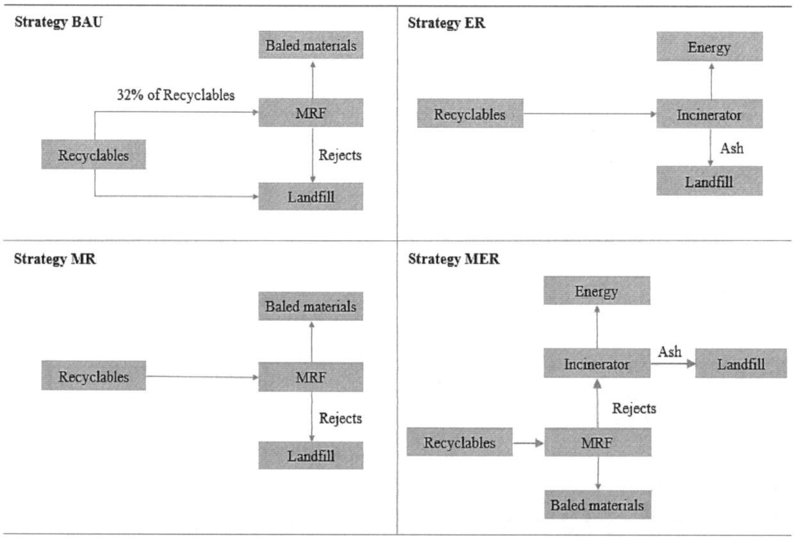

Figure 1. Proposed integrated waste management strategies

2.2. *Life Cycle Inventory*
The specific waste characteristics in UAE, such as composition and energy content, as well as the local market and operation parameters, particularly those related to energy consumption, are used in the LCI [5]. As reliable local data are limited, the Ecoinvent database was utilized to quantify the energy, resource use, and emissions of the involved processes and materials. The LCA software used in this study, WRATE, is specialized for waste management projects, and comprises an embedded database compiled from actual waste management facilities throughout the world.

2.3. *Life Cycle Impact Assessment*
The study covers the environmental impact categories which are considered significant and depict a wide range of environmental issues in the UAE. The environmental impact categories investigated are global warming potential (GWP), abiotic depletion potential (ADP), acidification potential (ACP), freshwater aquatic eco-toxicity potential (FAETP), human toxicity potential (HTP) and eutrophication potential (ETP). The environmental impacts are computed through the Problem Oriented Approach, known as the CML methodology [15]. The CML method has equivalence unit and normalization factor for each characterization impact assessment

2.4. *Interpretation*
The results of the environmental impact assessment for various waste management strategies are comparatively analysed and interpreted compared to the literature.

3. Results and Discussions
A comparative LCA is carried out for four alternative waste valorization strategies to evaluate the various material/energy streams and the overall environmental burdens in terms of different impact categories. Figure 2 illustrates the main technical aspects, including total energy recovered, various waste streams (recovered, recycled, and landfilled) and the land intake. Waste landfilled is a sum of biodegradable and non-biodegradable landfilled waste. The BAU strategy had the largest

biodegradable landfilled waste stream of 2.1 Mt, whereas the lowest amount occurred in two strategies: ER (zero due to complete incineration) and MER (zero when MRF rejects were incinerated and ~200 kt when rejects, particularly paper, were landfilled). The materials recovered through recycling offset the consumption of raw materials and associated extraction and production processes. As material recovery is essential in sustainable waste management, the amount of recyclable materials recovered in the two strategies involving MRFs was identical at the maximum possible amount (~5.2 Mt), compared to ~1.7 Mt recycled in the BAU strategy (Figure 2). The BAU and MR strategies did not include any controlled conversion process, i.e., no energy was recovered. Higher amounts of incinerated waste materials lead to the recovery of more energy with the waste management strategy. Figure 2 shows that ER strategy increased the energy produced by 90% compared to MER strategy due to the high energy yield of the recyclables, particularly paper & plastic materials. Strategy MER has the largest land intake since an inclusive land is required for the construction and operation of both MRF and incinerator facilities.

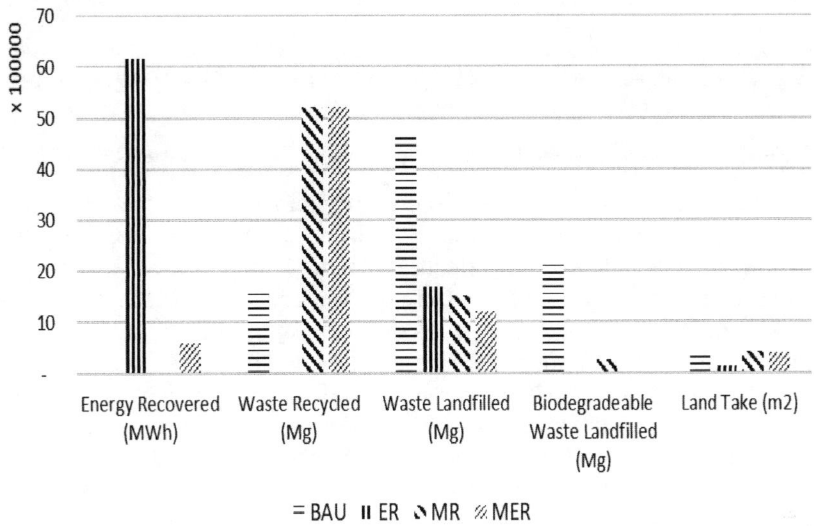

Figure 2. Main characteristics of the proposed WTE-based waste management strategies.

A comparative LCA was performed for four strategies to evaluate the various environmental burdens through six impact categories: GWP, ACP, ETP, FAETP, HTP and ADP. The GWP represents the combined effect of GHGs (CO_2, CH_4, N_2O, CO, CFCs, HCFCs, and HFCs) which can absorb heat radiation and consequently increase the atmospheric temperature. As shown in Figure 3, BAU strategy displays the highest GWP (~520 kg CO_2 eq.). This agrees with the main findings of LCA studies conducted in developed and developing countries, e.g., Lithuania, Thailand, Brazil, USA, and UK which specified that landfilling practices would constantly show a lower environmental performance unlike other conversion systems [16–20]. As burning plastics contributes to the GWP, the high calorific value results in more energy recovered, thus offset significant emissions from electricity that could have been generated in conventional power plants. The GWP caused by incineration depended on the feedstock fed to the incinerator; the GWP contribution was slightly higher in the ER strategy rather than MR and MER strategies, in which high-energy plastics, were combusted rather than recycled. A larger amount of recycled materials resulted in a lower GWP, as can be observed for MR and MER strategies.

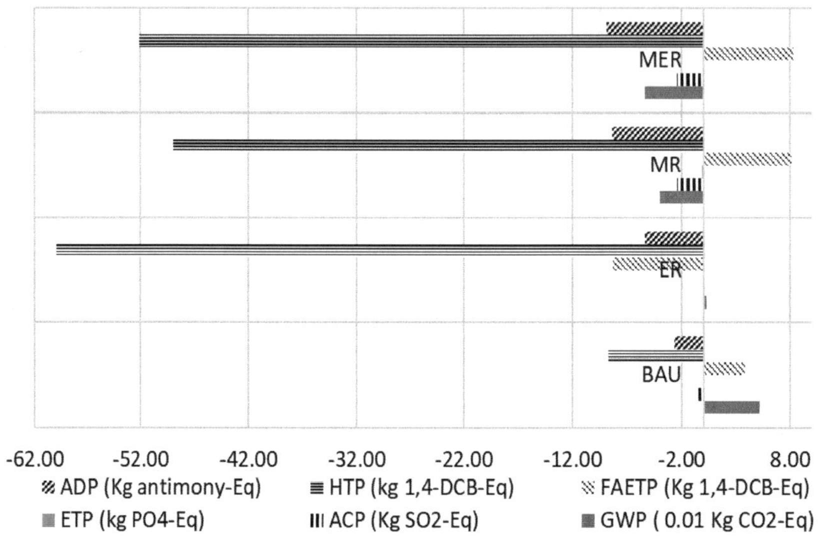

Figure 3. Environmental impact categories of the proposed strategies.

ACP is caused by the release of hydrogen ions to the environment through acidifying substances (SO_2, NO_x, HCl, and NH_3). The acidic gases released are washed out by atmospheric precipitation forming acid rain which has widespread noxious effects on plants, soil, and surface waters. ACP is also harmful for human health, especially on the respiratory systems, as well as the degradation of various assets and valuable landmarks. This later impact is of great importance to the UAE as the local non-oil economy is based on real estate and tourism. Acid gases typically emitted from incineration processes to some extent increased the ACP, which was not significantly offset by the energy recovery, and mainly related to the flue gas treatment process. The two strategies involving MRFs showed a positive impact on ACP mainly due to the recovery of materials and the avoided emissions during extraction and production. The ER strategy caused ACP impact ranging from -0.14 to -2.40 kg of SO_2 eq., primarily due to NH_3, NO_x, and SO_2 emissions.

ETP comprises all the potential impacts with high environmental levels of macronutrients, (mainly N and P), in aquatic and terrestrial ecosystems. This increases the production of aquatic plants and reduces water quality and oxygen depletion in the bottom layers. For incineration, gases such as NO_x from waste combustion contributed to much of the impact. ER strategy was the highest contributor for ETP by 0.05 kg of PO_4 eq. Strategies that involved MRF facilities showed low impact, and the lowest ETP impact was depicted in MRF strategy with a value of -0.10 kg of PO_4 eq. followed by MER strategy with half the value.

FAETP measures the relative impact of toxic substances on the freshwater aquatic environment due to the emissions to environmental compartments air, freshwater, seawater, agricultural, and industrial soil. As shown in Figure 3, a positive favourable FAETP impact on the environment was depicted in ER strategy by -8.34 kg 1,4-DCB-eq, mainly due to Barite, Zinc and Vanadium. Whereas the highest FAETP contributor was depicted in MER strategy by 8.32 kg 1, 4-DCB-eq. BAU strategy showed FAETP impact since the areas near landfills have a greater possibility of groundwater contamination because of the potential escape of metals in leachate discharge.

HTP is related to the negative effects of toxic substances (e.g., volatile organic compounds, particulate matter, heavy metals, NO_x, SO_2) from the waste management process affecting the biological human system (i.e., excluding workplace exposures). All strategies contributed to a positive HTP impacts on the environment, i.e., reduced negative effects of toxic substances. The highest reduced HTP impact was in ER strategy with a value of -60 kg 1, 4-DCB-eq and the lowest reduced impact was in the baseline strategy by -8.76 kg 1, 4-DCB-eq. This was contradicted with a LCA study that stated HTP emissions in incineration was 134 times greater than in landfill due to the air emission of dioxins

IOP Conf. Series: Earth and Environmental Science **897** (2021) 012024 doi:10.1088/1755-1315/897/1/012024

during incineration [20]. MRF and MER strategies presented reduction in HTP impact (~49-52 kg 1, 4-DCB-eq) due to Chromium, PAH, Arsenic, Nickel in air compartment and Barite, PAH, Vanadium ions in water compartment.

ADP is defined in terms of the annual rate of depletion of the stock of minerals and fossil fuels relative to ultimate reserves. Like HTP, all strategies contributed to positive ADP impacts on the environment, i.e., reduced negative effects of resource depletion. The main savings were due to the replacement of fossil fuels; thus, the strategies with more landfilling gave the worst results. Strategies that involved MRF facilities showed a lower ADP impact mostly due to coal, natural gas, crude oil, and molybdenum (0.11% in sulphide, Mo 0.41%, and Cu 0.36% in crude ore). The highest reduced ADP impact was in the MER strategy, with a value of -8.89 kg Antimony-eq. Also, the ER strategy had a positive ADP environmental impact due to the electricity generated by the process.

4. Conclusion

Material and energy recovery technologies can potentially make a substantial contribution to renewable energy production in the UAE while alleviating environmental impacts associated with landfilling. This study discussed the six environmental impacts: GWP, ADP, ACP, FAETP, HTP and ETP of selected MER-based strategies as alternatives to conventional landfilling. Combing MR and ER was the optimum environmental option for the management of recyclables in terms of GWP and ADP. The ER strategy presented a positive favourable FETP and HTP impacts yet showed the worst environmental impact in terms of ACP and ETP. On the other hand, the MR strategy showed positive ETP and ACP impacts, i.e., mainly due to the recovery of materials and the avoided emissions during extraction and production. The baseline landfilling strategy was the least environmentally favourable compared to the examined waste valorization systems.

5. References

[1] SCAD 2018 Waste statistics Abu Dhabi waste management centre
[2] EAD 2016 Waste and environment annual report
[3] Abdallah A, Shanableh A, Arab M, Shabib A, Adghim M and El-Sherbiny R 2019 *J. Environ. Manage*. **232** 58-65
[4] Shabib A and Abdallah M 2020 *Int. J. Environ. Stud.* **7233**
[5] Mendes M, Aramaki T and Hanaki K 2004 *Resour. Conserv. Recycl.* **41** 47-63
[6] Burnley S, Coleman T and Peirce A 2015 *Waste Manag.* **39** 295-304
[7] Buttol P, Masoni P, Bonoli A, Goldoni S, Belladonna V and Cavazzuti C 2007 *Waste Manag.* **27** 1059-1070
[8] Tunesi S 2011 *Waste Manag.* **31** 561-571
[9] Liamsanguan C, Gheewala S 2008 *J. Environ. Manage.* **87** 132-138
[10] Demetrious A, Crossin E 2019 *J. Mater. Cycles Waste Manag.* **21** 850-860
[11] Wang M, Li Z, Chen L, Liu P, Zhang R, Liu W, Ma Y, Wang Y and Li X 2020 *Environ. Sci. Pollut. Res.* **27** 42158-42167
[12] ISO 2006 Environmental management life cycle assessment-principles and framework *Environmental Management System Requirements*
[13] Abdallah M and Elfeky A 2021 *J. Environ. Manage.* **280** 111839
[14] Abdallah M, Shanableh A, Shabib A and Adghim M 2018 *Waste Manag.* **82** 207-219
[15] Tawatsin A 2014 Environmental assessment of waste to energy processes specifically incineration and anaerobic digestion using life cycle assessment
[16] Miliute J and Kazimieras Staniškis J 2010 *Waste Manag. Res.* **28** 298-308
[17] Cleary J 2009 *Environ. Int.* **35** 1256-1266
[18] Emery A, Davies A, Griffiths A and Williams K 2007 *Resour. Conserv. Recycl.* **49** 244-263
[19] Kaplan P. O, Decarolis J and Thorneloe S 2009 *Environ. Sci. Technol.* **43** 1711-1717
[20] Leme M. M. V, Rocha M. H, Lora E. E. S, Venturini O. J, Lopes B. M and Ferreira C. H 2014 *Resour. Conserv. Recycl.* **87** 8–20

AUTHOR INDEX